Recountings

Recountings

Conversations with MIT Mathematicians

Edited by Joel Segel

A K Peters Ltd.
Natick, Massachusetts

Editorial, Sales, and Customer Service Office

A K Peters, Ltd.
5 Commonwealth Road, Suite 2C
Natick, Massachusetts
www.akpeters.com

Library of Congress Cataloging-in-Publication Data

Recountings: conversations with MIT mathematicians/edited by Joel Segel.
p. cm.
Includes index.
ISBN 978-1-56881-713-2 (alk. paper)
1. Mathematicians—Interviews. 2. Mathematicians—Intellectual life—20th century.
3. Massachusetts Institute of Technology. I. Segel, Joel, 1961–
QA28.R43 2008
510—dc22

2008021795

First paperback printing, 2010.

Printed in the United States of America
14 13 12 11 10 — 10 9 8 7 6 5 4 3 2 1

Dedicated to

William "Ted" Martin (1911–2004)

&

Norman Levinson (1912–1975)

Contents

Introduction

This book is not about mathematics but about mathematicians, individuals who have made mathematics the pursuit of a lifetime. It also recounts the history of an academic department that wanted to learn about its most recent and successful half century from the faculty members with the longest memories. The project began with a thought. Professor Gilbert Strang was regretting the loss of Gian Carlo Rota, who died suddenly in his sleep at age 66. Rota had single-handedly put combinatorics on the map of mathematical respectability. He held the unique title of Professor of Applied Mathematics and Philosophy, had published a collection of thought-provoking essays called *Indiscrete Thoughts*, and was one of MIT's most popular teachers. What a pity his lectures were not preserved, Strang thought.

His wife, Jillian Strang, brought a historian's perspective to the problem. Why not interview some of the older generation? she suggested. Get their recollections down on paper before these memories, too, were lost. Strang mentioned the idea to I. M. (Is) Singer, who discovered the index theorem with Michael Atiyah and whose opinion is key to department enterprise. Singer liked the idea, as did department head Michael Sipser. The core mission: ask surviving members of today's senior generation what they remember of the department's history.

By any account, after all, the department that these mathematicians built is a success story. In the first half of the twentieth century, astonishing as this may seem for the home of Norbert Wiener, the department's primary mission was teaching MIT's engineering students the mathematics they needed to know. The generation interviewed here was hired as part of an ongoing effort to transform MIT mathematics from a service department to one focused on both teaching and research. Today MIT is ranked in the top handful of anyone's list of the best mathematics departments in the country. The collective memory of the department's oldest representatives—twelve men, 65 and older, all but two still working, as well as the widow of department godfather Norman Levinson—provides an important chapter in the story of that success.

The core idea of the project, once articulated, soon expanded to include not only memories of the department but also a focus on the individuals themselves. Anyone hired to a tenure-track position at MIT is exceptional. It therefore made sense to ask about the years before they came on board, beginning in childhood. How did bright youngsters first become attracted to mathematics, and how was their initial interest nurtured? Any educated person today will have a sense of mathematics as the fabric underlying our understanding of the physical world, a crucial element of the language we use to think about science and engineering. This realization was dawning only gradually in the years following World War II: most of the interviewees were only dimly aware, if at all, of a research mathematician as something real people became. In the 1930s this was to be expected: Norman Levinson (1912–1975), a formidable mathematician whose judgment and behind-the-scenes leadership profoundly shaped the mathematics department at MIT, assumed he would train as an actuary and get a job in insurance. Levinson knew jobs were scarce in those Depression-era days, even more so for the son of Jewish immigrants. Yet an entire generation later, an Estonian immigrant of practical bent like Alar Toomre still saw mathematics as "too much like opera. You got paid only if you were terrific." Even Michael Artin, a future leader in modern algebraic geometry, assumed he would get a teaching position at a small-town college somewhere.

The mathematics community has made great strides in identifying and nurturing mathematical talent and inclination. Fifty years ago, the process was much more haphazard. The paths these young men took to a career in mathematics were varied, and few had the final goal in mind. Some ambled up to the front door, set on their path by an early delight in number manipulation that deepened over time. Others clambered in through the back window, trying to understand the mathematics that undergirded an initial love of chemistry or physics. Bertram Kostant remembers being fascinated by the formulas of chemical reactions, as well as by "the magic and *power* of chemistry"—a dangerous attraction in the hands of a brilliant but rebellious youngster. But he was always good at mathematics as well, and he was entranced by the notion of statements that could be *proved*, in such contrast to the argumentative, fast-talking culture of Brooklyn where he'd grown up. Kenneth Hoffman attended Occidental College because of their track-and-field team. When told by his mathematics teacher, Mabel Barnes, that he should consider graduate school, his response was, "What's that?"

Then, as now, exceptional students will often gravitate toward—or attract—exceptional teachers, thesis supervisors, mentors. Fagi Levinson remembers how her husband's talent got the attention of Norbert Wiener, not a mathematician known for his attentiveness to the faces in front of him but a generous mentor once he realized that the young Levinson was worthy of his attention. Many here remember giants of the previous academic generation: Emil Artin, Alonzo Church, Paul Erdős, Alexander Grothendieck, Peter Lax, Solomon Lefschetz, James Lighthill, Marshall Stone, Oscar Zariski. They remember the workings and atmospheres of other top-flight institutions: Harvard, Chicago, Oxford, the Institute for Advanced Study, and the IHÉS. They remember teaching that worked for them—some appreciating precision, others a more improvisatory style that left room for their own contributions—and guidance that stayed with them decades later.

A working mathematician's first years in the field are often the most pressured. There's the holy grail of tenure, of course, but many young mathematicians are also haunted by G. H. Hardy's famous pronouncement, in *A Mathematician's Apology*, that "mathematics, more than any other art or science, is a young man's game." Fagi Levinson recalls John Nash as particularly anxious to secure his place in the pantheon before he reached the age at which, he believed, a mathematician's best days are over. "*Thirty* was a dirty word," she says. Youth conveys undoubted advantages in tackling certain kinds of problems. Yet Norman Levinson himself was an inspiring example of a person with the focus and determination necessary to make important contributions throughout his career, as are Is Singer and others in this volume.

The life of most mathematicians is more measured tread than glorious gallop, and a forty-year career often lies beyond those early years. The years after tenure bring other rewards as well, as these conversations testify. Careers can take surprising turns, for one thing. New areas of fresh inspiration can fall into one's lap almost by accident. Alar Toomre, though always interested in aviation and aerodynamics, wasn't planning to devote his life to the mathematics of spiral galaxies. A newcomer to the department with a PhD on wind tunnels and shear flows, he was converted by a series of let's-learn-this-together sessions organized by C. C. Lin, who had returned from a meeting at the IAS all fired up with enthusiasm for these "wonderful objects," in Toomre's words. A buzz among engineers about the finite element method persuaded Gilbert Strang to look into it further and, ultimately, to provide it with a rigorous mathematical foundation. A request for an article on Schubert calculus drew Steven Kleiman to enumerative geometry, where he became one of the central figures in the field. A PhD student's interest led Michael Artin into noncommutative algebraic geometry, "a nice, quiet pond where I could spend my declining years," Artin thought, though he allows that, due to recent interest shown by theoretical physicists, "the water [is] getting a little turbulent."

Many in this book have thought deeply about how mathematics is passed on to the next generation. MIT is known as an excellent place to get an education in mathematics, both at the undergraduate and graduate levels. How does a department earn such a reputation? Even senior department faculty commonly teach basic courses at MIT, and the faculty has written well over a hundred textbooks in the last half-century or so. But these conversations suggest that excellence in teaching is more than just a matter of books and blackboards. It is Arthur Mattuck's decades of devotion to undergraduate education, including his wry booklet on the art of recitation teaching, "The Torch or the Firehose," which has spread to other universities and been translated into several languages. It is the Research Science Institute (RSI) and the Summer Program for Undergrad Research (SPUR), programs that pair high school students and undergraduates with graduate students as mentors in a summer research project. (Both were run for many years by Hartley Rogers, who has now passed the reins to David Jerison.) It is Michael Artin's continued search, even in his third decade of teaching undergraduate algebra, for the perfect exercise. It is Steven Kleiman's focus on imparting the craft of setting out one's research findings in a coherent paper and Gilbert Strang striving for that trademark conversational tone in his latest textbook. It is Daniel Kleitman dreaming up Java applets to illustrate the concepts of calculus more viscerally. It is videoed lectures on MIT's OpenCourseWare, downloaded and viewed free of charge by students all over the world.

Administration is another key to excellence. Some see administrative service as a distraction or, worse, an admission that your work has dead-ended or your mathematical powers are not what they used to be. Others, as these narratives show, develop a genuine interest in the larger departmental or institutional view, the tactics and logistics that undergird scientific accomplishment. If a department is only as good as its faculty and students, how to secure the brightest of both? What new departmental structure might guarantee applied mathematicians the autonomy they need? How to chip away at budget cuts from above, obtain the next generation of desktop computers for the faculty, or conjure badly needed office space out of turf whose borders seem set in stone? Such initiatives are rarely the stuff of headlines, but they are no less necessary for that. Other contributions are more dramatic: helping the institution weather the student riots that rocked the MIT campus in 1969; arguing for more mathematics funding in Washington, D.C., and making sure that the arguments are not only made before the microphones but acted upon afterwards.

Frequent references to current events and social trends belie the perception that mathematics is a solitary and isolated pursuit. The interviews reflect a surrounding landscape that is always in flux, an institutional and societal context that continued to evolve throughout these men's lives and careers. World War II brought the G.I. Bill, a great physical and intellectual migration from Europe, and a dawning perception of scientists as heroes. Korean War deferments shaped decisions as to what degrees to pursue. Vietnam and the Sixties sent waves of protest through campuses across the country and around the world. The Cold War and a changing political climate led to seismic shifts in federal funding. All of these, in big ways and small, put their stamp on the stories these interviews tell.

Here, then, is the recent history of a university department, in the context of its times, through the eyes of some who worked within its walls. Rather than imposing a single voice and central organizing principle—a simple timeline, say, or the reign of successive department heads—we have retained a narrative form, distilled in each case from longer conversations, that allows each faculty member to give his own account. The historical picture emerges from a range of vantage points, "like triangulation in surveying," as Arthur Mattuck remarked.

A final note: The voices of women are largely missing from this account. Norman Levinson's widow, Fagi, recalls that during his earlier years on the math faculty, "women were allowed to be TAs but weren't allowed to teach, but he finally changed that, and of course they did very well." Today the proportion of women among mathematics faculty members at MIT is still low, but it rises steadily as one looks to successively younger groups—instructors, graduate students, and undergraduate mathematics majors, almost a third of whom are women, including some top students and national prize-winners. As these words are being written, the department and the National Science Foundation are sponsoring a conference, "MIT Women in Mathematics: A Celebration." Speakers include women faculty from eight different universities, all but one of whom received their PhDs in mathematics at MIT.

Joel Segel
Cambridge, Massachusetts
April, 2008

From the Department Head

The idea of compiling this collection of interviews came about several years ago with the recognition that many of the key figures in the mathematics department were nearing retirement. We decided to commission a series of interviews to preserve their recollections of the evolution of the department, to capture their anecdotes about mid-twentieth-century mathematical life and its intersections with the political and social climate of the times. As a newly appointed head of the department, I also felt it could be useful to learn how previous heads succeeded in building the department into the mathematical powerhouse it is today. We hoped the interviews would be informative, interesting, and perhaps even entertaining.

We recognized one thing: we would never manage to do it ourselves. I consulted with Joel Segel, a writer and editor with extensive experience in interviewing. Joel had grown up with an appreciation of our mysterious world and had a personal connection to the department as well: his father, Lee Segel, had been a well-known applied mathematician with a 1959 PhD in mathematics from MIT. Joel put together a detailed proposal for the project, and we decided that he would conduct and edit the interviews.

As the project started, it took form and grew. Early on it became clear that we should include Fagi Levinson, Norman Levinson's widow, who had been central to the social fabric of the department for decades. Joel interviewed her first. I polled various faculty to come up with lists of other possible interviewees. (One person we wish we could have interviewed is Ted Martin, head of the department from 1947 to 1968 and mentioned in so many of the interviews. He died shortly before the project began.) In some cases, faculty came forward and volunteered to be interviewed. Though we had originally planned to include all of the older faculty, the project's length and duration prevented us from doing so while keeping it to a manageable size and timeframe. For the same reasons, we decided not to interview any of the younger faculty. Our apologies to those who wanted to participate but didn't have the opportunity. We offer these recollections in the hope that current and future faculty will take up the story where we left off.

Michael Sipser
Cambridge, Massachusetts
March, 2008

Acknowledgments

Thanks first to all the interviewees for their enormous generosity and patience in the midst of busy schedules. Special thanks to Gilbert and Jill Strang, who conceived of this project, and to department head Michael Sipser, the driving force behind it. Thanks to David Buchsbaum, Fred Brauer, and Beverly Kleiman, for background help; Paula Duggins, Danforth Nicholas, Dennis Porche, and Claire Wallace for logistics, kindliness, and patience; Carolyn Artin, for additional help with photographs. Thanks to Klaus and Alice Peters and everyone at A K Peters. Thanks to all those whose labors make mathematical lore available on the Web and other public channels. Most of all, thanks to Arthur Mattuck for endless hours of his time and for his sage counsel, broad and balanced perspective, and gimlet eye: the project is incalculably better for his input. Finally, in loving memory of Lee A. Segel (1932–2005), whose lifelong devotion to mathematics and to mathematicians, past and future, inspired this project in turn.

Photo Credits

p. xxii	Zipporah (Fagi) Levinson: Courtesy of the Levinson family.
p. 1	Norman Levinson: MIT Archives.
p. 6	Norman and Fagi Levinson in Copenhagen, 1949: Courtesy of the Levinson family.
p. 8	Norbert Wiener lecturing: MIT Archives.
p. 11	Dick Shafer, George Whitehead, Kay Whitehead, Barbara Brauer, George Brauer, and Alice Schafer: Courtesy of the Levinson family.
p. 18	H. B. Phillips: MIT Archives.
p. 25	Isadore M. Singer: Courtesy of the MIT News Office.
p. 26	Is Singer lecturing: MIT Archives.
p. 27	Irving Segal: MIT Archives.
p. 30	George Whitehead: MIT Archives.
p. 33	Singer receiving the Abel prize: Photograph by Mark Germann.
p. 35	Is Singer: MIT Archives.
p. 40	Howard Johnson and Is Singer: Courtesy of the Singer family.
p. 42	Arthur Mattuck playing his cello, 2008: Photograph by Carolyn Artin.
p. 44	The Mattuck family: Courtesy of the Mattuck family.
p. 46	Arthur Mattuck, Dick Gross, Dana Fine, and Mike Artin: Photograph by Carolyn Artin.
p. 48	Emil Artin: Courtesy of the Artin family.
p. 55	Reunion of some Princeton PhD's: Courtesy of the Mattuck family.
p. 57	Miyanishi, Grothendiek, and Abyankhar in Montreal, 1970: Photograph by Studio O. Allard, Inc.; courtesy of the Artin family.
p. 61	Mattuck family in New York City: Courtesy of the Mattuck family.
p. 68	Arthur Mattuck, Monika and David Eisenbud: Courtesy of the Mattuck family.
p. 76	Arthur Mattuck, c. 1982: Courtesy of the Mattuck family.
p. 88	Arthur Mattuck, c. 1970: Courtesy of the Mattuck family.
p. 104	Hartley Rogers, 1986: Courtesy of the Rogers family.
p. 107	Hartley Rogers punting with his daughter Caroline: Courtesy of the Rogers family.

Zipporah (Fagi) Levinson

Editor's Note

Zipporah "Fagi" Levinson, wife of the late Professor Norman Levinson, is a vibrant and irreplaceable part of the Mathematics Department's institutional memory and perhaps the person best qualified to put a human face on the period when the department was coming into its own as a full-fledged research entity. Warm, down-to-earth, yet acutely conscious of the lofty endeavor that her husband and his colleagues were devoted to, she was the department's unquestioned "den mother" for decades, a sympathetic ear and helping hand to the likes of Norbert Wiener and John Nash. "I felt comfortable with mathematicians," she said. "They are an unconventional group, to be sure, but very decent people." Clearly devoted to her husband's memory, Fagi Levinson retains a humorous, no-nonsense view of his idiosyncrasies and those of the colleagues who surrounded him.

Norman Levinson.

Norman Levinson spent almost his entire professional life at MIT, from undergraduate student to Institute Professor and department chairman. A man of extraordinary talent and iron discipline, he was awarded the Bôcher Prize for his contributions to nonlinear differential equations and did some of his best work, on the Riemann zeta function, shortly before he died at the age of sixty-three. The preface to Levinson's Selected Papers summarizes his contributions as follows.

> The deep and original ideas of Norman Levinson have had a lasting impact on fields as diverse as differential and integral equations, harmonic, complex and stochastic analysis, and analytic number theory during more than half a century. Yet, the extent of his contributions has not always been fully recognized in the mathematics community. For example, the horseshoe mapping constructed by Stephen Smale in 1960 played a central role in the development of the modern theory of dynamical systems and chaos. The horseshoe map was directly stimulated by Levinson's research on forced periodic oscillations of the Van der Pol oscillator, and specifically by his seminal work initiated by Cartwright and Littlewood. In other topics, Levinson provided the foundation for a rigorous theory of singularly perturbed differential equations. He also made fundamental contributions to inverse scattering theory by showing the connection between scattering data and spectral data, thus relating the famous Gelfand-Levitan method to the inverse scattering problem for the Schrödinger equation. He was the first to analyze and make explicit use of wave functions, now widely known as the Jost functions. Near the end of his life, Levinson returned to research in analytic number theory and made profound progress on the resolution of the Riemann hypothesis.[1]

Levinson was the behind-the-scenes architect of the mathematics department in the days of its great expansion, lending counsel to department head Ted Martin on every important decision. He also had an impact on the teaching and communication of mathematics: his exposition of the Selberg-Erdős proof of the prime number theorem was honored with the MAA Chauvenet Prize, and fifty years after the publication of Coddington and Levinson's Theory of Ordinary Differential Equations, the book is still frequently cited and assigned in courses on differential equations, numerical analysis, and advanced calculus.

From Revere to Cambridge

J.S.: *How did Norman Levinson come to MIT?*

Z.L.: My husband came to MIT from Revere High School. His parents were poor immigrants. They didn't know anything about education here, but the high school

---

[1] Nohel, J. A. & Sattinger, D. H. (Eds.). (1998). *Selected papers of Norman Levinson* (2 Vols.). Boston: Birkhäuser.

principal told him to come to MIT and to apply for a scholarship. So he came, and he thought that the only thing to do was earn a living by getting a degree in electrical engineering. But Norbert Wiener soon spotted this kid and told him what courses to take, to stop his major in electrical engineering and to go into mathematics.

Wiener's lectures weren't very clear, and his classes were small. On the other hand, he gave every student an A. There was no bell curve. And if he had a student like my husband, who asked intelligent math questions, understood what he was talking about, and made suggestions, he would often make him his collaborator. My husband said he wanted to do his own work. He didn't want to feel he had to do what Wiener was doing.

By the time Norman was a senior, he'd written several papers that they figured were worthy of a PhD, though he wasn't registered in the graduate school. So Wiener sent him to England on a Proctor Fellowship to study with Hardy for a year, which was why Hardy knew about his work so well. He used to work all night and sleep in the daytime, but he had this colleague who would go to Hardy's lectures and bring back the notes and tell him the stuff, and Norman would send notes to Hardy, too. So Hardy and Littlewood knew him in this way. When he came back, he got his degree.[2]

---

[2] "Levinson entered the Massachusetts Institute of Technology in 1929 but he did not register for a mathematics degree, rather he studied for a degree in Electrical Engineering. In June 1934 he was awarded a Bachelor of Science degree and a Master of Science degree, both in electrical engineering. However he had by this time taken twenty graduate courses in mathematics at MIT, almost all the graduate courses the mathematics department offered. In fact by this time, while still an electrical engineering student, he had written a thesis with Norbert Wiener and, according to the Head of Mathematics, had:

... *results sufficient for a doctoral thesis of unusual excellence.*

"...After the award of his Master's degree in electrical engineering, Levinson applied to MIT to begin studying for his doctorate in mathematics. However the Mathematics Department were convinced that he had already done sufficient for the Ph.D. [sic] before starting the course! Instead Wiener together with Phillips, the Head of Mathematics, arranged for Levinson to receive an MIT Redfield Proctor Traveling Fellowship so that he could spend the year at Cambridge in England. He was assured that he would receive his doctorate on his return to MIT irrespective of any work he did in Cambridge."

See a more complete biography of Levinson in the MacTutor History of Mathematics archive of the School of Mathematics and Statistics, University of St Andrews, Scotland, at http://www-groups.dcs.st-and.ac.uk/~history/Biographies/Levinson.html.

Now, there was a *numerus clausus*,[3] not only for Jewish students, but for Jewish faculty. Most schools, if they had one Jew, figured that was more than enough. MIT began as an engineering school; they had a math department that was just a service department for the engineers, so they didn't care. They had Wiener, his brother-in-law [Philip] Franklin,[4] [Samuel] Zeldin, and Jesse Douglas—four Jews. And Jesse Douglas had had a nervous breakdown; he was in the hospital, and his classes had to be taught.[5] But when my husband was proposed by Wiener, the school said too many Jews, too many Jews. So he went to Princeton, along with all the other unemployed Jews, to the Institute for Advanced Study.

They'd get these little stipends, just barely enough to live on. My brother [Henry Wallman] was there with Hugh Dowker.[6] They shared a place. Hugh Dowker had a bed on the floor, and he had figured out the best diet, which was a head of raw cabbage every day. He claimed that this had the vitamins and the minerals and the carbohydrates—name it: proteins, everything. Warren Ambrose felt that he could afford to have his car only by not having a room and living in his car. All the young mathematicians would come to tea time at 4 o'clock and gobble up all the tea there was—the cookies, whatever they had there—because that represented a very important meal of the day.

If they ever wanted to come to New York, they obviously didn't have money for hotels or anything like that, so my brother would invite them to our house and they would stay overnight. So I met Norman there.

Norman Levinson was appointed as an instructor at MIT in 1937 after G. H. Hardy interceded on his behalf, at Norbert Wiener's request. Tell me how that happened.

---

3 Limited (lit., closed) number, or quota.

4 Philip Franklin, a number theorist who achieved the first real breakthrough in the four-color problem, married Norbert Wiener's sister Constance.

5 Douglas had just become one of the first two recipients of the Fields Medal.

6 Wallman and Dowker were both topologists who received their PhDs under Solomon Lefschetz at Princeton in the late 1930s. Wallman later wrote *Dimension Theory* with Witold Hurewicz (1941). Dowker was assistant to John von Neumann at the Institute for Advanced Study, worked at the MIT Radiation Laboratory, then was appointed associate professor at Tufts. During the McCarthy era, however, he left for Birkbeck College in London, where he eventually held a chair in Applied Mathematics until his retirement in 1970. He was the author of *Lectures on Sheaf Theory* (1956).

Well, you see, Hardy came to Princeton for the 200th anniversary. They went on boats, they didn't come on airplanes in those days, so you didn't stay for just a week; you stayed for a few months. He made his speech, he was a little bored, there was nobody to talk to in mathematics who knew his work except my husband, who'd spent a year in England, so he knew his work very well. So he came to Cambridge, and Vannevar Bush[7] took him around, and every time he showed him a new thing, Hardy said, "What a marvelous theological institution this is!"

Finally [Bush] said, "It's not a theological institution!"

And Hardy said, "Then why don't you hire Levinson?"

Hardy was an atheist, and he loved to make jokes about God anyway, so he enjoyed this. So Norman came in as an instructor and took over Jesse Douglas's classes.

And you were married about a year after that?

Yes, that's right. When Norman first got his job at MIT, he decided that he would like to come to New York for his first Christmas vacation, so he stayed with some friend's family, and he took me out. We were both Communists.[8] I joined when I was a graduate student—but actually my father was a Communist. You know, these Russian Jews were very proud of the fact that the corrupt Czarist government was overthrown, and at least in the earlier Communist days, they were tolerant of Jews—Trotsky and so on.

It was like a religion: you don't marry outside the religion, or even socialize. It was a very sectarian, dogmatic sort of thing. So one day we went to a Communist meeting, the next day we went to a Communist play, etc., etc.

---

7 Bush was a vice-president of MIT and dean of the School of Engineering at the time.

8 "[Norman] came into the job market at the height of the great depression when thirteen million people were unemployed. Anti-Semitism was widespread...The over-all misery of unemployment and anti-Semitism were disconcerting. The new government of the Soviet Union claimed to have eliminated unemployment. The American Communist Party had as its platform the recognition of the Soviet Union, the recognition of the labor unions and fighting to combat anti-Semitism and discrimination against blacks. This program appealed to Norman and he joined the Communist Party. When the excesses of Stalinism became known, Norman and other intellectuals left the Communist Party." (Levinson, Z. (1998). Remarks by Zipporah (Fagi) A. Levinson. In J. A. Nohel & D. H. Sattinger (Eds.), *Selected papers of Norman Levinson* (2 Vols.) (pp. xxix–xxxi). Boston: Birkhäuser.) Much more information on HUAC testimony, 1953—its effect on schools, lawyer Staurt Rand who helped Levinson keep his job, etc.—is available in Remarks, xxxi.

Norman and Fagi Levinson in Copenhagen, 1949.

By the time the week was over and he was going back, he said to me, "How would you like to come to Cambridge and see a football game?" That was the standard way of inviting a young woman.

I said, "Well, I'm not crazy about football."

He said, "Neither am I. Let's get married."

So we did. We married the next week, and I tell you, it scared the hell out of his parents, and it was quite a sensation in the mathematical community here. On the morning of the wedding, a special delivery package came to me, signed by all the mathematicians of the Thursday Forum, wishing me good luck. I still have the book.

Norbert Wiener

What was your husband's relationship with Wiener like?

Norbert Wiener was his mathematical father, and he was very, very devoted to Norman. For example, Wiener was very much impressed with Norman's parents, that they were these simple, uneducated Jewish people who managed to send a son to MIT and a daughter to Radcliffe. When my husband was going to England for a year, they were scared silly, because he was going to go across on a boat, and they remembered their coming across in steerage: it was filthy, it was diseased, lousy food. So Wiener went to their house in Revere—you know, the slums—and he told them all about Cambridge, about the meals and the High Table, Hardy's love of sports, all kinds of things like that.

Now, by the time my husband finished writing his book,[9] we were married and expecting our first child. The math department had one secretary at the time, so he brought the manuscript into her, and she said she was too busy, she couldn't type it for him: he would have to hire a typist, and it would cost three hundred bucks. It was a lot of money for us.

So that afternoon my husband and I were sitting in our tiny apartment, figuring out how we would raise the three hundred bucks. In walks Wiener with his Royal typewriter: they were very junky little portable typewriters in those days, but it had a few mathematical symbols on the keyboard. He plunked it down on my kitchen table, took off the cover, and showed us how, by back spacing and half spacing, you could make more mathematical symbols. Then he said to me, "Fagi, take the typewriter and type the book!" I told him I didn't know how to type—in those days, people didn't type except secretaries. So he said, "Don't worry, go to the Coop.[10] They have a Royal typing manual; it has pictures of the keyboard, and graded exercises." And he half sang, "'a-s-d-f-g-f, semi-l-k-j-h-j.'" He said, "Practice with the book for a couple of days; then type the manuscript." And that's what I did. I finished the typing a week before the baby was born.

Since there was only this one secretary in the math department, every time Wiener wrote a new paper—and he was very prolific—he would sit down at this typewriter and type it out—he was a hunt-and-peck typist—and send it off to a journal, fast, and that was much better than waiting for the secretary to get around to it. But he gave me the typewriter, and I had it for four, five months. He never once said, "Hurry up, give it

---

9 *Gap and Density Theorems,* which Levinson published in the American Mathematical Society's Colloquium Publication Series in 1940, at the age of 28. Being invited to write a book in this series was a rare honor for a young mathematician.

10 The Harvard Cooperative Society.

Norbert Wiener lecturing.

back." He was glad; he felt Norman was his baby, and he had to nurture him, to take care of him.[11]

Wiener had this manic-depressive pattern. When he was high he would go all over the school and tell people his latest theorem. I remember once he came to my house. Norman was away, I was feeding my kids; he couldn't wait, he's telling me his latest theorem. He knew perfectly well I didn't understand a word of mathematics, but he told me anyway. And when he was depressed, he would sit, and the tears would come down his face, and he'd say, "I'm going to *kill* myself," and he would do that [gestures across throat]. You know, he had this family history of schizophrenia. His brother Fritz was schizophrenic, and then their sister Constance's son became schizophrenic.

[11] This was not the first time Wiener had put his typewriter at his student's disposal. Ted Martin quoted Levinson on the subject as follows: "As soon as I displayed a slight comprehension of what he was doing, he handed me the manuscript of Paley-Wiener for revision. I found a gap in a proof and proved a lemma to set it right. Wiener there upon sat down at his typewriter, typed my lemma, affixed my name and sent it off to a journal. A prominent professor does not often act as secretary for a young student. He convinced me to change my course from electrical engineering to mathematics." (Martin, W. T. (1998). Remarks by William T Martin. In J. A. Nohel & D. H. Sattinger (Eds.), *Selected papers of Norman Levinson* (2 Vols.) (p. xxxii). Boston: Birkhäuser.)

And this was a deep, dark secret. In those days you didn't talk about these things. I mean, when Mrs. Leo Wiener went to visit Fritz in the institution, she disappeared for the day, and then she appeared back. It was a no-no; nobody talked about it.

Wiener would build up to a certain point of tension, but then he would sleep, and it would drop off. He was a sort of narcoleptic. He was famous; he'd sit in a classroom or lecture room, and he'd be snoring in the front row seat, and then he'd wake up, but somehow he would ask the right questions of the speaker, as if he had heard the whole thing through his sleep.

Wiener was very, very childish in many ways, very egocentric, but I think he was a very kind man. His kindness just rubbed off; everybody at MIT felt that his kindness made MIT a much, much nicer school to students than, say, Harvard or other schools. Norman was very loyal to him. At one point Wiener asked my husband if he would tutor one of [Wiener's daughters] to get her through some exams. Norman never did things like that, but of course he wouldn't refuse, and he did tutor her and got her through.

My father, Lee Segel, who was also a mathematician,[12] told me a story about Wiener that was corroborated by your brother, Henry Wallman. Here it is, in Dad's words: "Wiener and Levinson and Wallman were going to a meeting, and Wiener was driving, and they stopped for gas. And Wiener started talking with the kid that was pumping the gas. He said, 'We're all going to a meeting of the American Math Society in Charlottesville, Virginia, where the University of Virginia is, and these two young guys here are giving fifteen-minute contributed talks'—maybe it was twenty minutes, I don't remember—'and anybody can give a contributed talk who's a member of the American Math Society, and anybody can join the American Math Society if you just pay the dues. However, I am giving a one-hour invited talk, and people who are invited to give a talk of one hour, they are among the leading American mathematicians of the time.' And maybe he said what he was going to talk about. By this time the fueling was over, and Wiener paid this kid, and as he drove off, said to the other guys, 'I love to keep in touch with the common man.'"

[Laughs uproariously.] Oh, that's wonderful. I remember Wiener's mother, Bertha, said to Wiener, "I know President Pusey!"—the president at the time at Harvard. So Wiener answered, "But does President Pusey know *me*?"

12 Lee A. Segel received his PhD under C. C. Lin at MIT in 1959, and the two later wrote *Mathematics Applied to Deterministic Problems in the Natural Sciences* (1974). He went on to hold professorships at RPI and the Weizmann Institute of Science in Rehovot, Israel, switching from fluid and continuum mechanics to mathematical biology.

Life with a Mathematician

Do you have any sense of how Professor Levinson chose problems to work on, what appealed to him in a problem?

Well, he changed fields many, many times, and he was able to grasp problems in one field, and then in another, and then another. He liked hard problems and could work very hard at them.

My husband liked to get home from work and work in his study. Sometimes he would be so busy working, he didn't have time to get dressed; he would wear a bathrobe. You know how people would have a patch on the elbow? His bathrobe would wear out on the seat, because he'd sit on that damn chair in the study so long. When I'd see the bathrobe getting that way, that was when I would get him a new bathrobe.

He never went shopping for clothing and was never in the grocery store—although when he was a kid, he worked in a grocery store in Revere after school. I would buy him the clothes, the underwear, whatever was necessary. The only thing he really had to buy was shoes, because he had to try them on, and that was always a big problem; I had to trick him into getting shoes. It was a pain in the neck.

When the door of his study was closed, *nobody,* including the children—nobody opened that door. Once one of the children, the older one—she was a three- or four-year-old kid, a young kid—figured out what infinity was. She says, "No matter how you do it, there's always one more, there's always one more." And I got so excited I opened the door and I told him, and he was pleased. But in general, I didn't open the door.

My husband knew how to delegate work very, very well. So he would make me read proof, do editing of manuscripts, and go to the library for certain things that he didn't want to check up himself. I was always proofreading, typing mathematical manuscripts—he didn't want a graduate student to do it; they might make mistakes.

Was that satisfying to you, though? Did you ever want to pursue your own career?

Well, this was an era when women weren't emancipated. I felt that I had to take care of the children and run the house and make the meals and wash the clothes, and if I had spare time to do these things to help out my husband, why not?

I was very careful about money. I'm still careful about money. I always felt that I had money for the luxuries because I was so careful with the necessities. For example, I never took a taxi, unless it was a situation where I

Left to right: Dick Schafer, George Whitehead, Kay Whitehead, Barbara Brauer, George Brauer, and Alice Schafer.

couldn't drive or had to get somewhere fast, but Norman *always* took a taxi, because he didn't want to waste time, to get away from mathematics too long, and a taxi was a simple solution.

We were living in Concord, and I had to pay bills for dentistry, for piano lessons, all the things that involve children, and even my husband one day noticed that the kids weren't well dressed, compared to the other children. So I figured, "My God, if *he* notices, it must really be bad." Well, I was talking to a neighbor; she said there was an ad in the local paper for a converted Singer sewing machine, made electrified, for five bucks. So I bought that. Then the high school had a night school where they have little classes and things for adults, and one of them was sewing, for another five bucks. I took that course for eight weeks, and after that I was in business. I found out that I could go to the Woolworth's, and for a dollar a yard, I could get the most beautiful materials! For two yards, yard and a half, I could make my kids a most marvelous dress. So I started making my kids their clothes, my mother-in-law, and then even my own. Only in recent years did I start buying a coat occasionally, but I used to make everything.

A Beautiful Mind, Sylvia Nasar's book about John Nash, quotes you as follows: "These mathematicians are very exclusive. They occupy a very high terrain, from which they look down on everyone else. That makes their relationships with women quite problematic." [13]

Well, mathematicians certainly stick with mathematicians, because very few people understand them other than other mathematicians. But it's a very exclusive little club. They all got married, and they all believed in families and sex and all that, but when there was a mathematical party, the women sat at one side, and the men were at the other and discussed mathematics. The women were excluded.

The usual thing is there were these Thursday colloquiums at MIT or Harvard, and even the mathematicians from Brown would come. They would invite some speaker to give a mathematical talk. And afterward there was a dinner and a party in somebody's house. If it was in one guy's field, he would make the party. So that's how they socialized. They couldn't socialize with people who'd talk about tomatoes or clothes or something; they had to talk to people who understood the values they had, which were mathematical values. They're attracted to mathematics like a drug addict is attracted to drugs. They can't stay away from it.

Professor Levinson and His Students

I asked Fred Brauer [14] *what it was like to have Professor Levinson as his supervisor. Here's what he said:*

> He would start by suggesting a problem and some things to read. And then I would work on it and get stuck. He always taught Monday, Wednesday, Friday from 10 to 11. So at 11, there would actually be a line, because he had several students working with him, people waiting to ask him something, and when it was my turn I would go to the office and explain what I had done and where I was stuck. He would take another bite of his sandwich, suggest something that was obviously ridiculous—couldn't possibly work. And then I would go on and try what *I* thought was reasonable, and after a week or two I would get over the difficulty, and then afterwards I would look back and realize

---

[13] Nasar, S. (2001). *A beautiful mind.* New York: Touchstone, p. 172.

[14] Professor Fred Brauer received his PhD in 1956. He spent more than 35 years at the University of Minnesota and today is Professor Emeritus at the University of British Columbia.

> that, in a sense, what I had done was what he had suggested. I hadn't quite understood it that way. You really had to work at it until you saw it. That kept repeating.

Yes, he had about 35 students and he was very good to them, very patient with them and very kind. He wrote a book[15] with Redheffer, who had been his student. Norman treated these [younger men] as equals; he gave them full credit. When the publisher said he wanted to make the name [of *Theory of Ordinary Differential Equations*] Levinson and Coddington, he said, "No, have it alphabetical."[16]

Only one or two of Professor Levinson's thirty-odd students were women.[17] Can you say more about his attitude to women mathematicians?

Well, he had a few women students; Violet Haas was one of them. He was very kind to them. The women were allowed to be TAs but weren't allowed to teach, but he finally changed that, and of course they did very well. And the men had offices; the women didn't have offices. Finally the school found room. They pushed the janitor—who had a room with his brooms and mops and stuff like that—to one side of the room, *squashed* three desks in, and the women were so used to being treated badly that it never occurred to them to complain about having to share a room with the mops and the brooms.[18]

The McCarthy Years

Stanislaw Ulam said, "In many cases, mathematics is an escape from reality. The mathematician finds his own monastic niche and happiness in pursuits that are disconnected

---

15 Levinson, N. & Redheffer, R. M. (1970). *Complex variables.* Oakland, CA: Holden Day. Raymond M. Redheffer received his PhD in 1948.

16 Levinson worked with Earl Coddington on their book in the early fifties, when the junior mathematician was a C. L. E. Moore Instructor at MIT, circa 1951.

17 "At that time, there were three female graduate students in the department. The male graduate students received a stipend as teaching assistants, teaching freshmen. The women were not deemed qualified. Based on their records, Norman decided that the women were equally qualified and should be allowed to teach. This equal treatment of the women was unprecedented but the women proved more than equal to the task, and no freshman complained." (Levinson, Z. (1998), xxx.)

18 The site http://alumweb.mit.edu/groups/amita/esr.html is a website devoted to "125 Years of Women at MIT."

from external affairs." Yet one external affair, McCarthyism, affected not only your husband's work, but your family life as well. Professor Levinson was called to testify before the House Un-American Activities Committee (HUAC). Did you ever see the transcript of your husband's testimony?

Oh, yes. That was issued to the faculty, the whole MIT community. There was this thing called "guilt by association," so MIT felt very vulnerable. There were these three former Communists: my husband; [Ted] Martin, who was in the math department, and a man in the chemistry department, [Isadore] Amdur.[19] So they felt very, very vulnerable. The administration spent the whole time before these hearings consulting with lawyers, and the faculty debated and debated how best to remove this stain on its integrity, its patriotism. Guilt by association. So they all felt that the best thing would be if the three would be cooperative witnesses, that is, a witness who would name names. Two of them said yes; Norman said no. Most of the people—his own father—thought he was a damn fool for not being a cooperative witness. But Wiener understood this. He said Norman had to make a decision he could live with for the rest of his life.

In most cases people were fired. MIT kept them on. Compton was the president at the time,[20] and he was a very decent man, and he wanted these three people to have as good a chance as possible to get out of this. MIT got [Stuart Rand of Boston law firm] Choate, Hall and Stewart to defend these three people in Washington. The previous people who had been subpoenaed came with these junky little Communist or radical lawyers. This was the first time that a lawyer from a prestigious company came. So the people on the House Un-American Activities Committee in Washington were very, very pleased, and they were kowtowing to this Stuart Rand.

---

19 Another MIT mathematician who paid a price for his politics in the McCarthy era was Dirk Struik. "Dirk Struik was a member of the MIT faculty from 1928–1960, a highly respected analyst and geometer and an internationally acclaimed historian of mathematics, beginning with his book, *Concise History of Mathematics*, in 1948. Professor Struik maintained a life-long interest in social justice and Marxist Socialism, which, during the McCarthy period in 1951, led to an indictment 'to overthrow the governments of the United States and Massachusetts.' The indictment was dropped five years later, during which period Professor Struik was on paid leave from MIT; he was reinstated to the faculty by President James R. Killian, Jr. in 1956." See. http://web.mit.edu/annualreports/pres01/16.05.html, the Department of Mathematics' annual report to the president for 2000–2001.

20 Karl Taylor Compton (1887–1954) was president of MIT from July 1, 1930–October 15, 1948, but chairman of the Corporation at the time of the HUAC testimonies.

Norman said he wouldn't name names. So Rand had the other two fellows talk first. And they *did* name names. And then he said to Norman, "Why don't I ask the Committee to just ask you questions about yourself and not to name names?" And Norman did that. He told them quite frankly all the things he did in the Communist Party, but they didn't ask him to name names.

And as a matter of fact, one of the things they asked my husband—it turns out that there were three brothers at MIT, the sons of Earl Browder, the head of the American Communist Party.[21] And the older son, Felix Browder, was a mathematician, and he couldn't get a job because of this. They didn't want any trouble, you see, with the authorities. My husband told the Committee that it was wrong for them to condemn a fellow, that you don't blame the child for the sins of the father. And they retracted, and the result was Felix was able to get a job after that. He had two brothers; they were also mathematicians, and after Felix got a job, finally, they were able to get jobs. And when their father retired, what do you think he did? When he visited his sons, they would teach him mathematics!

Felix Browder had a hard time; he was put in the army, and because of his father he was put in a camp with criminals, and he not only had no affinity to these guys, but he had no privacy. He wrote my husband a letter once, and he said he's going crazy there; he can't stand it, the absence of privacy, and the absence of mathematics, and so on.

My husband spoke to Compton, and Compton, who'd been going to Washington as a sort of science advisor to the government and knew some of the generals, spoke to the general who was in charge of that camp, and as a result Felix was given a job pumping the gasoline for the tanks in the morning. There was a little office there where he pumped it, and when he was through with the pumping, he had the rest of the day to himself, and he could do some of his mathematics. Not only that, the general told him anytime he wanted to go to a mathematics meeting, that's okay. So Compton was a very, very nice man.

You must have been worried about your husband's testimony.

Well, the temper of the times was such that after I drove him to the airport to go to Washington, I didn't go home; I went to the superintendent of schools in Concord, where we lived, and I told the superintendent, "Tomorrow morning, in the newspapers, there's going to be this whole thing about my

---

21 Earl R. Browder was twice the presidential candidate for the U.S. Communist Party.

husband." I said, "You can think anything you want about me and my husband, but is there some way you can protect my children? I don't want them beaten up because of this."

This guy was a very sweet guy. Tears welled in his eyes, and he immediately called the teachers in for a meeting and told them what would happen, and he said, "You be careful. Protect their children. Protect the children." And they did. So it was a very, very difficult time. We lived in this community in Concord called Conantum; it was a community of over a hundred families, mostly MIT and Harvard, and some engineers, graduates from MIT. After the war, the housing situation was bad, and there was an economist at MIT named Rupert McLaurin who got the money to buy this land and an architect to design the houses. It was a very nice community, and I found also that during this whole thing, they were also very supportive of us, my children and me. Everything was very anti-Communist, but in the case of my neighbors, who knew me and knew my husband, they could assume that we had a certain amount of sincerity and integrity.

I remember when my husband was first subpoenaed by the House Un-American Activities Committee; my aunt came up from New York to be with me, and I took her around my neighborhood. She said, "What are you worried about? They must all be Communists or Communist sympathizers." This was her attitude, that this was an intellectual movement, that the more thinking people were Communists, or Socialists.

Was there a certain point at which you said, "We're not of the Communist Party any more"?

Oh, yes! The Roosevelt New Deal made a big difference. One of the attractions of the Communists was that there were 13 million unemployed, and the government wasn't doing anything about it. Then Roosevelt started doing things.

[At first] I didn't believe a lot of things that were reported back from the Soviet Union. I figured, "They're capitalists. Who's going to believe them?" I was really infused with that religion. But certainly with the Russo-German pact and all these excesses of Stalinism, these trials, persecution of Jews, everything like that, the intellectuals dropped out of the Communist Party.

The Communist Party had a brief rise and fall. It fell very fast. I tell you, over the years I have often thought about what a damn fool I was, what poor judgment it was to believe in Communism, and this haunted me for many, many years.

This so-called paradise was far from a paradise [laughs]. When we finally went to the Soviet Union for the International Congress of Mathematics, this was a very, very wonderful thing for the Soviet mathematicians, because they weren't allowed to travel out of the country the way the American ones were, to go to meetings all over and speak with other mathematicians and hear mathematical papers. The Russian ones, they were deprived of this. On the other hand, the spying system: the hotel, everything was bugged. The only reason they felt free to talk to you would be to talk to you in the street, where there was no bugging. They wouldn't want to talk on the phone except, "Let's have lunch together," or something like that, but never anything more than that.

Life in the Department

Professor Levinson received his entire university education, from age 17 onward, at MIT, and seems to have been the central figure in the department in his time. Ted Martin said that "during the time I was head of the department, I consulted with Norman on every major move." [22]

He did, yes. Norman was the brain that he needed for advice that was solid, about hiring people. He helped build up the math department to a very respectable research department.

On the other hand, A Beautiful Mind *referred to you as "Fagi Levinson, the department's den mother."* [23] *How did you come to fill that position in people's minds, to become the person the mathematicians turned to?*

Well, when Norman and I were married, as I told you, we had this one-week courtship. So he had been sharing an apartment with Ted in Cambridge. When we got married, Ted moved out, obviously; it was a small apartment. So they knew each other for a long, long time. And when Ted was courting Lucy, his wife, who came from Northampton, she would stay with us.

Ted Martin was from Arkansas. His father was a country doctor. He remembered sitting in his father's lap and driving their car as they went to visit the patients. If Ted entertained people, he would worry about how much the food and stuff would cost. I didn't give a damn if one more per-

---

22 Martin, W. T. (1998), xxxiv.

23 Nasar, S. (2001), 144.

son, two more people, a dozen people came to dinner. What the hell, you know? So if I couldn't afford a steak, we would have spaghetti, but there was enough for everybody. And I got to know their problems, and it was obvious that some things needed a solution that was easy, and I would suggest it, or I would help them. When Alicia Nash had to go to the hospital [to have a baby], I took her to the hospital. She didn't know how to drive. And the same thing's true with other people; I did what was necessary, as I saw. And so did Norman.

Norman took care of the important things as he saw it, and saw to it that I took care of the things that had to be done that he didn't want to do. I remember Nash would say to him, "How do I buy this?" or "How do I get that?" He'd say, "Ask my wife; she'll take care of it." And this continued even when Nash was in the hospital. Once he called me up—there was a pay phone in the ward there at McLean Hospital—and he said, "Come over right away, I have to have this and this." And I said, "Nash, I can't come now; I have to feed my kids, give 'em lunch," and I said, "It's hot as hell," it was July or something, "I can't come until I take a shower and get into some clean clothes; it'll take me two hours to come." And the reason I remember this is because, as nutty as he was, on some levels he was rational. He understood that I had to feed my kids, take a shower.

As you said, Norbert Wiener would come over when he was feeling high or suicidal. John Nash called from McLean's to tell you he needed a shower.[24] *When Alicia Nash was going into labor, she called you. What was it that motivated you to do all this?*

It was my duty! Just as I would feed my kids, I would dress my kids, I would buy my clothes—you know, this was part of life; you had to do these things.

Did the "Nash problem" really cast a shadow over your lives and the department's, or was it just something that flared up occasionally but wasn't really all that disruptive?

Well, Nash, we always hovered over him. I remember once that there was a mathmatician in Michigan, I think, who decided *he* had the best solu-

24 "Nash used the pay telephone in the lounge [of McLean's Belnap One building, the day he was committed]. He did not call a lawyer, but rang Fagi Levinson instead. 'John wanted to know how he could get out of there,' she said. 'He said he wanted a shower. "I stink," he said.'" Nasar, S. (2001), 255.

tion.[25] And his idea was to have Nash go to this very good mental hospital in Michigan, and the hospital would give him a little job doing statistics there, and they'd pay him a little something for it so he wouldn't feel that he was completely cut off from mathematics while he was getting treatment. When they told this to Nash, he said, "I don't want to do that; I'm not interested in statistics." Then another time, when he was on medication and in reasonably good health, Brandeis invited him for a year, no duties but to take part in the seminars and the mathematical meetings with the faculty, away from the tensions of MIT, and Nash liked that very much. The only trouble was, Brandeis didn't have any money to pay him. So my husband, who was still hovering over him, put Nash on his ONR contract—Office of Naval Research contract—so Nash had money. And then it was a question of where's he going to live? So I found out that Joe Kohn, one of the mathematicians at Brandeis, was going on sabbatical for a year. I said, "Joe, would you sublet your apartment to Nash?" He said, "Sure!" Everybody had this feeling that they wanted to help Nash, and this is the way it went.

When he was at McLean Hospital, they felt that the best for Nash would be if all the mathematicians came and visited him. I was the one who had to engineer this visiting, and I said, "My God, they *hate* him, such a egocentric guy"—he was obnoxious, and nobody liked him. But I was wrong. They didn't, but they knew he'd done wonderful mathematics, and if by coming to visit him they could hasten his recovery so he could do more mathematics, of *course* they'd come. They all came. Even the ones at Brown came. Nobody liked him; he was too egocentric, too secretive, too selfish, but he did wonderful mathematics, and that was the important thing, the *sine qua non.*

You've talked about Norbert Wiener, John Nash, Ted Martin. Who were some of the other colorful people in the department?

Well, [Paul] Erdős wasn't in the department, but he was always coming, and he never had a home; he only had a suitcase and would come and stay with this one or that one. Erdős was very peculiar in the way he lived, in the way he would talk about children as little epsilons, but he was very sweet, very humanitarian, very kind. [Gian-Carlo] Rota was a very good expositor [of mathematics] and he was very likable. [Warren] Ambrose liked jazz, Charlie

---

[25] A number of people were involved in this plan, including Robert Winters, "a Harvard-trained economist who [was] the business manager of the [Princeton] physics department at the time," mathematicians Donald Spencer and Albert Tucker, James Miller of the University of Michigan psychiatry department, and others. Nasar, S. (2001), 293, 303.

H. B. Phillips.

Parker or something, and he would take people to hear this sort of music. He got involved with some sort of Latin American group and got beaten up because he opposed the government.[26]

The chairman of the math department in the early days was a man named H. B. Phillips.[27] Phillips came from South Carolina. Very backward area: his attitude was, "You never call a doctor until you're ready to call the undertaker." So his wife died of something or other because he didn't call the doctor. He told me this himself, once when *I* got sick. He said, "You sure you had the doctor? You know, I was foolish."

---

[26] Warren Ambrose's MIT obituary reports that he "was known for his commitment to political and social causes, particularly in Argentina and Chile in South America when those countries were ruled by military regimes.

"He came to international attention in the summer of 1966 while a visiting teacher at the University of Buenos Aires in Argentina, when he was severely beaten along with other faculty members and students by military police. This occurred shortly after a military regime, which had taken over the government in a bloodless coup, had ordered a government takeover of public universities." Ambrose died on December 4, 1995, at the age of 81. See http://web.mit.edu/newsoffice/tt/1996/jan10/41416html for further details.

[27] Henry B. Phillips was head of the department from 1935–1947.

Well, anyway, he had a child by this wife, and he believed that [his daughter] had no right to get an education, that women don't need an education. So she secretly went to study to become a nurse. When he found out, he was just furious. But nevertheless, when he married the second time, he married a woman who was a PhD in physics. He had two kids by this second wife, and of course, they had a very good education.

And he—very strange man—he bought a piece of land at the top of a mountain in Lincoln, Massachusetts, and to get up to the house you had to make three hairpin turns up the road to the house. The mailbox was at the bottom of the road, of course; the groceries that were delivered—anything was down there. He had a Jeep, and that was the only way he could get up the mountain. I remember they had a huge garden, he had fruit trees and just any vegetable you could think of, in *vast* quantities, and he and his wife worked on it very, very hard; they froze stuff, they canned stuff. The only things they bought were flour and milk products, meat, but everything else—she baked her own bread. They were health nuts.

A Mind That Didn't Grow Old

Your husband was preoccupied with his health as well, wasn't he?

Oh, *terrible*, terrible! Norman was a hypochondriac. He was always complaining about one thing or another. When he was a kid, his mother was very poor, you see, and she took him to the clinic at the Mass General [Hospital], and some intern there said he had rheumatic fever. So he never learned to throw a ball, and for his whole life he never played any sort of sports.

He was a nut about medical things, and he was always interested in following up certain things, so he used to make me go to this medical library along the Fenway—this was the days before Xerox machines and stuff, so I would have to get these articles and make briefs on papers and bring them all back. And boy-oh-boy, it had to be clear, because he was surveying the thing afterward and would ask me many questions about this medical research or report.

He invented a cholesterol-free diet long before it was ever publicized. He was a pain in the neck about his food, so much so that when we were going to dinner, to people's houses, they would call me up in advance and say, "What will he eat for dinner? Will he eat lobster, or chicken, or beef?" And I would have to tell them, no, he could eat that, but he couldn't eat this.

I used to make yogurt. This was before you could buy fat-free yogurt: I used to buy fifty-pound vats of skim-milk powder, because then he was sure it was not whole milk, and make the yogurt. The way you do it was get some sort of infusion, a culture, once a month, that came flown in from California, and make the yogurt in sterilized bottles, put them in the oven between 105 and 115 degrees—this was an electric oven; you couldn't do this with a gas oven—and after two hours it would thicken up, you'd take it out and refrigerate it, and that would be a week's supply. And then for the next week, take part of the last bottle, this was part of the infusion still, and repeat the process, and this went on all month until the infusion was too weak, and you'd get a new thing flown in. So I had all this nonsense about food.

A Beautiful Mind said your husband suffered "wide mood swings, long manic periods of intense creative activity followed by months, sometimes years, of depression in which nothing interested him."[28]

Yes, yes. When he would get depressed he wouldn't do mathematics. That's right. He could still give his lectures; he knew mathematics, he knew the faculty. These things had nothing to do with being high or low. Those were intermediate things. The highs or lows had to do with mathematics, but not human relations, family relations, stuff like that, you see. He loved his children, he loved his family, but he wasn't interested in doing mathematics. And this bothered him, because this was the milk that mothers live on, for him. It would last a long time. But he would come out of it.

He tried to contain it a lot. You know, he died of a brain tumor, and for years he had these terrible, terrible headaches, and the only thing I knew about the headaches is that he would eat up bottle after bottle after bottle of aspirin.

He was an Institute Professor towards the end. I remember at his memorial service, there were tremendous tributes to him. This was not fake stuff, the usual baloney, but heartfelt stuff from people who valued his assistance, valued his mental ability. Many young mathematicians dry up soon, and they're no longer creative. *Thirty* was a dirty word. John Nash always felt this way, and got very nervous as he approached thirty. But my husband was very lucky in this regard: he wrote his most brilliant work on the Riemann zeta function when he was in his sixties, shortly before he died. He was very disciplined and very bright, and his mind didn't go old in mathematics.

~ January 18, 2005

---

[28] Nasar, S. (2001), 137.

Isadore M. Singer

Beginnings

J.S.: *How did you first get into math?*

I.M.S.: I was interested in science when I was in high school and won a scholarship to the University of Michigan. Freshman year at Michigan was not very challenging, disappointing and routine, until, in calculus, our instructor explained how Newton derived the elliptic orbits of planets from his inverse-square law. I felt something special had happened, something intellectual and wonderful. I was at the right place! I still feel this way and tell many laymen, "Just think how remarkable the human mind is; you can scribble on a sheet of paper and compute the orbits of planets!" It's incredible, really—*awesome* is the right word. I was thrilled by that derivation; it reinforced my decision to go into science.

I entered the University of Michigan in September 1941, just before the Second World War began for the United States. I majored in physics and got my degree in January 1944 before entering the Signal Corps.

Two physics courses were especially interesting, but also puzzling—quantum mechanics and relativity. Though I could do the problems, I felt that I didn't really understand these subjects. I wanted to take some math books with me in the army that would give me a better background. Luckily, the University of Chicago had two correspondence courses that suited me—one in differential geometry and one in algebra—which I completed when I was stationed in the Philippines.

And you did your graduate work at Chicago.

Yes. The textbooks [from the correspondence courses] were both from the University of Chicago, so I thought it a good idea to go to the University of

Is Singer lecturing.

Chicago. I applied to the math department, telling them I wanted to spend a year learning more mathematics and then go back to physics. Later, Paul Halmos, a professor there, told me they almost didn't take me because of that sentence! [Laughs.] Paul was very kind to graduate students, particularly to me. It wasn't long before we became close friends.

Luckily they did admit me, and in the first year I sat in on some physics courses, particularly quantum mechanics. Again I found the course unsatisfactory from my point of view, which by then meant a mathematical point of view. On the other hand, the math courses were very beautiful, and I was much taken by them. So I didn't move into physics; I decided to continue as a math grad student in Chicago, where I got my degree.[1]

Can you give a sense of what you mean when you say the math was very beautiful?

Yes. [Pause.] That's a very good question, and it is difficult to answer. To some extent, it is like explaining why music is beautiful to someone who is hard of hearing.

Mathematics has a logical structure to it, a coherence to it, that I find very appealing. That's what I mean by "beautiful." To be more specific: in Galois theory, for example, groups—symmetry groups—allow you to answer ques-

[1] Singer received his PhD on "Lie Algebras of Unbounded Operators" in 1950 at the University of Chicago.

Irving Segal.

tions you wouldn't be able to otherwise. Group symmetry is an example of elegance in physics as well as mathematics. By the way, Newton's inverse-square law can be derived from symmetry considerations.

At Chicago, you did your PhD with Irving Segal, who later joined the faculty at MIT.[2] *What was he like to work with, and what were the important things you learned from him?*

He was a first-class analyst, interested in differential equations and physics. He taught us functional analysis. He was a terrible lecturer, but a wonderful teacher in this respect: he was always doing advanced stuff, which we students loved; but he would get stuck halfway through his lecture. He would spend the rest of the lecture trying to complete a proof. I would go home and try and figure out how to get unstuck, and I would come to class very interested in what Segal would do. Either I had solved the problem, and I could compare my solution to his; or I couldn't solve the problem, and I would see what he did. The way he ran the course made it good training to be a research mathematician. I learned a great deal.

[2] Segal joined the MIT Mathematics Department in 1960 and retired in 1989.

The other nice thing about Segal was that he was always available, even at midnight. His door was always open; you could always walk in and talk.

And working with him one-on-one, as you later did? How did his abilities or tendencies translate into a closer relationship?

I didn't work with him one-on-one; I wrote a dissertation under him. I would just come in, show him what I could do, and sometimes where I was stuck. His door being open was a general invitation to talk about mathematics.

Building a Great Department

I gather that Chicago was the preeminent math department in its early years. It got a lot of funding, and it was never regarded as a service department: all the faculty were expected to do research. Then things kind of stagnated for a while, but Marshall Stone was in the process of rebuilding the department when you were there. Is it possible to compare the atmosphere and how things were done at Chicago and MIT? How do you build a great department?

There was a real difference, and that gives me an opportunity to talk about our department. Marshall Stone managed to induce some of the best mathematicians in the world to come to Chicago: for example, Andre Weil, Saunders Mac Lane, and S. S. Chern. Some younger members in the department were Paul Halmos and Ed Spanier. All of them were at the forefront of their research fields.

Undergraduate teaching was separate. I think it still is. Chicago was a graduate department with its focus on research. Faculty taught the basic grad courses in order to bring students to the cutting edge of research; they also gave courses in their own fields. We graduate students thrived in that exciting environment.

MIT is different. I first came here in 1950, as a Moore Instructor. Ted Martin was chairman, and he, Norman Levinson, and MIT had the aim of building mathematics into a first-rate department. They succeeded. The structure of our department is different than Chicago in that we have the serious responsibility of teaching mathematics to undergraduates whose interests are largely science and technology. In fact, I think the MIT math department is unique; it carries out its teaching obligations with real aplomb and enthusiasm and at the same time is ranked first in the country for research.[3]

---

[3] *U. S. News and World Report.* (2007). America's best colleges.

How that came about is very interesting. Ted persuaded the university to have special instructors called Moore Instructors. They were originally sponsored by either the Air Force or Office of Naval Research, "sponsored" meaning that instructors didn't have full-time teaching. We taught half or two-thirds time; the rest of our time was devoted to research.

The Instructorship program started in 1948. Felix Browder and F. I. Mautner were the first two instructors; Earl Coddington and Tom Apostol came in 1949. I began in 1950 and John Nash in 1951. The programs expanded with time. In 2006 we expect twenty-six Moore instructors.[4]

Over the years, the department has always succeeded in recruiting new PhDs excited about research. These bright young people stay two or three years and help with teaching. They participate in seminars, sometimes *running* seminars. Permanent faculty are mentors, but also learn a great deal *from* them. For the most part, instructors come from major universities, with different skills and different points of view. Their presence has created remarkable *esprit de corps* here.

The math department celebrated the fiftieth anniversary of the Instructorship program in 2000. We invited all former instructors, and many came; it was a very special occasion. When you looked around, you saw major prize winners, former presidents of the American Math Society, and senior mathematicians from around the world. They got their basic post-graduate mentorship here. The instructorship program impacted not only MIT but also the world-wide mathematical community. Wherever I go, someone will remind me of their happy days at MIT as an instructor.

I might add that one of the most important people in our department is Arthur Mattuck. He is a gifted and witty lecturer. For decades students have enjoyed learning calculus from Arthur. In the early 1960s, he took over the calculus program, making one large lecture section with accompanying recitations, revising it from time to time to fit the needs of students in other departments. I was a recitation teacher at the time of the changeover. Since we TAs no longer had anything to do with quizzes or the final exam, it made for a more cooperative effort in learning between student and instructor. A year earlier students felt I was someone who was judging them. That was no longer true; I was there to help them.

Another example of the department's focus on undergraduates is Mike Artin, a world-class algebraist and former president of the American Math

---

[4] In the following year the number rose to thirty.

Society, who was in charge of the undergraduate math program for seven years.

What brought you to MIT in particular? Why not stay in Chicago, for example?

When I got my degree, I could have gone to the Institute for Advanced Study and been von Neumann's assistant. But I had family, and the MIT salary was $500 more a year. Salaries were very low at the time. The difference of $500 meant a lot to us.

What was MIT like when you got here? What were your first impressions?

I'll repeat what I have described elsewhere. My first day at MIT was in early July 1950. We had to teach summer school then. I was staying in Back Bay and walked across the bridge to Building Two and the chairman's office. Ruth Goodwin was the only secretary; I introduced myself to her. A guy sitting opposite her reading the *Globe* lowered his paper and said, "Singer, I'm Ambrose. There is a seminar in Lie groups in five minutes. Let's go."

I said, "But—but I have to meet Professor Martin."

He said, "Oh, that can wait. Do that later! Let's go to the seminar."

George Whitehead.

We went to the seminar. George Whitehead, Barrett O'Neill, and John Moore, among others, were there. After the seminar, Ambrose took me back to Martin's office and said, "We meet for coffee at midnight at Hayes-Bickford. I'll pick you up at 11:30?"

That night we went to Hayes-Bickford, a dismal cafeteria in downtown Boston. There I found the seminar plus Kay Whitehead, George's wife. We talked a lot. And then Ambrose said, "Let's go!" We spent an hour driving around, as Ambrose introduced me to Boston and Cambridge at night. By the time he dropped me off at two in the morning, I felt that MIT was my new home and that Ambrose and I were close friends.

My strongest impression of the department in the 1950s was the enormous number of seminars, as we absorbed as rapidly as we could the remarkable developments that were taking place in mathematics. In topology, cohomology theory was codified. Sheaf theory and spectral sequences needed to be absorbed. George Whitehead explained all this to us. George was very special. He had great geometric insight that guided us through the weighty algebra in cohomology and homotopy theory. I'm delighted but not surprised that we now have one of the best topology groups in the country.

Seminars in Lie groups, like the one I attended my first day, incorporated the global view of Lie groups. Representation theory under Gelfand and his school was just beginning, followed by the seminal work of Harish-Chandra. Bert Kostant was instrumental in making MIT a center of groups and their representations.

The most exciting developments for me were in differential geometry. S. S. Chern had come to the University of Chicago in 1949 and given a course in differential geometry. His exposition emphasized E. Cartan's work, the frame bundle point of view and differential forms. He also explained his remarkable global results: what are now known as the Chern-Gauss-Bonnet theorem and Chern classes.

As a graduate student I had listened to Chern and taken notes. At MIT, at Ambrose's urging, I explained what Chern said. Ambrose and I struggled to make Chern's viewpoint available to other mathematicians. I think we and our students succeeded in that.

We reorganized the geometry curriculum and courses in analysis that were needed for it. We had many enthusiastic students who wrote excellent textbooks based on Ambrose's and my lectures. So MIT became one of the first centers of differential geometry in the United States and this continues to be one of our strengths.

Our first seminar in analysis was on the Hodge theorem and the Weyl lemma, based on the Princeton Kodaira-Spencer notes. It's amazing how many first-rate analysts spent some time at MIT. I'm afraid to list them because I'm sure I'll forget one or two. But I will always remember the day I met Alberto Calderón in the hall. He said, "We've just shown that singular integral operators are invariant under coordinate change," meaning the remarkable thesis of Bob Seeley.[5] I replied, "Manifold theory will never be the same."

Indeed, that turned out to be the case: singular integral operators became a tool in the study of the geometry and topology of manifolds. I didn't know it at the time, but in 1963, using Seeley's work, Atiyah and I extended the index theorem for geometric operators to general elliptic pseudodifferential operators.

I want to emphasize Ambrose's role. He and I worked together and were very close friends; but he was the one who supplied the energy for so many of our activities: having seminars, carrying them out, changing courses. I provided a lot of support, but Ambrose was *the* leader in bringing modern mathematics to the pure mathematics department.

What other figures stand out?

Norbert Wiener, of course. All the books about him bear out that he was quite a character. He was hard to talk to mathematically, because he was always telling you what he was doing and wouldn't listen at all to what you were doing. But he was an enormous influence on MIT. He had an evening seminar for a while in which he talked about communications, computers, cybernetics. Jerry Wiesner and Walter Rosenblith were among the participants and were much taken by Norbert's vision. Jerry was head of the RLE Lab at that time and later became dean, provost, and president of MIT.[6] Walter ultimately became provost.[7]

---

5 Seeley's thesis (under Alberto Calderón) was "Singular Integrals on Compact Manifolds" (1959). He went on to become Professor of Mathematics at Brandeis University (1962–1972), later moving to the University of Massachusetts, Boston (1972–2002).

6 Jerome B. Wiesner was director (1952–1961) of the Research Laboratory of Electronics (RLE), successor to the WWII Radiation Lab. He spent the next three years as chairman of the President's Science Advisory Committee and special assistant to President Kennedy for Science and Technology, returned to MIT in 1964 as the dean of the School of Science, became provost in 1968, and served as president from 1971 until he retired in 1980.

7 Walter Rosenblith came to MIT in 1951. He was chairman of the faculty from 1967 to 1969, associate provost from 1969 to 1971, and provost from 1971 to 1980.

Is Singer receiving the Abel Prize with Sir Michael Atiyah from King Harald V of Norway.

Wiener's view of the role of electrical engineering was pervasive. Electrical engineering changed somewhere in the forties and fifties, from the focus on power to a focus on communication. Now the department is called Electrical Engineering and Computer Science. I have always found that our administration has a great deal of respect for core mathematics. I think that goes back to Wiener's insight and foresight.[8]

Norman Levinson also stands out. He was a world-famous analyst whom the administration listened to when he described the needs of the department. As chair, he ensured that pure and applied mathematics remained one department by giving each considerable autonomy.

---

[8] Wiesner referred to Norbert Wiener's weekly dinner seminars, which began in the spring of 1948 and went on for several years, as "a seminal experience which introduced us to both a world of new ideas and new friends, many of whom became collaborators in later years." (Singer, I. M., Stroock, D. W., Jerison, D., and Wiener, N. (Eds.). (1994). *The legacy of Norbert Wiener: A centennial symposium* (p.19). Cambridge, MA: MIT Press). For a brief appreciation of Wiener by Walter Rosenblith and Jerome Wiesner, see "From Philosophy to Mathematics to Biology" at http://ic.media.mit.edu/projects/JBW/ARTICLES/WIENER1.HTML.

The Index Theorem

Your most influential and wide-ranging work has been on the index theorem and its developments with Michael Atiyah, for which you recently received the Abel Prize.[9] *How did you come to work together?*

I was on sabbatical leave, academic year 1961–1962. I called Michael, whom I had gotten to know in 1955 at the Institute for Advanced Study, and asked whether there was room for me at Oxford for the second semester. He simply said, "Come."

I arrived in January 1962. Michael came up to my little office when I was unpacking the next day to check on how I was doing. Then he asked me, "Why is the A-roof genus an integer for a spin manifold?"

I replied, "Michael, you know the proof. Why are you asking me this question?"

He simply said, "There is a deeper reason."

When you said to him, "Why are you asking me a question that you know the answer to," what did you mean by that?

I was puzzled. Why is someone so well versed in topology asking me this question? I was really asking, "What's motivating you? Why is the proof unsatisfactory to you?"

And when he said, "there's a deeper reason," how do you understand that?

What Michael had in mind are what are now known as integrality theorems, explaining why certain integrals are integers. Examples are Hirzebruch's signature theorem, Hirzebruch's generalization of the Riemann-Roch theorem to all algebraic manifolds, and the Atiyah-Hirzebruch formula for what turned out to be the index of a spin-C Dirac operator.

---

[9] From the mathematics department's Report to the President of MIT for 2003–2004: "Institute Professor Isadore Singer shared with Sir Michael Atiyah the 2004 Abel Prize, given by King Harald of Norway in a ceremony at the University Aula in Oslo, Norway, on May 25, 2004. This is the highest distinction for landmark work in mathematics, likened to the Nobel Prize. This is the second year of the award. The citation of the Abel Committee reads in part, 'for the discovery and proof of the index theorem, bringing together topology, geometry and analysis, and their outstanding role in building new bridges between mathematics and theoretical physics.'"

Is Singer.

Our exchange took place in January. I was working on a manuscript with Shlomo Sternberg on the infinite groups of Élie Cartan.[10] But as I sat on the benches in the gardens of Oxford, I couldn't help thinking about Michael's question. In March I thought I had the answer. Surprisingly, it involved analysis and the generalization of the Dirac operator to Riemannian spin manifolds.

I explained to Michael what I thought was going on and how most of the known integrality theorems fit together using the Dirac operator. It was not hard to conjecture the index formula; we obtained the proof of the index theorem in September.

The popular explanation for the original index theorem is that it tells you how many solutions there are to a differential equation. From a lay person's point of view, that seems a little like having a car mechanic tell you that there is a solution to why your car won't start

---

[10] This manuscript was later published as Singer, I. M. & Sternberg, S. (1965). The infinite groups of Lie and Cartan. Part I (the transitive groups). *Journal d'Analyse Mathématique, 15,* 1–114.

and then leaving. Besides giving you hope, what is the value to knowing the existence of solutions without knowing what they are?

For a differential equation, one wants to have a *unique* solution. Index theory was stimulated by the Fredholm alternative: if the index is zero, then existence implies uniqueness and uniqueness implies existence. That observation is important, for it is usually much easier to prove uniqueness than existence; when the index is zero, existence comes free of charge. Knowing that a unique solution exists gives a great impetus to finding it. But that doesn't always help.

To answer your question more broadly, pure mathematics isn't immediately practical but often gives insight into problems others work on. Over the years, people in different fields have come to my office and asked my advice on a problem. I usually can't solve it, but I often suggest a different approach or a book that would provide some needed background. Many leave disappointed, but later I learn I've put some on the right track.

Every important theory has a certain trajectory in time, in terms of its effect. Sometimes it's sort of a sleeper and lies around for a while, but this one seemed to explode onto the scene. Yet evidently the full import of the theory took a while to unfold. The Daily Telegraph apparently quoted Michael Atiyah as saying, "The index theory is a Trojan Horse that mathematicians have used to get into physics, and vice-versa. When we first did it, we had no inkling that this would follow." [11] *It has been more than forty years now. Can you plot the ripples, the mathematical and scientific effects, over that time?*

The original index theorem expressed the index, an analytic object in geometric and topological terms. The theorem brought some parts of analysis, geometry, and topology closer together in a new way. That's why it was instantly famous.

In the late sixties, with the help of Graeme Segal, we extended the index theorem from one operator to a family of them. We also allowed for symmetries in the geometry.

In the mid-seventies, with Patodi, we extended the index theorem to manifolds with boundary. To do so, we introduced an important new invariant, the eta invariant.

In the late seventies, with the discovery of instantons, self-dual gauge fields, index theory became important in physics. It counted the number of

[11] Devlin, K. (April 2004). *Devlin's angle. MAA Online.* The complete article is available at http://www.maa.org/devlin/devlin_04_04.html.

instantons, and it provided a formula for the axial anomaly. It even clarified a problem in physics, the 't Hooft problem. Physicists observed eight light particles of a certain kind, but their theory predicted nine. The index theorem explained why the ninth was not light.

In 1984 Atiyah and I wrote our last paper together to date, applying the general index theorem to the problem of chiral anomalies in string theory and gauge theories.[12] That application turned out to be very important, so much so that young physicists are now experts in index theory and K-theory.

Index theory has been robust over a forty-year period—very impressive, and, of course, gratifying.

Was the work that emerged from the index theorem weighted more heavily toward the analytical side or toward the algebraic geometry or topology side?

I think both subjects were stimulated by the index theorem. On the analytic side, pseudodifferential operators became an integral part of operator theory; some work we did later, the mod 2 index and real K-theory, illuminated parts of topology. K-theory in operator algebras and noncommutative geometry were both influenced by the index theorem.

Is it possible to foresee new ways in which the index theorem is going to be used in the future?

I wouldn't hazard a guess on future uses. New things are happening. For example, in the string theory explanation of black holes, the index theorem is used to compute entropy.

The Washington Period

You have been prominent in representing the importance and interests of mathematicians and of science as a whole in Washington. Can you describe some of your activity? What strategies did you use in representing math and science to lawmakers trained in neither?

My first experience in Washington was service on the David Committee, which explained the needs of mathematics to our representatives in D.C.[13]

---

[12] Atiyah, M. F. & Singer, I. M. (1984). Dirac operators coupled to vector potentials. *Proceedings of the National Academy of Sciences, USA, 81,* 2597–2600.

[13] The ad hoc committee of the National Research Council on Resources for the Mathematical Sciences, informally known as the David Committee, was convened after the

Ken Hoffman, who had recently finished as the MIT math chair and was on leave in D.C. to play a role in the national mathematics scene, organized the committee and persuaded Ed David to serve as chairman.[14] We wrote a report, the David Report, which demonstrated the need for a fellowship program at the postdoctoral level and more funds for graduate students. Ed taught us that writing the report was just the first step. The next step was to visit key administrators and explain the importance of our recommendations. We met with the Director of the National Science Foundation and the Science Advisor to the President, and our report was effective in getting more resources for mathematics.

From 1982 to 1988 I served on the White House Science Council in the Reagan administration. We gave broad advice on matters of science to the president's science advisor, Jay Keyworth.

Later I served on the Council of the National Academy of Sciences and was chairman of its Committee on Science, Engineering, and Public Policy (COSEPUP). We supported the Academy's efforts to represent the science community and explain the needs of science and engineering to our elected representatives in Washington.

But again, how does one advocate on behalf of mathematics or science to a lay audience? Does one point to significant public benefits? "We wouldn't have space travel and heart transplants if it weren't for...."?

Rarely does a direct appeal to Congress for support of a specific scientific effort work. Our role was to help the president of the National Academy of Sciences represent science in Washington. What fields should he/she emphasize? What projects should be singled out?

Occasionally we appeared before a congressional committee or spoke with the head of the National Science Foundation, or talked with people who make up the science budget.

National Research Council's Office of Mathematical Sciences, chaired by William Browder, presented evidence suggesting an alarming deterioration in the federal support of mathematics research in the United States. The committee's report, published as Renewing U.S. Mathematics: Critical Resource for the Future (1984) (Washington, DC: National Academy Press), confirmed those suspicions.

[14] Edward E. David, Jr., was then president of Exxon Research and Engineering Company, and formerly science advisor to President Nixon and director of the White House Office of Science and Technology (1970–1973).

You served on a committee on strengthening the linkages between the sciences and mathematics for the National Academy of Sciences. The committee's report said, "It is particularly important to develop effective criteria for the evaluation of cross-disciplinary research," [15] *to make sure that the promotion and tenure, the reward mechanisms, are there. How is MIT doing, do you think, in promoting, evaluating, and rewarding research across disciplines?*

It is a difficult problem. I think that cross-disciplinary research is very exciting these days. MIT is doing a good job in encouraging students to combine biology and engineering. If I were starting my career now, I would choose theoretical biology; biology is replete with interesting mathematical problems. But I don't know of any effective criteria for the evaluation of cross-disciplinary research. There is no substitute for shrewd, unbiased judgment.

Did your service at the national level give you a different perspective, and how did you bring that back to MIT? You're an Institute Professor, so I gather you're consulted somewhat on the running of MIT. You were on the committee to choose the next MIT president, for example.

I do have a different perspective. At the National Academy and on the White House Science Council, I met brilliant people in different fields, with different insights about the problems we were confronting. I learned a great deal from them. It broadened my entire view about science and academia.

In fact, when I became chairman of COSEPUP, I went to see Jerry Wiesner, then MIT's president. I said, "Jerry, I've just gotten involved in mathematical physics; I want to teach the subject differently—outside the constraints of the math department."

He replied, "Whatever you want, that's okay with me, Is."

And then he said, "By the way, Is, I understand you are chairman of COSEPUP. Congratulations!"

Well, I cringed, feeling guilty, as it would be taking me away from my research. I said, "Yes, I—I did agree to do that," very hesitantly, and he replied, "*Is*. Don't worry about it. It'll do you *so* much good for your research later on!" Jerry was a terrific guy; he understood something I hadn't—that I would have a different perspective about my work in *mathematics* because of my experience in Washington.

That broadening was reinforced by my service on the selection committee for the MacArthur Foundation. All committee members were involved

[15] Committee on Strengthening the Linkages Between the Sciences and the Mathematical Sciences, National Research Council. (2000). *Strengthening the linkages between the sciences and the mathematical sciences.* Washington, D.C.: National Academies Press, 2–3.

Howard Johnson and Is Singer.

in the choice of fellows, whatever their field. I got an entirely different perspective about human creativity there.

As to MIT, I think I do see MIT differently than my mathematical colleagues. We have a wonderful administration. It is so much more creative than most university administrations. It is sensitive to the needs of faculty as well as students. Our administration tries hard to give faculty free time to be creative.

I don't know that I was chosen to be on the search committee because I am an Institute Professor. More likely the chairs of the search committee thought I could bring something to the search because of my broad experience.

U.S. News & World Report recently ranked the MIT math department first in the country, as you mentioned. What does MIT need to do to stay at the top?

Continue doing what it has been doing: compete for the best mathematicians. It might expand its search to fields poorly represented here. The competition is getting stiffer. We need more resources to maintain our position.

Back to the Center

Can you say anything more about how the interaction with top people in different fields, even art and literature, really made a difference in your own work as a mathematician?

[Slowly and thoughtfully] Having seen what deep, original work in other fields is like, I have to ask myself whether I am doing that in mathematics. I can't articulate this well, but these outside viewpoints are compelling. They make you look at mathematics as a whole and ask whether you couldn't be doing something much more penetrating and valuable than you presently are.

I have only known one mathematician who does that, and does it to me. That's Gelfand.[16] I. M. Gelfand and I are very good friends. He is in his nineties. He comes to mind because we are just publishing his ninetieth birthday-party workshop.

And he always asks me: "What are you doing? *Why* are you doing it? What about such and such?" One of the reasons we are such good friends, since I met him in Moscow in 1963, is he *always*, in his gentle way, attacks me in this very constructive way. It's Gelfand who keeps bringing me back to the center; and all my experiences with brilliant people, whose work I admire in other fields, bring me back to the center.

~ September 9, 2005

[16] Israel M. Gelfand (b. 1913 in Odessa, Ukraine) is a mathematician of enormous range and prodigious output (some five hundred papers), famous for his work in group representations and cell biology alike. He emigrated to the U.S. from Russia in 1990.

Arthur P. Mattuck

Beginnings

J.S.: *How did you decide to get into mathematics?*

A.P.M.: In high school I liked math, and I read standard books like Eric Temple Bell's *Men of Mathematics*—chapter biographies of famous mathematicians, well written, but a lot of it is considered suspect; it's not scholarly work, but very enthusiastic—and Courant and Robbins, *What Is Mathematics?* Stuff like that. And I was on the math team. This was in Midwood High School in Brooklyn, and the New York City schools had inter–high school competitions. So we would meet for the year or the semester with a particular school, like Brooklyn Tech, where my father was the chemistry chairman, or Madison, Erasmus—good nearby Brooklyn schools.

The curriculum was standard throughout the city. Everybody who was interested in math took the same eight courses, and so did many people who were not particularly interested. I didn't know any school that taught calculus, for example. There were no advanced placement courses or anything like that.

When I got to college—I went to Swarthmore—I majored in chemistry. I suppose I was sort of following in my father's footsteps; it wasn't any deep commitment. But I took mathematics all the way, in every term. At that time honors at Swarthmore meant you took only seminars, two per term, for your last two years, so each semester I took a math seminar and a something-else seminar, chemistry or physics.

Then I met my Waterloo in organic chemistry in a laboratory. I guess it must have occurred in the first semester of my senior year. I was on about

The Mattuck family; from left to right: Rae, Arthur, Jack, and Richard.

the tenth step of an organic synthesis that had already taken two months. The vacuum system was an old-fashioned aspirator: you put a funnel over your little beaker of stuff, boiling away, so the fumes would go through some tubing to an aspirator thing, through which water flowed into a sink, and the flow of the water would draw the fumes down the sink. I hadn't tightened the funnel support enough, and I jarred the table slightly: the funnel fell down into the beaker, and in an instant, my two months of synthesis got sucked into the sink and disappeared.

I'd already had intimations that I could not do laboratory work: no matter how hard I tried, others got better results. I was very good on paper, but at Swarthmore, theoretical chemistry wasn't a well-known activity at that time; if you were a chemist, you were in a lab, mixing things. So the next day I became a math major, just in time to apply to graduate school. I got into Princeton, largely on the basis of a recommendation by Arnold Dresden, who taught me what mathematics was about and encouraged me in my stumblings; he also introduced me to chamber music through Monday night sessions at his house, where he could pull music from off his shelves for whatever combination of students and faculty showed up with instruments, wanting to play.[1] I will always be grateful to him.[2] From that point I decided on mathematics and never looked back. Surely the right decision.

I got to Princeton in September '51. Those were the days when you got a PhD in three years, not seven. The ones who took longer were the square-dancers. The Princeton class was filled with all the Putnam teams from Harvard and Toronto—those were the schools that regularly won the inter-collegiate Putnam math exam. They had taken mostly graduate courses while undergraduates. When they got to Princeton, they said, "Hey, I know all this," and they square-danced. But I was scared, like many of the other first-year graduate students, so we worked! When I left, they were still square-dancing. It took some of them a long time to get a PhD. They were brilliant, but somehow they couldn't work.

You studied with algebraist Emil Artin. What were his lectures like?

Well, you didn't have to go to lectures, but he was the best lecturer there. I liked algebra. With my summer roommate Lou Howard, whom I followed from Swarthmore to Princeton to MIT, I'd studied Van der Waerden's *Algebra* the summer before I'd entered Princeton, so I already knew some, and I liked it: it's a clean subject; you know when something's proved. I went for a while to Steenrod's lectures in topology and Bochner's in analysis, but Artin was the only one whose lectures I never missed.

This sounds idiotic, but in a sense I didn't learn much from his lectures. They were very polished, but by and large he wouldn't work out any examples, so it was hard to get a foothold in them. Each year he would figure out an even

---

1 An article by David E. Zitarelli of Temple University chose Arnold Dresden as one of five "rank-and-file" mathematicians, known primarily for teaching, whose "lives and works will reveal numerous contributions to the American mathematical community," and who thus "deserve to be rescued from obscurity." Dresden (1882–1954) designed an innovative mathematics honors course for juniors and seniors at Swarthmore College, where he taught from 1927 to 1951. Zitarelli writes that he "was one of the most respected and effective leaders in both the AMS, and the MAA," serving as president of the MAA in 1933–34. He was the author of the popular *An Invitation to Mathematics* (New York: Henry Holt, 1936). Zitarelli adds, "Dresden was known as much for his musical talent and interests as for his mathematics; his Monday evening chamber music sessions at Swarthmore were celebrated." See Zitarelli, D. E. (2001). Rank and file American mathematicians. In Amy Shell-Gellasch (Ed.). *Proceedings of the history of undergraduate mathematics in America* (pp. 179–193). United States Military Academy: West Point.

2 Arthur Mattuck has played cello in an amateur string quartet with other Boston-area faculty for the past fifty years. See page 355.

Arthur Mattuck and Dick Gross (Harvard) waiting for their entries in the opening fugue of Opus 131. Dana Fine (University of Massachusetts) and Michael Artin complete the quartet, July 2008.

cleverer way of doing Galois theory, and by my year, it was incredibly slick, but you didn't get much feeling for what problems it was designed to solve.

That echoes something that Gian-Carlo Rota said about the lectures at Princeton of Alonzo Church. The lectures were word-for-word repetitions of a typewritten text that was obtainable from the Fine Hall library, but the point of actually attending Church's lectures was the opportunity of seeing pure logic incarnate.[3]

Well [laughs], he certainly had a great reputation. I went to just one of Church's lectures and did not go back, because they were monstrously slllowwww. Church was an enormous man, maybe six foot six, and broad: he must have weighed over 250 pounds, and he had enormous hands. He would always appear at tea at 3:30 in the Common Room. The cookies would be placed on the tray standing on edge, just as they were in the box. He would do the same thing every day: he would open his hand as wide as possible, take all the cookies that would fit between these two [shows outstretched thumb and middle finger]; and then he'd turn them upright, so they'd be a

[3] Rota, G.-C. (1997). *Indiscrete thoughts* (p. 5). Boston: Birkhäuser.

stack of cookies, and then he'd retire to a corner of the Common Room, speak to no one, and just stand there, methodically eating one after the other, and when they were all gone he would leave.

The blackboard in his class was perfectly erased with horizontal strokes, and then erased vertically to remove the horizontal erasure marks, and then he would write in perfect handwriting, verrrry, very slowly, leaving nothing out. It was not an animated performance, but it was possible that Rota could have learned something just from seeing the phenomenon: this is a logician, this is the way a logician thinks. He gets to the bottom of things, is systematic, leaves no stone unturned.

[Emil] Artin's lectures were too smooth. If a lecturer makes occasional mistakes, or some things are inelegant, or if he writes down a messy example, or indicates there *is* an example, it gives you a place to put your foot in and try to fix it up or make it look better, and then that's the way you start learning how the theory is put together. In Artin's lectures, there tended not to be footholds, or anyway, not for me—maybe for somebody with a different kind of mind. You admired the way it was put together, but you didn't feel there was anything that you yourself could do there to improve it. Ideally we should have been smart enough to make up our own examples, but most of us didn't really understand things well enough to do that. When Artin gave lectures, he worked very hard polishing them. (Once I knocked on his door a half hour before class and found the board covered with the lecture for that day.) It wasn't that he was trying to hide the sources of where these things came from; I guess he felt that you really understood best by seeing how the different pieces of the theory fit elegantly together. But he himself had studied as a student many, many examples, and I think he did not appreciate that without those examples, it's harder to appreciate an elegant theory.

It's the same in algebraic geometry. Take the Italians: their proofs might not have been airtight, but they had studied lots of examples, so they were fairly sure of what the theorems were, even if their proofs were defective. A major theorem of Enriques—a great Italian geometer who relied heavily on his intuition—was discovered after several years of heavy use to have a defective proof. A correct proof only came much later, using newer methods. Severi[4] said the faulty proof was really a good thing for the theory of surfaces, since it let geometers use the theorem much earlier than if they had had to wait for a correct proof!

---

[4] Francesco Severi (1879–1961) published hundreds of papers, most importantly in algebraic geometry.

Emil Artin, c. 1960.

I remember once, just once, Artin took time out of a lecture on Galois theory to work out an example, and everybody's reaction was "Oh! Is that what he's been talking about all this time!" Galois theory talks about number fields; there's a group that operates on the number field; its subgroups and certain subfields bear a relationship to each other expressed in the theorems. Artin took a *specific* number field and a *specific* group acting on it, and calculated the subgroups and corresponding subfields explicitly, showing the relation between them.

The only one I've ever known who could work without examples was Grothendieck.[5] He did algebraic geometry seemingly uninterested in specific curves and surfaces. He constructed his own view of algebraic geometry, building it up from scratch, just using algebra. For a time it completely dominated algebraic geometry. Now people are more eclectic: they use Grothendieck when it's necessary. When it's not necessary to use Grothendieck's theory because the problem doesn't require that degree of sophistication, they use more classical ideas. The wars, so to speak, between Weil's and Grothendieck's views of algebraic geometry

[5] Alexander Grothendieck (b. 1928) made major contributions to many fields of mathematics, including algebraic geometry, before retiring at age 60. He won the Fields Medal in 1966.

were settled by the following generation, which uses what it finds useful to prove its theorems. They take what they need; they're not interested in the wars.

How did Emil Artin become your thesis advisor, and how did you find a topic?

I think 1951 was the first year that the NSF pre-doctoral fellowships were announced, and I decided to apply for one. Somebody had to write a recommendation for you, so there was only Artin, since he was the only one whose class I was going to regularly [laughs]. At the same time, I had gotten friendly with a couple of his more advanced graduate students and was busy trying to study notes from the seminar two years earlier on local class field theory that one of them, Iain Adamson, had written. And so very hesitantly I gave Artin the NSF recommendation form; he looked and said, "Ach, vat iss thees?" and he wrote something, but he really didn't know me from Adam.

I studied that summer in the heat of D.C. where I was working, took my orals as soon as I got back to Princeton in the fall, and passed them. So as far as I was concerned, I didn't see any reason why I had to take any more courses, and I started looking around for a thesis topic. I can't remember whether I even asked him for anything at that point, I sort of decided to study algebraic geometry, because André Weil's book had just come out—a peculiarly written book, but at that point it was the only rigorous book in the subject. Grothendieck later claimed that its point of view set the field back ten years. Serge Lang[6] had graduated the June before I entered and was now spending time at the Institute [for Advanced Study at Princeton], and he decided to run a seminar to go through that book. So he studied it, and what he learned he gave a lecture on, and I went to that seminar and picked it up, and then I started looking for a thesis topic.

Earlier in the fall I had gone to Artin with a stupid idea for a topic, which he dismissed, somewhat puzzled. Then he suggested something, but I didn't like it, and he said "Vell, you're on your own then!" I remember exiting prematurely and precipitously, and him shrugging his shoulders. [John] Tate[7] was in the office at the time and told me much later how funny the whole scene was.

So I ended up spending most of the rest of my second year thrashing around by myself for a thesis topic. It didn't help that I had very little idea

[6] Serge Lang also did his dissertation with Emil Artin.

[7] John Torrence Tate, Jr., another Artin student, received his PhD in 1950.

of what math research was about. But in an odd way, it was something Artin said later on about research that helped me the most.

Word had gotten down to us—I guess from Adamson, but he had already graduated and was off teaching in Belfast, so he couldn't help us—that Artin's PhD students (real ones and hopeful ones) were supposed to invite him sometime during the year to dinner at the Graduate College, where we lived. There were three of us—Tim O'Meara, in his third year and working in quadratic forms, plus Dick Semple and me, in our second year, both just hopefuls. We were really nervous about asking him and didn't know exactly how such things were supposed to go, but we had some wine with Camembert and crackers in our dorm room, before going down to dinner, and he pronounced it to be his favorite cheese—dumb luck, but we started breathing a little easier.

Anyway, later the four of us were sitting across from each other in the middle of one of the GC's long dinner tables waiting for dessert. We talked about the number theory seminar Artin was running that we were all going to, and I blurted out something like, "Boy, some of those proofs are really hard—it must have taken months to find them."

Artin turns and stares at me, eyes opened wide. "Months?? [long pause] *Years!!*"

"I tell you a little story, about the Reciprocity Law. After my thesis, I had the idea to define L-series for non-abelian extensions. But for them to agree with the L-series for abelian extensions, a certain isomorphism had to be true. I could show it implied all the standard reciprocity laws. So I called it the General Reciprocity Law and tried to prove it but couldn't, even after many tries. Then I showed it to the other number theorists, but they all laughed at it, and I remember Hasse in particular telling me it couldn't possibly be true.

"Still, I kept at it, but nothing I tried worked. Not a week went by—*for three years!*—that I did not try to prove the Reciprocity Law. It was discouraging, and meanwhile I turned to other things. Then one afternoon I had nothing special to do, so I said, 'Well, I try to prove the Reciprocity Law again.' So I went out and sat down in the garden. You see, from the very beginning I had the idea to use the cyclotomic fields, but they never worked, and now I suddenly saw that all this time I had been using them in the wrong way—and in half an hour I had it."

This wasn't a story you forgot—it gave me a better idea of what might lie ahead, so it was in a way actually comforting. Finally, after a few more months of beating the bushes, I came upon André Weil's lecture to the International Congress in 1950, in which he casually suggested some things that were worth studying, so I picked one of those. See, the purpose of Weil's writing the algebraic geometry book was because there was a conjecture called the Riemann hypothesis for function fields—formulated for the first time by Artin in his thesis—which was in a sense the analog of the usual Riemann hypothesis for number fields. Artin had proved it in some special cases; André Weil had proved the general case using algebraic geometry during the months he was imprisoned at the beginning of the war for refusing to serve in the French army or something like that.[8] A colleague finally secured his release, after which I believe he emigrated first to Brazil and then to the United States.

Anyway, he proved the Riemann hypothesis for function fields by using the inequality of Castelnuovo-Severi, which he found in a book by Severi. Now, as I said, the Italians had a *terrible* reputation—Castelnuovo, Enriques, and Severi were great mathematicians, but you couldn't trust anything in their papers, certainly not the proofs.[9] If Weil had just quoted the inequality in his proof, he couldn't really have claimed to have solved the problem. It was not just its doubtful proof; he needed to know it continued to be true even over a different kind of ground field than Severi had worked with. So he then went on to a long labor of putting classical algebraic geometry on a firm logical foundation, proving everything. This essentially required writing two or three books. The first book is *Foundations of Algebraic Geometry*—that's the one that set the field back ten years, because it used clumsy methods that he'd adapted from the Italians. The second book proved the Riemann hypothesis for function fields, based on his rigorous proof of the inequality and involving the Italian notion of correspondences between algebraic curves.

The third book went on to generalize the theory to what are called abelian varieties, a certain kind of manifold. A curve you know. A surface you know.

---

[8] Weil was in Finland when World War II broke out and decided to remain there rather than return to France and be drafted into the army. When the Russians invaded Finland, he was arrested as a spy. Thanks to the intervention of Rolf Nevanlinna, Finland's most prominent mathematician, Weil was saved from execution and deported to Sweden. The Swedes returned him to France, where he was sent to prison for failing to report for military service. He sketched out his proof while in prison, and it was published soon after, but only several years later was he able to complete the work necessary to put the proof on firm footing.

[9] Zariski's biographer excludes Castelnuovo from this list: "Although he was a central figure in the development of the Italian School, [Castelnuovo's] commitment to reason and discipline made less than rigorous proofs distasteful to him, and he watched with dismay as his colleagues became increasingly dependent on 'intuition.'" Parikh, C. (1991). *The unreal life of Oscar Zariski* (p. 19). New York: Academic Press.

What do you call something of higher dimension? If it's smooth, that is, has no singularities—points like the vertex of a cone—it would be called a manifold, but in algebraic geometry, typically, there would be singularities, or places where the thing folds over on itself, and so the generic word for it is *variety*. Hypersurfaces, they were often called, but that has a different meaning today.

Anyway, it was the abelian varieties in that book and their classical predecessors—the Riemann matrices and complex tori—that really attracted me. I was fascinated by them. My thesis was on abelian varieties over p-adic ground fields. I made progress on it by myself, and got stuck on a foundational point in p-adic geometry, and went to Artin. "Vell," he said, "it seems to me you would have to prove thus-and-such," and he was right. Then the next time I saw him, I said, "Well, I solved *this* now." And he said, "Good!" And then I went to him a few months later, and I said, "Well, I solved *that*," and he said, "Vell, there is your thesis!"[10]

But he never really read it [laughs]. He told me he trusted me. Weil refereed the paper, so it did get read. Artin wrote recommendations for an NSF post-doctoral fellowship and for what was then called a Peirce Instructorship at Harvard. But just to Harvard. I didn't ask him for other recommendations. I didn't want to bother him. Recommendations were written by hand in those days. You went to an office with him; he just wrote down a few sentences. I didn't really get along with him; we essentially had no contact at all, and he was so impressed by that—you know, a trouble-free graduate student—he must have written me a decent recommendation.

I got the NSF post-doctoral fellowship and took it at Harvard for a year. Before that, in my third and last year, there had been a conference at Princeton in May '54 in honor of Lefschetz's seventieth birthday.[11] André Weil had sent a little paper to the Lefschetz conference that I was asked to referee, because Princeton was going to publish the conference proceedings and there was no one in the math department familiar with Weil's language. That was my first referee job.

---

10 Mattuck, A. (1955). Abelian varieties over p-adic ground fields. *Annals of Mathematics 2*(62), 92–119.

11 Solomon Lefschetz (1884–1972). His contribution is best described by his famous quote, "It was my lot to plant the harpoon of algebraic topology into the body of the whale of algebraic geometry." The word *topology* comes from the title of a monograph he wrote in 1930. Lefschetz was the author of *Algebraic Topology* (1942) and won the Prix Bordin, the Bôcher Memorial Prize, and the National Medal of Science. He served as editor of the *Annals of Mathematics* for 30 years and as president of the American Mathematical Society.

Zariski[12] came to the conference. He was at that time one of the greatest algebraic geometers in the world, perhaps the greatest. And Zariski was at Harvard, so I took my post-doc at Harvard. Zariski gave the most exciting lecture at the Lefschetz conference, announcing, in effect, that he had essentially solved the problem of the resolution of singularities of an algebraic variety.

Can you give some sense of what made Zariski's lecture the most exciting?

As I said, an algebraic variety, an *n*-dimensional object, can have a singularity, like the vertex of a cone, where it doesn't look smooth, or like the point where a curve in the plane crosses itself or has a pinch point. And there's a theorem that had been conjectured for a long time, that by repetitions of a certain type of algebraic transformation, all singularities could be resolved. The simplest example is if you have a curve in the plane that crosses itself—well, if I sort of pull it apart, by going into the third dimension, that double point will have separated into two separate points on the pulled-apart curve. For the vertex of a cone, there's a certain type of transformation, more complicated but well studied, where you essentially pull it apart, so the cone becomes a cylinder. You can do something similar in higher dimensions. And the theorem was that by doing these standard transformations, you could resolve the singularities of any variety.

That was *the* problem in algebraic geometry at the time. The Italians had worked on the problem and gotten only as far as doing it for surfaces; for curves it had been known classically. There were four different proofs that they'd given. Zariski wrote a book on algebraic surfaces in the thirties, when he was in *his* thirties, in which he comments on each one of the proofs in turn, pointing out that this was no good for this reason, this next one has a very serious gap, this one has some incomprehensible reasoning, this last one has *some* possibility of being fixed [laughs].

And then Derwidué, a Belgian mathematician, sent in a proof of the resolution of singularities, and Zariski devoted two entire pages of *Math Reviews*—that's a lot of copy; nothing ever gets a review much longer than

12 Algebraic geometer Oscar Zariski (1899–1986) won the Steele and Cole Prizes and was awarded the National Medal of Science. He was the author (with P. Samuel) of *Commutative Algebra*, editor of the *American Mathematical Journal*, and president of the American Mathematical Society.

that—taking apart Derwidué's proof.[13] First mistake; second mistake; nonsense; this is impossible—and so on and so on. It would have been enough to just say, Look, this paper is riddled with errors—here, for example, in the beginning, which vitiates the rest of it—but he continues for several columns.

So at the lecture, Zariski said, "I decided that if I was going to have to review such papers, the only way of avoiding this labor in the future was to prove the resolution of singularities myself." He announced he was going to give the idea of his argument; he said he had finished about 90 percent of the details, but he was confident that it was going to go through, confident enough that he would give the lecture at a meeting with so many well-known algebraic geometers, including Lefschetz.

Anyway, I saw Lefschetz out in the hall afterwards and asked him what he thought of Zariski's 90 percent proof. You know, Lefschetz was an algebraic geometer. And he said, [abrupt, squeaky voice], "Let me tell you something, Mattuck! In the theory of resolution of singularities, 99 percent equals zero!" A mistake can be on the last line. It's either perfect or it's nothing.

Then a half year later, Shreeram Abhyankar, a student of Zariski's at Harvard, decided he was going to fill in the proof, finish it. And within, I don't know, a few days, he observed what should have been clear from the beginning, namely, that the method could not possibly work. The method required the theorem to be proved in a certain strong form, and there was a simple counterexample to the strong form. Then a few years later, another Zariski student, [Heisuke] Hironaka, by a completely different method, succeeded in proving it in a long paper, for which he won the Fields medal.

Did you have much contact with Lefschetz at Princeton?

Just very casual, but he and [Albert] Tucker were the ones at Princeton I liked most. In fact, he gave me a good idea—I made no use of it, to my regret later; but there was something about him, he sort of liked me for some reason, and I liked him. I went to a few of his lectures when he talked about nonlinear differential equations. That was in my first year. But Lefschetz was an algebraic geometer of the old school. He had done fun-

13 Zariski, O. (1951). Review of L. Derwidué: Le probléme de la reduction des singularities d'une variété algébrique [The problem of the reduction of singularities of an algebraic variety]. *Math Reviews 13,* 67–70. *Mathematische Annalen 123,* 302–33.

A reunion of some Princeton PhD's (1954, 1955); front row: Frank Peterson, Sigurdur Helgason; back row: Arthur Mattuck, Gian-Carlo Rota (all MIT), Walter Baily (University of Chicago).

damental work in the twenties and thirties, applying topology to algebraic geometry, and before that in abelian varieties. And so he called me into his office during my second year: "Mattuck!" [imitates clipped, squeaky voice] "I'm going to give you a good idea here!" He had a funny way of talking. "I've been thinking that all that theta function stuff, that's purely algebraic! That's really just algebraic! You hear what I'm saying?" It was related to the abelian varieties and actually relevant to my thesis work, but I went off in a different direction.

I remember one very long walk at Princeton; he was walking home and I fell in and kept him company, and we just talked, partly about mathematics and partly about other things. But as we got closer and closer to his house, he got more and more nervous [laughs]. I certainly didn't expect to be invited in, but he found some way of making sure that I didn't get too near the house. He didn't have children and, as far as I know, didn't entertain, so he wasn't sociable in the usual sense, but as I said, there was something about him.

Do you remember anything that you talked about on that walk with him?

Well, I had been reading the long paper on abelian varieties that made him famous in the early twenties; part of it talks about relations among the minors of a determinant, some of which were wrong, so I asked him about that. "Oh, I'm not surprised; there are probably a million errors, but that doesn't matter, because the end results are right." So I guess he had thoroughly absorbed the Italian point of view.

I asked him how he had gotten interested in abelian varieties during the time he spent at Kansas, which was very isolated at the time. He began by talking about how he had been an electrical engineer, but lost his hands in an accident in his mid-twenties, and turned to getting a math PhD at Clark University in Worcester, at the time mainly known for its geography deparment. "You know, Mattuck, that hot wire was the luckiest thing that ever happened to me, because without that I never would have become a great mathematician.[14]

"When I finished with Clark, Kansas was the only place that offered me a job. And there was no one within a thousand miles that ever heard of algebraic geometry. But it was a great place to go and think and not be bothered, if you had something to think about. When I got there, I thought I would study Laguerre polynomials, so I did that for a couple of years, but then I got totally disgusted with them and looked around for something else. Picard's book had just come out, so I thought I would try that, to see what it was all about. Boy, that book was tough as the dickens! And there was no one to explain anything to me, so I had to try to figure it out myself. I could see the proofs were full of holes. But I kept at it.

"Then after a few years they announced in Paris that the Bordin prize for that year would be for a memoir on abelian integrals. Well, that was just what I had been studying. So I worked really hard and sent in my paper. You didn't put your name on it: it was in a sealed envelope attached. Well, they all read my paper and couldn't figure out who wrote it—some thought maybe Picard himself—but they gave it the prize, and then they opened the envelope. Well, do you know, their jaws dropped open. They were completely flabbergasted. None of them had ever heard of Kansas, let alone me; they hardly knew there was any mathematics in America, except for Birkhoff, because he had solved Poincaré's problem. They thought it might

---

[14] Lefschetz lost both his hands and forearms in a transformer explosion while working for the Westinghouse Electric Company in 1907. He worked with the aid of prosthetic hands.

Miyanishi, Grothendiek, and Abyankhar in Montreal, 1970.

be somebody's idea of a joke. But I got my prize, and after that things were easy—[James W.] Alexander invited me to Princeton and I've been here since."

Early Years at MIT

What brought you to MIT from Harvard?

Oh, that was very simple. I'd come within something like a week of being drafted for the Korean war while I was working in D.C during the summer of '52, and I only got a last-minute deferment by appealing directly to the Presidential Appeal Board. Now it was spring 1955, but my education was officially over and I was again subject to the draft. I guess the Korean War was over at that point, but the draft still continued.

I had an appointment as a Peirce Instructor starting in fall '55, and also at that point an offer of a Moore Instructorship at MIT. Harvard said that they did not write letters requesting deferment for their instructors; MIT had extensive experience with the military and of course did. So John Tate—we both lived in the same apartment house, he lived over me; I think he was an associate professor at Harvard at that point—saw Van Vleck, who was a dean at Harvard,[15] at a party and mentioned the situation to him. Van Vleck said, "Send him to me!" So the next day I went to see Van Vleck. He didn't remember anything from the night before, but after I described my situation, he shook his finger angrily at me and said, "MIT doesn't do ANYTHING that Harvard doesn't do." I disliked him so much that I immediately accepted the offer from MIT [laughs]. A couple of years later my brother Dick had a similar run-in with him when the results in his MIT physics thesis contradicted one of Van Vleck's theories, and he emerged with a similar opinion of him.

That first year I had hung out both at Harvard and MIT. I had an office in the Harvard math building—really just the second floor of 2 Divinity Avenue—and started the Princeton tradition of daily afternoon tea and cookies in my office, mostly for the algebraic geometers, but others were welcome, too. It continued after I left. We had for a while a little study group on Weil's book on abelian varieties, to which Zariski came and contributed. I saw some of the graduate students and the younger faculty—Tate and Gleason especially were friendly. But the older ones were more formal and distant; and it wasn't a large department. I had studied Zariski's paper on holomorphic functions, but it was a poor choice: I encountered difficulties, and anyway it was superseded a few years later by Grothendieck's version, one of the early triumphs of his approach.

Meanwhile, during that year I went to seminars at MIT and spoke in some, and I liked many of the faculty and graduate students at MIT a lot; we went out to dinner and there were evening gatherings. As far as the general student atmosphere went, Harvard seemed to me a snobby place, like Princeton had been, with its carefully ranked eating clubs. In both places it mattered whether you had gone to a public or a private secondary school—and it was important which private school it was.

Whereas MIT was a meritocracy, which I was used to from Brooklyn. The good guys were the smart ones. It didn't matter how much money you

[15] John Hasbrouck Van Vleck (1899–1980) served as dean of engineering and applied physics at Harvard from 1951 to 1957. Van Vleck won the 1977 Nobel Prize for Physics with Philip W. Anderson and Nevill F. Mott.

had, or where you came from, or anything like that. It didn't matter to MIT that 35 percent of the students were Jewish. If you were smart, or creative, or passionate about something, that was it. That was what was respected at MIT, and it's still to a great extent the same. At Harvard or Princeton it was perfectly okay to have a C average if you had social connections. The only acceptable excuse at MIT to have a C average was if you were spending all your time inventing something great, or pursuing some fixed interest with such intensity that you couldn't be bothered with all the other stuff. So it was really very different.

Was the math department still officially a service department then?

We were a real department, with a Roman numeral, XVIII, like the other science and engineering departments. We gave undergraduate and graduate degrees and had been since the thirties. The only reminder of our earlier servitude was that the old subject designation was still used—math subjects weren't prefaced by our number 18, but by the letter M. Only the humanities courses (with an H) and the language courses (L) were like this.

This was so for my first few years at MIT. Then in 1959 all the math subjects became 18.xx in the catalog, and later when there got to be too many of them to be numbered by just 1 to 99, we went to a three digit system, 18.xxx, and renumbered them all using the digits to indicate the general area and the subtopic. But somebody's always trying to give their subject a number which violates this.

And the department was outstanding. We had Wiener and Levinson, Whitehead, Hurewicz, Salem when I first came, though I don't remember him being here more than a year more, and Hurewicz died in 1956, in the summer, so he wasn't around much longer, either.[16] I. S. Cohen was a very fine algebraist, but he died in my second or third year here while spending a leave at Columbia. There were some applied people, too. Eric Reissner, for example, was a well-known applied mathematician in elasticity theory. C. C. Lin in fluid mechanics and astrophysics. Iwasawa, of course, was here when I came, I believe, and Warren Ambrose and Is Singer. I'm probably leaving out a lot.

Can you talk a little bit about the people who were here?

---

[16] Witold Hurewicz (1904–1956) was at MIT from 1945 until his death. Hurewicz was killed in a fall from a ziggurat while at a conference at the International Symposium on algebraic topology in Mexico.

They were very friendly. I was a Moore Instructor, and at that point there were still only one or two Moore Instructors. So essentially we were part of the family, taken in just like any faculty. Now there are maybe 25 Moore and Applied Instructors, and socially it's totally different, but at that point it was nice.

My office mate was another Moore Instructor—Kobayashi, a genial differential geometer from Japan who tried to teach me how to pronounce in Japanese the haiku I was reading in English. He was determined to stay in the U.S. and did ultimately end up in Seattle. Once I asked him why he liked the U.S. so much. At the time, Akizuki was visiting Cambridge for the year from Kyoto—he was one of the two most influential mathematicians in Japan, and an art-lover who spent many hours at the oriental collection in Museum of Fine Arts. Kobayashi said, "Back home, when I see Akizuki on the street, I bow very low and say, 'Good day, Professor Akizuki.' Here I just wave and say, 'Hi!'"

Warren Ambrose was in some ways the social center. Ambrose always liked to hang around with the younger faculty, the young instructors and so on. He immediately took them under his wing. Singer tells the story of how as soon as he walked for the first time into the department office, Ambrose grabbed him before he got his coat off and dragged him off to a seminar.

He was very sociable, and he'd gather people toward him. "Let's do this! Come on, come on, come on! *No?* Don't say no, I'm not interested in no's! Come on!" That sort of person. "Okay, everybody's going to Pritchett!" The cafeteria on the first floor of Walker was only open at meal times, but on the second floor was Pritchett, a lounge serving sandwiches and desserts, which was always open. So we'd go there in the afternoon and get some ice cream or something, and sit around a table chatting about mathematics.

I remember being one evening with Ambrose (nobody ever called him Warren) and Lee, his wife at the time—they got divorced a few years later. "Let's go visit Hurewicz; he has a cold!" So we piled into a car and drove to his place near the Harvard Observatory, and Hurewicz came to the door in his pajamas, "Ohhh!" he said, "Thank you!" We stopped and chatted for a while, and then Ambrose said, "Is there anything you need?"

Hurewicz said, "Oh, I wish I had a *New Yorker*."

Okay, eleven o'clock at night, where do you get a *New Yorker*? There was only one all-night drugstore, at the Boston end of Longfellow Bridge—it's

A Mattuck family visit in New York City, c. 1978; Arthur Mattuck in rear; on couch, from left to right: nephews Robin and Allan, daughter Rosemary, mother Rae, brother Richard.

still there. We got the *New Yorker* and then drove back to Cambridge to deliver the *New Yorker* to Hurewicz a half-hour later. That was the sort of thing that happened.

Or it's midnight, so just time to go to Jack and Marian's, a popular late-night restaurant in Brookline. And if it was too late for Jack and Marian's, you went to the Hayes-Bickford's nearby, a place just one step above the Automat in New York City, but as a mathematician friend, Joe Sampson, said, "There at two o'clock in the morning you get the cream of Brookline society!"—it being the only place that was still open.

And I hung around with [John] Nash, and some of the graduate students were part of that group also. Nash was also sociable, in a distant, funny way, but not like Ambrose. Nash was a good talker, and often it was fun to hear Nash and Ambrose square off at each other. Ambrose could make sarcastic remarks. Nash wasn't quick that way. He didn't take offense at sarcastic remarks; he was pretty thick-skinned. He wasn't sensitive to anything except what impugned his abilities. He and D. J. Newman were always trying to put each other down, but it was all just one-upmanship.

Many of us were New Yorkers and used to that sort of sarcasm. Jack Bricker, Gus Solomon, and I were from Brooklyn; D. J. Newman was from New York somewhere. New Yorkers have a way of talking so almost every word that comes out is slightly sarcastic. And since Nash was from

West Virginia, I guess we teased him; he was very egotistical, had a very good impression of his own abilities, and so there was naturally a certain number of sparks. The others had different senses of humor. Nash had almost none at all, in the sense of anything that New York Jews would recognize as humor. You know, there are big cultural differences in these things. Others would laugh, and Nash would just look puzzled. Or he'd try to analyze it.

So it was a sociable time. We did mathematics, but we also hung out together. We had a colloquium once a week, and there'd always be a colloquium party at somebody's house, and of course everybody was expected to come. It was—I don't know, we did things. I don't think that happens now anymore.

Several people have mentioned that. Why is that, do you think?

Size, primarily. The department's much, much bigger now. It's more spread out, and the difference between having two instructors and having twenty—you know, there's a constant turnover; people have gotten older and sort of overwhelmed by the number of new faces they don't know. It was small and cohesive then. We went to each other's seminars, even if we didn't quite understand what was going on.

Mathematics itself, in fifty years, has changed a lot. Someone, I forget his name, not a mathematician but a scientist who knew a lot of mathematics, wrote that in looking through *Math Reviews* of fifty years ago, he would have some idea of almost every topic—have at least a feeling for what it was about—but that now it would be *totally* impossible. Individual fields have gotten so deep and technical that it requires all your energy just to learn and keep up with your own field. You know, I shopped around at Princeton a little bit, heard some lectures. I wouldn't have any idea of probability theory or deeper parts of analysis, but I had some idea what was going on in adjacent fields—fields involving geometry in some sense, topology, or differential geometry, complex manifolds. As mathematics itself has grown, it's gotten more compartmentalized. It's harder for people to talk to each other now or follow the average over-technical colloquium lecture.

Another reason why we have much less social glue than we did then is probably that the department lacks a couple of central figures, people like Ambrose, who made it their business to draw everybody in and try to get people talking to each other. Or not deliberately, but just from their natural personality, people for whom it was natural to invite people to their house. Ambrose was a constant colloquium party giver. Whitehead was very shy by contrast, but nonetheless he and Kay liked to give parties. He would not be the life of the party, but he would be there, clearly enjoying the company, enjoying having people around.

And Ted Martin, too, the chairman, did it very deliberately. Once a month, Ted would have an open house in Lexington for the MIT math faculty, but instructors were treated just like faculty members, and a lot of people trooped out there and would stand around eating. Then a certain number of people would leave, and the rest of us would sit around in a circle, and Lucy—that's his wife—invariably said, "Oh! Well! Now comes the best part!" and we'd sit around talking, and just shooting the breeze, sometimes talking about problems the department had. Whatever anybody was interested in at the time.

So those were the principal entertainers, but they were steady about doing it. You met people in a social context as well as a mathematical one. You know, for a long while in the sixties and seventies, people gave dinner parties. Now—I don't know, maybe it's just me, but they seem by and large to have disappeared. It's very unusual to do that now. The times have changed.

Even the colloquium parties have disappeared. It used to be, MIT had a weekly colloquium, Harvard had a weekly colloquium, and that meant that there would be two parties every week. That was while Brandeis was still a fledgling university. Then Brandeis started having its own colloquium every few weeks. At that point it was decided that it was too much; people were spending all their time going to parties [laughs]. So they got together and decided they would have only one colloquium a week, which would rotate among the three universities. Northeastern has joined in, too.

Meanwhile at MIT the colloquium party has been replaced by a fruit-and-cheese spread in the Common Room immediately after the lecture, and then the speaker gets taken out to dinner by a group of people, and that's the end of it.

That reminds me of the time Lefschetz came up and gave one of the colloquiums at MIT, and I took charge of him to deliver him back to the airport. This was when people didn't automatically hop in taxis to go to the airport, so we took the subway. And the thing I remember most vividly, it was so perfectly in character, was—let me give you an example of Lefschetz at Princeton. We're standing, a group of us, standing around in a circle in the Common Room, talking. It's pouring cats and dogs outside. Lefschetz enters: this is a man almost 70, wearing a very elaborate mackintosh raincoat, boots on because it's raining very hard, and one of those floppy fisherman's hats. He walks in, and he sees this group of mostly graduate students, maybe a professor or two, standing around in a circle talking at tea time, and he breaks in, walks dripping into the middle of the circle, turns around, slowly, just like a little kid, like, "See me in my mackintosh! And all new, too!" [Laughs.] The meaning was perfectly clear. Big smile on his face, you know: Admire me, admire me!

Anyway, he and I were on the subway, and somehow the conversation turned to Norbert Wiener, who was still in evidence at MIT at that time. Lefschetz clearly did not like Wiener, and he said [imitates high, excitable, irritable voice], "You know! I was at a conference in Mexico,[17] and I *found the key to Wiener's character.* He had to give a lecture. So he came up! And he gave the lecture in *Spanish!*"

I said, "That's the key to his character?"

"Yes!" he said. "At that point, his character was completely clear to me. He was simply a *complete, self-centered egotist!*" [Laughs]

It was the pot calling the kettle black, but he said it with such conviction. Lefschetz was gifted in languages and very proud to be able to talk in them, but so was Wiener. Lefschetz would automatically try to address everybody in their own language; he would try to lisp a little bit in Romanian—which Wiener also would do—tried to talk Chinese to the Chinese. They were definitely birds of a feather. But Lefschetz was more—he was aware of people. Conversations with him were two-way.

Wiener was different. I was never able to talk to Wiener. Wiener frequently would try to talk to me; my office was directly across the hall from his. But he didn't know who I was—he would talk to everyone the same way. He would stop and, with no preliminary, just start talking about whatever was on his mind. But all he wanted was a body there, he didn't expect you to say anything, it was just—I don't know. I couldn't understand him.

Levinson wrote a biography of Wiener, on which he worked quite hard. I guess it must have appeared in the *Bulletin of the American Math Society,* which is where obituaries of important mathematicians appear.[18] It's a long obituary, and part of it is personal reminiscence. Of course, a lot of

---

17 Lefschetz visited Mexico many times, beginning in 1944, and helped establish a school there.

18 Levinson, N. (1966). Wiener's life. *Bulletin of the American Math Society, 72*(1.2), 1–32.

it is mathematical. But according to Levinson, Wiener would constantly come in and say, to all his friends [anxious tone]: "Am I slipping?! Tell me frankly, am I slipping?" And then, of course, they would always assure him not, definitely not. "Oh, no, no, you're as great as ever." But, you know....

Levinson might have been worried at the time about failing powers as well, the way Wiener was. He was an older figure to me; Ambrose was older, too. I was, say, twenty-six, twenty-seven, and Ambrose must have been in his early forties, energetic and youthful. Levinson must have been close to that, but he *seemed* older. Levinson was the only one in the department who struck me as constantly slightly depressed. He didn't communicate *joie de vivre,* the way Ambrose did, for example, or Is Singer, and many of the others. Since we were in different fields, I didn't see him much, but he and Fagi liked to entertain occasionally, so sometimes I saw him at his house.

A couple of times Levinson spoke to me when he thought he'd done something interesting. He wrote a little paper, just a few pages long, proving you could turn an analytic singularity into an algebraic one by making a change of variables.[19] Not his own regular field, but something in two complex variables that he'd done just from scratch, by thinking about it. He said, "Yeah, I think I have a little something here." That's the way he would describe it. You know, a typical Jewish understatement, where something is terrific, you're saying [casually], "Yeah, not too bad." It's a way of talking that New Yorkers understand very well; I don't know how he picked it up in Revere.

So Levinson came in and he said, in just that tone of voice, "Yeah," he said, "I did a little something. I think somebody might find it interesting. A little problem about singularities—the equivalence of complex singularities. You know, I softened it up with the Weierstrass Preparation Theorem. That was it. As soon as I saw that, then it was just, you know, nothing. That was all you had to do."

I said, "Hey! That's a good theorem! People have been trying to prove—"

"Oh," he said. "Yeah, I guess." Or, "Well, we'll see. See if somebody likes it, if they'll print it"—stuff like that, you know, all in that way of talking. Still sounding depressed, but trying to find a little pleasure in it in spite of that.

---

[19] Levinson, N. (1960). A polynomial canonical form for certain analytic functions of two variables at a critical point. *AMS 66,* 366–368.

And at some point you decided to devote most of your time to teaching.

Yes. You know, I did standard things before that. I published some formulas which turned out to be useful. I worked in a particular area of algebraic geometry that I happened to like. It was quiet, not a lot of other people were doing it, and I liked that, too. For me, anyway, it's better to have the security of knowing that if you do something, sixty other people aren't going to be doing it at the same time. So in terms of algebraic geometry I didn't really have company here, which didn't bother me at all.

I'm a loner, mathematically. I haven't written many papers, but of the ones I've written, only two were in collaboration, and then purely by accident. One was with Tate—I was downstairs and he was upstairs, and he asked me something or I asked him something, and then we both went back to our apartments, and then half an hour later one of us said, "Hey! What about this?" and then the paper was born.[20] It was a little short paper, but it got well known because Grothendieck found out what was really going on and wrote *Sur une note de Mattuck-Tate.*[21] Grothendieck analyzed it and made a real theorem out of it. We had just done something cute. Tate said "Hey, we really missed the boat on that!"

A similar thing happened long-distance with Alan Mayer, but outside of that I worked by myself. I was on NSF review panels and the NSF Advisory Committee in mathematics and went down for meetings for that, and I was editor of one of the journals and, you know, did what mathematicians are supposed to do.

But all through this time, I also put in a lot of time teaching. I always liked teaching, starting with a calculus class when Lou Howard and I took over from Arnold Dresden at Swarthmore while he was being treated for cancer. At Princeton, I taught a lot my first year. I had two calculus classes all by myself, and the next semester Al Tucker picked me to run the four discussion sections—preceptorials, they called them—for his general-education culture elective "Introduction to Geometry." After that I didn't teach any more, because I got the pre-doctoral NSF fellowship, and then I got a post-doctoral NSF fellowship, so I didn't have to teach at Harvard, but I did anyway for free, just for the sake of teaching a course: a graduate course in Riemann surfaces.

---

20 Mattuck, A. & Tate, J. (1958). On the inequality of Castelnuovo-Severi. *Abhandlungen aus dem Mathematischen Seminar der Universität Hamburg, 22,* 295–299.

21 Grothendieck, A. (1958). Sur une note de Mattuck-Tate [On a note of Mattuck-Tate]. *Journal für die reine und angewandte mathematik, 200,* 208–215.

In 1960, MIT went to the lecture system for its elementary courses. Before then I just had an individual class, like everybody else—a class of 25 or 30 students in calculus, or whatever I was teaching. But the lecture course, you have to prepare a little more carefully, and there are other little problems—exams, stuff like that—that go with it. I remember the first lectures I ever gave to a roomful of students, maybe a couple of hundred, in calculus. Kennedy was running for president, so it must have been fall 1960. MIT was becoming more and more committed to the lecture courses, so first one-variable calculus, then multivariable calculus, and then differential equations became lecture courses. The core mathematics, which everyone took—there was still very little advanced placement—lasted two years. They needed lecturers, and they didn't have so many, so I ended up doing a fair amount of lecturing.

Notes on Teaching Calculus

You've taught calculus for a long time and seen the teaching of it evolve a good deal. Can you talk about your part in that evolution?

Almost from the beginning I started fooling around with calculus. At first it was mostly with the common exams that all the sections took, since it was largely me who made them up. Egbert and his kid brother Oswald put in regular appearances, always in need of calculus to get out of one predicament or another. On one exam, students had to use Simpson's rule to estimate the volume of Mrs. Simpson from several of her measurements. Students didn't do any better on these than on more sober tests, but at least they got entertained for their Cs. One mathematician confided to me many years later that he had gotten his college spending money by selling these MIT exams as practice calculus exams, with solutions, to his fellow-students. His brother was one of our grad student TAs at the time, a gifted teacher himself.

In the early sixties, after lecturing a couple of times, I started more serious activities with the single- and multivariable-calculus courses, writing notes and problems. I had various projects.

At that time, the multivariable course ended, like all the textbooks for it, with double and triple integrals. But the core physics program needed vector integrals—line and surface integrals—to calculate work and flux in gravitational and electromagnetic fields and to frame the physical laws governing them. This was hard on the students; I thought to help them we

Hiking in New Hampshire, c. 2003; from left to right: Arthur Mattuck, Monika and David Eisenbud.

should teach them some vector integral calculus, too. I wrote some notes; George Thomas sat in on my lectures one semester and produced his own written version for a new edition of his calculus text. Competition made the other popular calculus texts include the material, which is now standard in the multivariable texts, though taught principally in the more technically oriented courses and schools.

Another project was to put qualitative work into the beginning calculus, by giving problems requiring qualitative rather than calculated answers. One way of doing this is to emphasize estimations and approximations. For example, they have to know that $\sin x$ is approximately equal to x when x is near zero: that's used very early in the physics course. So that was in the first calculus lecture. But the reaction was always uncertainty and mild panic. Calculus is about getting answers, exact answers, and now he's talking about something where it's not an equality sign but a little double-wiggle sign. So when they produce an answer, it's always accompanied by [anxious tone] "Is this what you wanted?"

Another panic-inducer was to ask qualitative questions about functions—their relative rate of growth as x becomes large, for instance, or whether one property of a function implies another: does a continuous function have to be bounded? What if it is also periodic? Many of the students did not like qualitative mathematics. I'd say about a third thought it was great, another third was neutral but could learn to do it, and a bottom third really disliked it: "That's not what calculus is about—calculus is about getting answers."

At one point Charlie Townes, the laser physicist who was then MIT's provost, chaired a dinner meeting of some representative faculty to discuss what was happening to calculus at MIT. It all seemed pleasant enough, lots of questions and meaningful communication, and at the end, he wrapped things up by saying, "So we're agreed, then, that we're going back to the traditional course." I guess my antennas must have been down during the discussion, because I was really surprised and said, "Of course not!" The meeting ended in confusion, and an economist came over afterwards and said, "My, you're quite an infighter!" but it was simply my sheer obliviousness to the cues they had apparently been dropping all evening.

All this was recognized in due course by the students. A Big Screw contest was announced: by depositing money in a parade of jars in the main building lobby, students could vote for the most-deserving teacher to be awarded a three-foot, wooden, left-handed, home-made screw. For the first two years I was the winner. I still have it: for many years it was the centerpiece on my office chalk tray. The third year, they switched the prize to an aluminum screw, which the winner only kept for the year. I won that one, too, a while later, after which, like any good show dog, I retired from competition.[22]

Over the years, exactly what "deserving" meant has been left up to the students who hang around the money jars seconds before the contest ends, ready to deposit enough checks to make sure their candidate wins. One hugely popular calculus lecturer, Ted Shifrin, actively campaigned for the prize (and won) by giving half his lecture in French the day before the contest closed, but a more typical winner might be the administrator who started the pile-driving for a new campus building adjacent to two crowded dorms on the first day of final exams.

One of the things which ultimately made me more sympathetic to the students' nervousness about qualitative mathematics was a study made by

[22] The Big Screw contest is a fundraiser sponsored by the service fraternity Alpha Phi Omega. The money raised by each candidate goes to a charity of their choosing. Originally called the Institute Screw, the contest began in 1967 as a spring semester version of an older fundraiser called Ugliest Man on Campus, sponsored by the same fraternity.

William Perry, a dean at Harvard, which put this in much sharper focus.[23] It was about students in the humanities, but it had a direct parallel to what I was observing in mathematics. It involved many Harvard students, from freshmen to seniors. The conclusion was that questions requiring judgment—questions where students had to tell, not who was king of England from 1630 to 1650, but instead to analyze a qualitative question: which was more important, this or that in the development of Europe?—represented a step up in maturity. And that it was reasonable to expect that as students advanced at Harvard, they would be able to answer such questions, but not so reasonable to expect beginning students to be comfortable with questions like that.

So I finally decided that I was asking for a certain mathematical maturity that was unreasonable to expect of average beginning students. I teach a calculus course now for students who've had what they call AB Calculus. That's the slower version of the Advanced Placement course; we give it partial credit by offering the AB students a six-week accelerated calculus course, which covers what they didn't have in high school. They come in expecting to breeze along for a while, but they are greeted the first day by approximations. I can do that because they've had a year of calculus, of getting exact answers in high school, and therefore it's fair to ask them to take the next step up. So exact versus approximate remains something that we continue to emphasize. But for the students beginning calculus, I would not give them the sixties program; it assumes too much maturity.

A related issue is the proper role of proofs in the first-year course. I do not believe that students, even smart students, need to study mathematical proofs as part of their fundamental training. They would be better off studying more mathematics of a traditional kind: how to calculate different things, how to solve real-world problems, by mathematical modeling—translating the physical problem into mathematical equations to be solved, then using standard mathematical techniques to solve them. And I think that people agree with this. I've discussed this several times with Singer, for example. He remembered very vividly how excited he was to be able to calculate the volume of a sphere, using calculus—you know, just to *use* these mathematical techniques.

---

23 William G. Perry's decade-long study identified nine stages through which undergraduates progressed in their perceptions of the world. He reported these findings in 1968 in *Forms of Intellectual and Ethical Development in the College Years* (New York: Holt, Rinehart and Winston, 1968, 1970).

Newton's work was mathematical modeling. The planets went around in ellipses. Why? How to model the motion with equations whose solution would be ellipses? He assumed that $F = ma$ and that the force F was an inverse-square force directed toward the sun, and that gave ellipses. For 150 years they used calculus to solve problems; they didn't worry about the proofs. They used infinitesimals and all sorts of things that you would consider unacceptable today and that were even made fun of at the time. "What kind of mysticism is this?" "Well, we got the right answers using it!" They had procedures that produced answers, and they wanted to extend those procedures as much as possible to solve more and more real-world problems. The ones who straightened out the mess came afterwards; the ones who said, "Oh, no, no, we have to have exact definitions. We have to be able to prove theorems convincingly." That didn't start until 1810, 1820 or so, when people were running into contradictions and paradoxes.

So the first-year courses place much more of an emphasis on the practical applications, without the rigor.

Absolutely. The rigor is taken as intuitive. In other words, I draw a curve. You have some idea of what the "area under the curve" means. But there are relations involving the area that are not so obvious. In general, I'd give a proof or a partial proof only of the very few things that really would not be believable without proof. There's something called the fundamental theorem of calculus, which, without some argument, is not believable. That was one of Newton and Leibniz's basic discoveries. It took two thousand years to come up with what they discovered, and even now, if you tell somebody, you have to explain why it's true. It's not obvious as a formula: you have to draw pictures and show them why it's true. So I do that, of course.

But we don't teach the mathematical rigor—that was introduced in around 1820 or so by Cauchy, a French mathematician.[24] They were running into problems, and he was the one who wrote down in their modern form the definitions that plague some first-year courses now—limits, derivatives, continuity; he wrote down definitions for these using Greek

---

24 In *Men of Mathematics,* E. T. Bell wrote of Cauchy, "The methods he introduced, his whole program of inaugurating the first period of modern rigor, and his almost unequaled inventiveness have made a mark on mathematics that is, so far as we can now see, destined to be visible for many years to come."

deltas and epsilons. So the older calculus textbooks would begin with these rigorous definitions, and the high school advanced placement course would spend a month on these things. When I spoke once to the advanced placement teachers, telling them we didn't bother with the rigor—we just took it as intuitive and didn't spend time on it—some told me, "You just changed my entire course!"

They went overboard with the rigor. They *still* tell their students, "You should retake calculus when you get to MIT, because it'll be much more rigorous than what we've done here." But it isn't! The students tell me in September, "This is what my high school teacher said." I always tell them, "Forget it. If you passed the AP test, please take the next course. Don't repeat what you've had. Learn new mathematics." It's true that some will need rigor later on: there are certain science and engineering subjects that use such sophisticated analysis that you can't really study them unless you're comfortable with exact definitions and rigor. For them, you have to go back and start learning how to read and make proofs, but most users of calculus will not need that.

Calculus has always been constrained, in my view, by the advanced placement program. Cracks in that are starting to appear now, but for forty years we've supported the advanced placement program, which means that there's a standard advanced placement test whose content is agreed on by committees of mathematicians, basically from the MAA: this will be the standard course, this will be the standard test; each college decides what grade it will accept for AP credit. A whole industry has grown up around advanced placement. *Newsweek* ran a cover story, "The Top 100 High Schools."[25] What are the criteria for judging the top 100 high schools? Each high school is assigned a single number, which consists of the total number of advanced placement examinations taken divided by the number of students in the graduating senior class. So you boost your ranking by giving advanced placement courses and making sure that your kids take the exams: the tail wagging the dog.

But it's not just *Newsweek* doing the wagging. I remember vividly once, talking to a group of the Boston area's AP math teachers and being asked: how important is it, in getting into MIT, to have advanced placement courses under your belt? I had inquired about that, and I knew the official answer: You didn't have to have advanced placement courses; it was not a significant factor. Rather, it was letters from teachers saying you were a bright student,

---

25 Published in May, 2005.

grade point average, stuff like that. So that's the answer I gave, and the entire room burst into laughter. "That's what you may think, but it's not true!"

And so I wrote afterwards to the MIT Admissions Office, "My impression has been the following; was that wrong?" And I got a very craftily-worded answer back: "Yes, well, of course, naturally, we...." So I never said that again. But I don't know if they actually factor it in. They make two numbers, an academic number and another that tries to give an overall measure to nonacademic things. I don't know how or if the number of AP courses weights the academic number, but they see it on the record, and mentally it operates—it's taken as a sign of how ambitious you are, how anxious to get ahead, proof that you can do college-level work.

So you said calculus had been constrained by this.

Right. Once you agree that you're going to cooperate with the advanced placement program, then there's a standard curriculum, and that's what you need to cover. The only difference at MIT is speed. We cover most of the AP course in one semester: what's left out is done in the differential equations course. Most schools take two, or a little less than two, to cover that material. Not all schools: the last time I checked, for example, Caltech was giving its first-semester course differently, spending the middle third of the year on differential equations and teaching the calculus much more theoretically than we do. In other words, Caltech says, "We teach our own version of calculus. We can give you some credit, but you have to take our calculus, unless you can pass our own AP exam on it." It's a decision: it's up to them.

But AP credit is important to many students. AP credits give them flexibility in their program, allow more elective time, maybe let them get two degrees or graduate early. That's very important, I think. After all, math is math. If we think the AP course is deficient in some ways, we can make up for it in the next course.

Why is calculus so fundamental to everything these students are going to do?

Well, it's not a given. There are schools that insist that discrete mathematics is what students should take in their first year, or at least share fifty-fifty between discrete mathematics and calculus. For future computer scientists or management students, discrete mathematics or linear algebra is probably more important than calculus.

But answering for MIT: our freshman science core includes one year of chemistry and biology and a year of physics. The first semester is basically mechanics—the laws of moving bodies—while the second is electricity and magnetism. And it's really impossible to study either of those subjects without calculus. One-variable calculus is not quite enough for mechanics, but they have various theorems that enable you to get around that. For electricity and magnetism, you really must know something about multivariable calculus.

MIT's mission has always been, historically, to educate engineers; engineers must study physics; and anybody who studies physics must study mathematics. And so at MIT, at any rate, there's been over the years no challenge to that.

There's also another reason. Freshmen here sign up for a major, most of them, in the middle of their second semester. In other words, they register for their freshman courses without officially having a major. Many of them have an idea what they want to study, but many do not. It's different now. Fifty years ago, kids knew they wanted to be engineers or space scientists. Now they're just bright kids who got into a good school: my impression is that many are less focused on something when they enter. If you ask them, many will say they don't know. Or sometimes they'll say electrical engineering, just because you can study almost anything in the electrical engineering department; it's so large. You could be a computer scientist, you could study cryogenics, design music keyboards [laughs]; almost everything seems to have some connection with electrical engineering.

There's a strong tradition at MIT that students do not have to select a major before the middle of their second semester. They're not to be channeled at the beginning; any of the core courses they take should be adequate to whatever major they want to take. That's the principle.

So for example when a group of materials scientists said, many years ago, "We want to teach core chemistry, too; but we want to teach it *our* way," MIT let them do it, but it didn't ultimately work out exactly as expected. They had anticipated teaching primarily the chemistry that materials scientists and electrical engineers needed to know, and then they found that this was considered a violation of the principle. Students with too narrow a focus could not then decide to major in something else—could not become chemistry majors, for example, because they hadn't learned certain basic things that the chemistry department felt they had to know.

In many other schools, a whole set of alternative first-year courses has developed. If a lot of their students go on to study management or business or even biology and do not have to take physics, they have courses like minimal calculus for biology and management majors. In molecular biology, you're analyzing DNA sequences, comparing, for example: Are these two organisms related? How much DNA do they have in common? That's a mixture of combinatorics and statistics, so they might start off with a semester of discrete mathematics, rather than calculus.

Self-Paced Teaching

Another teaching innovation the department experimented with for a while was allowing students to study calculus at their own pace.

Yes. In the seventies, self-pacing was coming in, and that struck me as a very good idea, so I put some energy into that. Students could study calculus at their own rate, taking an exam whenever they're ready, repeating it as often as necessary if they failed it. The students liked it a lot, because MIT was viewed as a pressure cooker, and studying at your own pace and being able to repeat an exam you didn't do well on the first time was viewed as taking off some of the pressure. It showed them that the mathematics department cares.

In the beginning it was an enormous amount of work. There were weekly exams that students had to pass on their own. These weekly exams had to exist in several versions, so it was a lot of exams to make up! And there was a lot of individualized tutoring to get students to pass it, and integrating this with running a course in general—it was a lot of work.

But the basic flaw in the system, which should have been obvious *a priori* but was not, is that if you allow exams to be repeatable, a student studies for a repeatable exam, which means that some present themselves at the exam without studying, just to see what it's like. Well, you try to head that off by giving a sample exam. Okay, they look over the sample exam, "It doesn't look so hard; I'll still try it." It's like the viruses and the humans: the human makes a move with a drug, and the virus makes a countermove by developing an immunity to it. And so it turns passing a course into a game, and that can get very discouraging for a teacher. Tutors are there to answer questions, but nothing prevented a student from saying, in effect, "Hey, I didn't bother studying any of this. Teach me what I have to know to pass the exam." Some tutors acquiesced, but others didn't, and I don't blame them. You know, "God helps those who help themselves": do some studying first. So the system had flaws.

Arthur Mattuck, c. 1982.

On the other hand, MIT has always been characterized as a school where the pressure is more intense than other places. You're expected to learn faster, you're expected to work harder. Over 60 years ago, as a thirteen-year-old Boy Scout, I knew MIT was a pressure cooker. That was the reputation it had in Brooklyn. "Ohhh, you're going to MIT? Oh *boy,* I pity you." Courses went faster there, you worked all the time, it was terrible.

The MIT administration worried about "pace and pressure." A committee was appointed to make recommendations. Paul Gray, as he rose through the administration, made it one of his chief goals to somehow reduce the pace and pressure at MIT, but by the time he finished his term as president, his conclusion was, it's impossible: it's part of the culture of the place.

Why is that a problem? If you're really good, you come here. If you're not, there are plenty of other schools to choose from.

Well, you don't want nervous breakdowns. Education isn't supposed to be a time of suffering. Harvard and Princeton alumni have nostalgic memories of the wonderful time they had as students and leave huge bequests to their alma mater. At MIT that's less common, by and large. Many felt under stress while they were at MIT. Sure, they learned a lot—but they don't harbor sentimental memories of the place. Lowering the pace and pressure would make the MIT experience a more positive one.

I think the reason the self-paced system did not disappear at MIT entirely, the way it did at most other schools that tried it, was that from the beginning it was hailed as a very concrete way of lowering the pressure on students. Okay, they failed the exam; you would give them another chance at it, and still another chance, as many chances as they wanted. And so they were never completely at sea in the course; they would never have to drop the course if they were willing to just keep learning at their own speed. The math department got credit for that.

It seemed like a great idea at the time, but it's now pretty much disappeared from the curriculum, and the schools that tried it virtually all dropped it, as far as I know. I don't keep in close touch. Alan Guth tried it recently for a couple of years in freshman physics, but it got supplanted by a different teaching system. For many years it made the math department distinctive in its large freshman courses. But when we saw students were taking unfair advantage of it, we started limiting the number of repeat exams a student could take. Then we started, bit by bit, cutting down on the number of exams. They weren't weekly or biweekly exams any more; they went down to five a semester, and I cut this down to a more manageable four when I recently taught first-semester calculus.

And now they can repeat the exam only once. So if they take an exam on Thursday, they're told by Friday, the next day, if they've passed it or not, and they have any one of four days the following week to take a repeat exam, but they can only take it once. So it's cut down a lot, but it's still not hopeless: if they're willing to study over the weekend and make a serious effort, most of them do pass it, or come close to passing it, the second time. So it's a compromise. It still reduces the pressure somewhat and gives them some hope. Self-pacing disappeared, but its pressure-lowering aspect has persisted in some form now for thirty-five years.

The Pure/Applied Split

To backtrack a little bit, one of the things that happened at the end of the sixties was the pure/applied split, surely one of the forces that most profoundly shaped the department. What are your recollections of that?

Well, Ted Martin kept the department together with a fairly firm hand, but after he retired[26] there was a question of how the department would

26 Martin retired in 1968, having served as department head for 21 years.

be run, and some committee met to try to decide this. There were pure and applied mathematicians before then, of course, but I don't know the process by which faculty were appointed, or maybe it never really became a problem. But the essential reason for the split was the different views of the pure and applied mathematicians as to what constituted outstanding work in their area. The groups had different criteria and different mathematical goals. Pure mathematicians want to prove a great theorem that solves a difficult and well-known problem, or that turns out to be important elsewhere in mathematics, or that starts up an important new area of research. Applied mathematicians are less interested in proving theorems; they want techniques that give mathematical insight into physical or real-world problems, and they place a high value on interaction with scientists and engineers.

Classically, when this happens, the groups do not agree on the caliber of proposed new appointments or promotions. The solution is to become separate departments. So for example, at Harvard, the mathematics department is the home of pure mathematics, while the applied mathematicians are in the Applied Science Department. It's only by being a separate department that you get control over your budget and appointments. But the MIT administration, I believe, was flatly unwilling to do this. They saw no reason to have two separate departments.

So the solution that was worked out here was that there would be a pure group and an applied group, which could use their own standards for proposing appointments. Then there would be a Departmental Council, composed of a certain number of people from each group, plus the chairman, plus others *ex officio,* who would then look critically at the two groups' proposals for appointments and promotion. That would give the two groups a place to confront each other face-to-face, as it were. Cases that got through the Departmental Council would still have to be presented afterwards to the tough Science Council, composed of the science department heads and the dean of science, so the Departmental Council wouldn't be the last word. As for the budget, it would be split between the two groups, each of which would have therefore its own budget for faculty, for instructors, and for graduate students; I think 2:1 was the ratio decided upon for the pure/applied groups.

At that time MIT had permanent department heads, by and large. Somebody was chairman until the dean of science decided he should be replaced; it wasn't a fixed term. But after Martin stepped down and the two groups were created, it had to be a limited term, because of the necessity of alternating the chairmanship between pure and applied people. Nothing else would have been acceptable.

So who would be the first chairman under the new arrangement—would he be pure or applied? Well, Levinson had credentials in both camps. He was certainly viewed as a pure mathematician with his work in complex variables. On the other hand, some of his work in differential equations and complex variables had applications, and the applied mathematicians viewed him as somebody who appreciated the importance of applications. He was a student of Wiener, who had, I think, later in life gotten more interested himself in looking at real-world problems and trying to devise mathematics to solve them.

Levinson had both the status and the broad appeal, so he was the natural choice to be chairman. I think Levinson only served as chairman for three years. It was purely an interim arrangement: he didn't want to do administration, that was clear, but he offered himself just for a limited time, for the department to get things settled down. The new agreement also required chairmen for each of the two groups, chosen by the overall head, with the advice and consent of the corresponding group. The first pure chairman was Ken Hoffman, and Harvey Greenspan was the applied chairman; there was never any question about that, since it was Harvey who had been the group leader and had the strongest ambitions for it.

After Levinson stepped down as chairman, it was Hoffman from the pure side who took his place; Hoffman was chairman from '71 to '79—quite a long time, because I remember there was some grumbling at Hoffman's five-year term being extended by the dean of science. And then from '79–'84 it was Dan Kleitman, '84–'89 me, '89–'99 Dave Benney, who served for two terms, then Dave Vogan,'99–'04, and now Mike Sipser, starting in '04.

How did this new separation work for the department?

I think overall it has worked out pretty well, but several things have operated over the years to blur the lines somewhat. For instance, graduate students apply to and are admitted separately by the two groups, according to that group's budget, but they are free to switch to the other group, if their interests change. The physical separation—pure on the first two floors, applied on the top floor—is breaking down, since available offices have to be used for whoever needs one at the time. The graduate students' desks are getting mixed in together for the same reason.

Perhaps more fundamental is the enormous growth in the importance of discrete math over the last forty years. At the time of the split, applied mathematics was envisioned largely in terms of what would now be called continuous applied mathematics. At the time this was principally fluid mechanics, with Lin, Greenspan, Benney, and Orszag; elasticity theory, Reissner and Wan; astrophysics, Toomre and Lin; and theoretical physics, Cheng and Bender. There was also Wadsworth in probability and statistics. I think those were the principal areas.

Many of the applied mathematicians upstairs now are discrete, but at the time, it was not yet thought of as a significant part of applied mathematics. One of the first examples was coding theory, which essentially started with Shannon, the founder of information theory, who had been on the MIT faculty.[27] Well, information theory was partly probability theory, but also partly discrete mathematics, using ideas of algebra, combinatorics, and so on. So all of a sudden, discrete mathematics was needed for information theory; it was also needed for computers, which had to do everything discretely. So appointments started being made in these applied areas.

Levinson tried to educate Harvey [Greenspan] in this [laughs]. He wrote an odd paper, not in his field, which he published in the *American Mathematical Monthly*. The *Monthly* is a college-oriented journal; it's the principal magazine of the Math Association of America, which is more teaching oriented and less research oriented than the American Math Society. Levinson's was a lead article, giving an exposition of how the theory of finite fields and modern algebra was being applied in coding theory, building redundancies into the transmission of information so that if lightning strikes the line and alters a few digits, the message will still be readable.[28] It was just a simple, routine exposition of a few different codes, and how their construction and workings used the theory of finite fields, in words that he, Levinson, could understand, since he was not an algebraist.

And I asked him about it, I said, "How come??" And he said, "Well, to be honest, it was mostly for Harvey's benefit," a signal that there

---

[27] Shannon received a Master's in electrical engineering and a PhD in mathematics at MIT, both in 1940. He returned as visiting professor of electrical communication (1956), then served as professor of communications sciences and mathematics (1957) and as Donner Professor of Science from 1958 until his retirement in 1978.

[28] Levinson, N. (1970). Coding theory: A counterexample to G. H. Hardy's conception of applied mathematics. *American Mathematical Monthly 77*, 249–258.

was getting to be more in applied mathematics than the classical continuous applied math. I think he reported that Harvey's response was something equivalent to "One swallow doesn't make a summer." But that might have been just his usual truculent response to someone urging on him something he had already been thinking about, since while he was the applied chairman the group did acquire some great faculty in discrete math, both combinatorics and theoretical computer science. I think Rota's moving from analysis to combinatorics—in MIT math that meant moving from the second floor to the third floor, pure to applied—helped, too.

And there got to be more applications. During the time that we had Len Adleman and Adi Shamir, in joint work with Ron Rivest, a theoretical computer scientist in MIT electrical engineering, there was a breakthrough in cryptography, the Rivest-Shamir-Adleman algorithm,[29] which I think is the most popular one used now to securely encrypt information. The basis for the RSA method was a fundamental but easy little theorem in number theory, for which probably nobody except algebraists and number theorists had ever found any use. It's called Fermat's little theorem; you can prove it in a few lines. I was always amused by that, because I used to give this course that everybody had to take, "Intro to Algebra for Computer Scientists," something like that. Similar courses were given everywhere, and a lot of books were written for them, giving all the discrete mathematics, all the theorems of algebra and number theory, that they thought computer scientists ought to know. Before the RSA method appeared, not one of them mentioned Fermat's little theorem, even though it was in *every* book and course on number theory, but after the RSA method appeared, all those textbooks were quickly revised [laughs].

This is what pure mathematicians, or pure scientists, are always claiming, right? "You might think that it has no applications now, but we should push this boundary, because..."

That's right! Because if something is sufficiently fundamental, sooner or later it's *bound* to be used somewhere. You know, the tensor calculus was invented *well* before Einstein's general relativity, which was lucky for him because he could not have developed general relativity without it.

---

29 See Rivest, R., Shamir, A. & Adleman, L. (1978). A method for obtaining digital signatures and public-key cryptosystems. *Communications of the ACM, 21*(2), 120–126.

And tensor calculus had not had a lot of practical applications before then?

Well, it was only around twenty years old at the time. It was developed for studying curves and surfaces and higher-dimensional manifolds by what's called differential geometry: measuring distance along geodesics—shortest paths, like the great circles on a sphere—studying curvature, geometric ideas like that. Nowadays it is widely used. Engineers use it, for instance, to describe and study stresses and strains on their constructions; it's important in elasticity theory. Conceivably other topics in differential geometry could one day be of importance to these or similar applied fields.

How do applications of discrete and continuous math differ, broadly speaking?

I think it would be right to say that the way discrete mathematics is used to solve practical problems in coding or encryption is very different from the way continuous mathematics is used to solve problems in, say, fluid mechanics. That's modeling, and modeling is a genuine art, requiring a physical feeling and experience; the first is more like cleverness—knowing the right fact and seeing how it can be used.

Modeling is different. You look at the real-world problem; it's about quantities. There are these numbers involved, so there are variables, and basically, you have to find relations between the variables. Now, there are standard ways of doing that, but normally, they're too complicated, and the art consists of the approximation: okay, this variable isn't important; let's at first neglect that and see what happens. Or something interesting happens when x is near zero, so let's study that case first; maybe from that we can learn how the rest is going to behave. In other words, a tremendous amount of art is involved, of experience, of using what you see in the real world, having a physical feeling for what's important and what's not important.

For example, Lorenz, the mathematical meteorologist who discovered the phenomenon called chaos, was studying a problem of convection in the atmosphere. There were seven variables, and he had to decide that three of these were more important than the other four, and that he would probably be able to understand the phenomenon if he just focused on these three, and for these three he could modify the equations thus and so. That's experience. Experience, judgment, and something that I imagine one could be intrinsically gifted at. I know I'm *not,* personally. I could never do applied mathematics.

So what role did the growth of discrete mathematics in the applied field play in the relations between pure and applied?

Well, in a sense it blurred the distinction between the two. Actually, I think that's good for the health of the department. Graduate students who were admitted to the applied math group to study combinatorics, say, might end up writing a thesis with the pure group in Lie algebras, and vice versa. Two of the leaders of the discrete applied group, Rota and Stanley, in combinatorics, would have been called pure mathematicians anywhere else. In the opposite direction, Dudley, a probabilist and statistician in the pure group, for a while joined the applied group. Strang, in the pure group, was elected president of SIAM, the Society for Industrial and Applied Mathematics. Go figure.

Which brings up another facet to the split. Exactly the problem that had caused applied mathematics to want a split—namely, the need for being able to make their own appointments, on their own budget, using their own criteria for who was a good applied mathematician—that same problem was presented to them by statistics.

Statistics is part of applied mathematics; nobody had ever suggested anything else at MIT. But the statisticians said that applied mathematicians did not understand what made a good statistician; that the criteria they used were not the right criteria by which to judge a statistician; and therefore, that they wanted either, at a minimum, to be a separate subgroup in applied mathematics, with a separate budget, or better, to be a separate department, as in some good schools: Stanford for instance. Harvard has a small but separate group, not part of the math department.

The issue came up because MIT had succeeded in attracting from Stanford a top-flight statistician, Herman Chernoff, who would attract younger statisticians and build up a good, solid statistics group. But he met with total frustration, because the younger statisticians could not get tenure within the applied math group, so they didn't stay.

Herman came to MIT because he dreamt of making statistics a force in engineering, to make engineers aware of the importance of statistics in their work and of designing the things they did according to good statistical principles. A statistician's standard complaint is that after the experiment is all set up and run, then they're called in to analyze the statistics. They find all sorts of flaws in the way the experiment was designed, and they complain, "You should have called us in when you were designing the experiment, rather than asking us to justify your inadequate data."

But he found that engineers tend to regard mathematics as trivial and statistics as *very* trivial. You know, you just look at a statistics book, apply a few tests, and that's really all you need to do; that's what their culture says, and therefore, the answer to him was "We can do our own statistics, thanks. We don't need help from mathematicians." So the only people who would talk to him seriously were the people in the Sloan school. I think Herman considered trying to have the statistics group under the aegis of the Sloan School, but that wasn't his dream, as it were.

So after a number of years of this two-fold discouragement in the promotions and with the engineers, Herman finally left MIT for Harvard.[30] He wanted to stay in the neighborhood, and Harvard was happy to have a great statistician.

Thus my five years as chairman were plagued with the statistics problem, how to attract and retain statisticians in the face of these difficulties. I remember spending a fair amount of time trying to get names of possible tenure candidates who might be interested in visiting. Finally one did come, a top-flight statistician who was unhappy at Harvard, and so it ended up an exchange—Chernoff from MIT for Peter Huber from Harvard. Unfortunately, Huber stayed only a few years and then returned to his native Switzerland.

The statistics problem has never been resolved. We still, I think it's fair to say, are limping along. Usually there are one or two younger people in statistics who are good but don't stay. The administration refused to even consider having statistics as a separate department. They had a strong belief that to guarantee its excellence, it had to be part of a strong mathematics department. They wouldn't trust any other arrangement.

So the hiring and tenure process was really the main thing that strained relations between the pure and applied groups. Were there other major problems?

Well, outside of how to divide up the overall budget, that's all that matters, really. What makes a group is who you appoint to it and in what field you will appoint them. You know, pure mathematicians fight over the fields, too: "Look at what's happening in number theory. Stark has left. We have nobody in number theory now!" "Yeah, but Lie algebra is more important for MIT!" Arguments like that.

In general, when you have a couple of people in the field, they want others to talk to, and they yell quite loudly on the committee, "We gotta go after

---

[30] Herman Chernoff came to MIT from Stanford in 1974, then moved to the Department of Statistics at Harvard in 1985.

this guy, he's really great! Then we'll have a *really* solid group in topology"—or differential geometry, or partial differential equations. It's hard to get an orphan field into consideration unless the candidate has done prize-winning work and the field is viewed as one that we really have to cover in the future.

For example, there was a time when the theory of finite groups was hot. They were aiming to classify all the simple groups, a major open problem in mathematics and a huge effort. But the pure mathematicians did not appoint anyone in that area, because they considered the field difficult and specialized: group theorists talked only to each other. And since there was no resident group theorist who wanted company [laughs], we had nobody in group theory.

That was also for many years a criticism of logic. There's a well-defined field, mathematical logic, but logicians talked a language that nobody else could understand. They had relatively little contact with the rest of mathematics. That's not true anymore, but it was, forty, fifty years ago.

Like that old poem about Boston, "where the Lowells talk only to Cabots, and the Cabots talk only to God."

Yeah, that's right, but it wasn't snobbishness; it was just that they had developed a certain language and a certain set of problems that interested them but didn't seem to impact on the rest of mathematics.

That's changed to a great extent because of a field in logic called model theory, in which some problems in regular mathematics were solved by applying methods of logic. The first incredible success for that was in 1964—a thirty-year-old problem in pure math first formulated by Emil Artin, halfway between algebra and number theory, which was solved essentially by logic. It was a collaboration between an algebraist and a logician, but when it was written up,[31] all the logic was removed from their proof, because the algebraist felt that otherwise nobody would trust it! Of course, somebody who understood logic well enough would be able to see through the subterfuge.

That was only the first such success. There have been others since, and now there are serious efforts to prove major open problems using model theory. So that's at least made model theorists socially accept-

[31] Ax, J. & Kochen, S. (1965). *Diophantine problems over local fields. American Journal of Mathematics, 87*, 605–648. Parts I and II. Ax, J. & Kochen, S. (1966). Diophantine problems over local fields Part III. *Annals of Mathematics, 83*, 437–456.

able [laughs]. In the mathematics department we had a fine one, Ehud Hrushovski, but he went back to Israel. Too bad: that was a major disappointment.

So the hope of a wider diffusion of knowledge plays a part in the hiring process as well.

That's right. In other words, that as many people as possible, both inside and outside the department, should be interested in or have some mathematical overlap with the area in which an appointment is being considered.

I think at MIT the fact that somebody should be able to talk to other scientists and engineers is of more importance than at an average school. From their point of view, math ought to support fields that have relevance to engineering, to what the rest of MIT does. This can create difficulties for us, though. One of our best—but too obliging—applied professors, Lou Howard, had to start coming in daily at 6 AM, to get three hours quiet time before the phone calls from the MIT engineers started. He ended up retiring early and moving to Florida State University to have a little peace and tranquility for his own work.

Undergraduate Chairman and Class of 1922 Professorship

In the early 1970s you started to get more involved in administration yourself, right?

Yes. I became the undergraduate chairman, a position that Hoffman created when he became chairman in 1971. He changed the administration: the pure committee and the applied committee already existed, but he also created a graduate chairman and an undergraduate chairman, positions which didn't exist before. He felt that we should have administration this way [gesturing horizontally] and that way [gesturing vertically]. Otherwise, who worried about undergraduate courses? Nobody. If some faculty member was willing to take an interest in a course or create a new one, fine, but it was left to chance. If we wanted to change the requirements for majors, there was no mechanism for doing that. The pure and applied committees primarily made faculty and instructor appointments in their areas; but the courses and teaching were left mostly to individual enterprise. Teaching assignments for the upper-level and graduate applied courses were decided at a fall meeting of the group, but the lower-level and service teaching tended to fall between the cracks. There was no supervision of teaching.

The Math Undergraduate Office was perhaps the first in the Institute, though most departments have them now. I had my own office upstairs, but I felt I couldn't run the Undergraduate Office without spending a lot of time in it, so I had a desk there too. I was busy getting things organized and trying to do administration, which I was totally unfamiliar with. I do things myself, and I don't understand how to get other people to do things that need to be done. There was a good secretary, Pat Sprissler, who kept the records of majors, an earlier innovation. But the secretaries whose job it was to make sure the large lecture courses ran smoothly required closer supervision and changed frequently.

I assembled a group of interested pure and applied mathematicians who agreed to form an Undergraduate Committee; one of its first actions was to change the requirements for majoring in mathematics to what they are today for the general math major. I also started to supervise teaching, primarily the large-enrollment courses, observing lectures and recitations in person and videotaping.

We centralized most of the undergraduate teaching assignments in the Undergraduate Office. During Levinson's administration, the applied math group had met by itself to vote on who should teach what it considered to be its courses, graduate and undergraduate. So this led to conflicts with the Undergraduate Office over staffing elementary courses like differential equations, advanced calculus, and complex variables, which could be taught by pure as well as applied mathematicians. At some point we were able to reach an accommodation.

We also got some new courses going, stuff that we hadn't given before. There was a one-semester probability and statistics, oriented toward non-math majors—biologists and others. George Thomas volunteered to give that, and I think he had over a hundred students—more, in its heyday. Gil Strang started an Applied Linear Algebra course emphasizing matrices, writing a text that spread the point of view everywhere; that became a large-enrollment lecture course. Mike Artin started a new, year-long algebra course, using group theory to teach linear algebra and vice-versa. He has taught it more or less continuously since the mid-seventies, developing it, changing it each year—just like his father did at Princeton—and finally producing a book. It's a big book that is by no means easy, but it was picked one year as a science book club selection.

We had some excellent teachers, but unfortunately that didn't always help them getting tenure. One was Frank Morgan, an extremely energetic and well-loved lecturer, who by mid-semester would know the names of most of the 200 students in his calculus lecture, partly from having invited them in small groups to dinner at his apartment. He had been an MIT undergraduate and wanted to create a greater sense of community and give students something to talk about at dorm dinners besides problem sets, so he started an All-Institute Colloquium—a well-known speaker, a faculty panel, and questions from the student audience. He loved to draw students into his mathematical specialty, minimal surfaces, by entertaining lectures featuring wire outlines dipped into soap solutions. He left for Williams College, now known nationally for its undergraduate math research program in minimal surfaces.

Another was Carl Bender, a theoretical physicist in the applied math group. He was a really outstanding teacher, a very gifted lecturer who played a big part in revising and revitalizing the large differential equations course. Since MIT he's had a good career at Washington University in St. Louis. He had his revenge: for many years its team frequently won the intercollegiate Putnam math exam, and he was its secret weapon—their coach.

So the message at that time was that if you want tenure, we need major research accomplishments. If you have time and energy left over for teaching, great.

Arthur Mattuck, Class of 1922 Professor, c. 1970.

That's right. As undergraduate chairman, I was *ex officio* on the Departmental Council, which reviewed the recommendations for tenure of the pure and applied committees. In those days, if the discussion turned to the candidate's teaching, it was interpreted by the naysayers as meaning the research wasn't sufficiently ground-breaking to support tenure. The same interpretation was placed on letters that talked about the teaching.

The situation seems to be better now; I hear that by and large someone who is a poor teacher will not be considered for tenure, unless they've proved the Riemann hypothesis. Short of that, they have to be at least adequate teachers. But frankly, for the big lectures, the department needs more than "adequate." It has to have people who are *good* at it, and it's a persistent worry. Undergraduates pay a lot of money now, and it's a problem. Fortunately, over the years some of our tenured faculty have turned out to be fine lecturers, even if that's not why they were tenured. So I think we are in pretty good shape now on the teaching front.

A year after becoming undergraduate chairman, I was made MIT's Class of 1922 Professor, which at the time was the most heavily endowed chair at MIT, a recent fiftieth reunion gift of that class. It was designated for distinguished teaching, and it removed me from the department budget, allowing the use of my salary for other purposes, money that was sorely needed. I think it was supposed to be just for five years, but they kept renewing it all the time. I guess I held it for twenty years—'72 to '92, something like that.

I'm sure it was Ken Hoffman who was behind this. He was the department head. Rosenblith was provost at the time,[32] so the chair was his decision. I think he'd been harsh on the department's budget: it was math; it didn't need lots of money for labs and stuff that would drain MIT's resources—but it also didn't bring in a lot through huge government contracts, no chances of lucrative patents, either. So the math department was sort of—not the poor relative, but it didn't get first dibs on money. Later Rosenblith said to me casually at one point, "Well, we felt we should do something for Ken." Ken had spent his spare time for the previous two years chairing the MIT Commission, developing a response to the student unrest of the late sixties. So I assume that they decided that if there was somebody in the math department who would not embarrass the new chair, they would give it to the math department as a way of helping out on its budget.

---

32 Walter A. Rosenblith served as associate provost from 1969 to 1971 and provost from 1971 to 1980.

With the chair, I had a strong feeling of obligation. To MIT first of all, for singling me out; to the math department, for providing a congenial home. And also to the expectations of the Class of 1922, then in their early seventies. They had me fitted for a red jacket to wear at their reunions. I was invited to some of their homes and got a sense of how proud they were of their chair.

Since it was a teaching chair, I felt I should try to improve MIT's teaching in what area I could, both in the math department and in things which might be picked up elsewhere if possible: that would be doing something not just for the students but for the math department as well. We were one of the only departments that didn't get bad press. Students complained about some of the core physics and chemistry teaching. We didn't get so many complaints.

Videotaping

Could you say something about the videotaping program?

That was my first effort at improving teaching. It started in the early seventies. By this time we had a lot more instructors. Graduate students were doing a lot of teaching, and there was no program for helping them do it. So I got involved with the problem of teaching: how to teach, how to critique teaching, and how to get others to teach better. It helped that one of my father's important duties as a high-school chemistry chairman was to observe each of the thirty-odd department teachers twice a year and submit a written report for each of them. Watching his conflict between what he observed and what he felt he could say was very illuminating!

Videotaping was just coming in as a possibility for ordinary people to do, so we bought a system and started videotaping, and I experimented with various ways of communicating the results.

In the beginning, I ran little teaching seminars for the TAs where they would comment on each other's teaching. First I would review the videotapes. I would take snippets of each, and then I would invite them in, four or five at a time, and we would sit for a couple of hours watching ten minutes of one tape and fifteen minutes of another, stuff which I thought was revelatory and gave something to discuss, and discussing what they thought, how it could have been done better, what their reaction was to it, and try to give them the idea that it mattered, that it was interesting; they could learn something from this.

But those seminars were just too time-consuming. Now I do something that's equally time-consuming, but the hours are more flexible. I look at the videotape myself, making extensive notes, and then I email comments, which take an hour or two to write. If I think they'll profit from it, I ask them to look at the tape and send me their reactions to my comments, but often they won't watch the tape. So sometimes I don't send my comments until the teacher has watched the tape and sent me *his* comments. And that generally produces results. The students like that it's being done, and the teachers, in general, learn something about themselves and become better.

Some people can have their lecturing improved a lot, and others can't. TAs who say "okay?" five times in every sentence, or—there are a million bad habits you can have, but those usually disappear as soon as you see yourself doing them on the videotape. Other things—people who are insensitive or cut students off, or don't listen to questions they're being asked but just answer their own questions—those are harder. For a sensitive teacher, I wouldn't have to write anything. I've gotten criticisms from the teachers of their own videotapes that were much more detailed and more perceptive than anything I could have written. Though I still find something to add, or just to congratulate them.

Before we videotaped, I used to do personal observing only, and the poorer teachers would always be in denial. They'd say, "*No*, I don't do that. You must have misunderstood." Now, with the videotape, I can write, "At 23 minutes"—and I quote what they say. "Listen to the effect of that on the class." Or, "A student asks a question, but your answer has nothing to do with what the student asked"—very specific things. So they can see the evidence for themselves. I don't have a good personality; if somebody's being difficult, I tend to get into an argument. If somebody says, "I *don't* do that," I say, "No, you do!" Sometimes we've gotten to words, and videotaping is much better. Some people are good at communicating criticism, and I'm not. Good teachers have told me years later how traumatic for them my comments were. So it's much better if they can see things for themselves.

Every first-time teacher gets videotaped at some point, a month into their teaching, say, when they've gotten comfortable with the class. Also, all new faculty members when they're appointed here, no matter how experienced; even visitors just teaching for the semester, or undergraduate recitation teachers—all get videotaped.

It's had various formats over the years. In the beginning it was a graduate TA who was the cameraman. Now it's done professionally, so the class has to be moved for that hour to a room in Building 9, which has concealed cameras all over so that different angles can be shown. The videotape is made for the hour; the students know they're being videotaped. We pay for it, I think it's about $100 an hour. It's more professional and it's in color, but the results are the same as when we had a graduate student and just had a moveable camera up in the room.

These days our program to improve teaching is formal and structured: this started while Mike Artin was Undergraduate Chairman. TAs who haven't taught in the department attend a three-day workshop before term starts, where they are videotaped presenting the solution of a calculus problem to the "class," which consists of the other TAs there. Then everyone gets to comment on the presentation, and the TA gets the videotape to look at afterwards. This goes also for undergraduates who are being considered for recitation teaching.

Then, the semester before the one they will teach in, they go for a couple of weeks to a recitation being given by an experienced teacher, observe, discuss it with him, and get to teach it themselves for part of a class period, with the students and regular teacher critiquing it afterwards. After this they are ready to teach a class by themselves for a semester, and they get videotaped and critiqued as I just described it.

What about the videotapes of your lectures that are on the Web?

That's a totally different videotaping. There's this OpenCourseWare initiative at MIT. Haynes Miller has been involved with the large differential equations course for a number of years, and he wanted to include it in OpenCourseWare and give it a comprehensive Web site. So in spring 2003, when I gave the lectures—I've been giving them frequently over the years—he asked me if I would be willing to be videotaped for the entire series of lectures, to put them on the Web site. In the beginning I prepared very carefully, figured out where I was going to put everything on the board. After the first three lectures I no longer bothered; it came out just as well—or just as unpredictably well or badly. So that was for the OpenCourseWare, and those lectures are up on the Web now, edited somewhat. For me it's a mixed blessing: now if I mess up a lecture, I can tell students to look at it on the Web site instead, but it also means I have to avoid jokes I made on the Web, except I can't remember what they were, and I don't have time to look.

The Torch or the Firehose

How did your guide to section teaching, The Torch or the Firehose, come about?

I think that must have been about 1980; it couldn't have happened while I was head. We had been working to improve the teaching throughout the seventies. Peggy Richardson (Enders, now) from the dean's office would usually be in attendance at meetings of the core lecturers, where we'd describe what was happening in our courses. I mentioned the math videotaping program, and she said it would be useful for the rest of MIT to have a little booklet summarizing what we had learned. There had been one written 40 or 50 years earlier, I think, by Professor Robley Evans, "You and Your Students," but it was somewhat dated and no longer published; attempts to update it hadn't been successful.

An energetic math graduate student, Susan Colley, who was an outstanding teacher herself, said, "Yes! We have to do this!" So we formed a little committee that met during IAP.[33] It had some faculty, administration, undergraduates, TAs, and of course Peggy and Sue. I ended up doing most of the writing, based on what I had observed over the previous ten years. I'd bring in drafts and we would discuss them, others contributed small sections, and finally a little pamphlet emerged. And it had cartoons by two undergraduate cartoonists, which made it a lot livelier.

It stayed unchanged until about 15 years later, when Peggy said, "This *has* to be revised, it doesn't treat women properly, it's sexist in the use of pronouns," various things like that, "and haven't other problems arisen?" So I said all right, I would try to revise it, and I spent a *very* unhappy summer. It had been written in sort of a joking style, and meanwhile in the interim, I had been department head [laughs], I wasn't feeling very funny any more. I was also a lot older, and I was blocked for about a month or so. I'm not a very fluent writer.

Finally, I've forgotten what got me started—desperation, probably, I had only the summer to work on this, and I wanted to go off somewhere—so I started scribbling and finally got it done. And there were new cartoons: Peggy recommended Pawan Sinha, a graduate student who drew a strip for *The Tech*—he's now a professor at MIT, actually, in the Department of Brain and Cognitive Sciences, and quite amused that his cartoons are still cur-

---

[33] The Independent Activities Period of special and varied activities that takes place during the month of January at MIT. For more complete information, see http://web.mit.edu/iap/overview.html.

rent.[34] We used also another cartoonist from *The Tech*—Chris Doerr, whose strips had a sharp insight into undergraduate life at MIT.

The booklet had been sold for many years to other schools by the graduate school office at MIT, by an energetic administrator there who used the money to travel to conferences. She was less energetic about distributing it free to the MIT graduate students, which was the original purpose. Anyway, as a result of that, it got a certain distribution, though I never heard a word about where she was selling it or how many copies.

But quite unexpectedly, this past year, Richard Olivo, of the Derek Bok Center for Teaching and Learning at Harvard, said he wanted to publish a Harvard version of it. He is a Smith professor with a strong interest in teaching improvement who is at the Center one day a week. I knew the Center had been distributing their own teaching manual for many years, but he said he liked this booklet better, because it was more fun to read. So he changed a few words here and there to make it adaptable to the Harvard culture, but otherwise reprinted it verbatim: *The Torch or the Firehose* with "Harvard edition" at the bottom. So it got distributed at Harvard. I thought that was nice. A more recent Harvard version adds some material on teaching humanities sections, with added cartoons by Sinha. There's a German edition and a Chinese edition, and a professor in Chile wants to make a Spanish edition.

Department Head

How did your appointment as department head come about?

They didn't have anybody. To find the department head, the science dean forms a committee, and the committee goes around talking to people and making lists. Since Kleitman in the applied group had just been head, I guess it was to be a pure person this time. John Deutch was the science dean, so it was his responsibility to appoint somebody. I was on the selection committee. I don't remember it very well, but we talked to people, and I tried to urge them, "Come on, it won't be that bad. You'll get help." I guess we recommended a list, Artin and Melrose the two principal candidates, and maybe somebody else. But these natural choices didn't work out, for different reasons.

A couple of people volunteered, but they were not on the committee's list. So Deutch said, "People on the search committee don't have immunity

[34] Pawan Sinha was the author of an award-winning comic strip *Tumbleweed Garden* for the MIT student newspaper.

any more" [laughs]. That had been a condition of our service, and I think it finally came down to Dave Benney and me, and we ended up offering ourselves, mostly because we didn't know what else could be done, there being nobody else. We both agreed to do it if asked.

Deutch ultimately picked me, probably since I was pure and Dave applied, or maybe because he already knew me and wanted to see a familiar face on the Science Council.

So there you were, head of the department—what did you do?

Well, there was a lot to learn. In the beginning, it was all trying to understand the department budget. Ken Hoffman had told me it was all done "with smoke and mirrors": the money for all our two-year instructors wasn't actually in our budget, because the administration didn't really understand what an instructor was—some sort of post-doc who for some reason also taught? As far as I recall, only the math department had instructorships. In other departments, post-docs were purely research people; post-docs who taught were appointed to the faculty as assistant professors. Having instructorships meant that math could give temporary two-year positions to promising new PhDs without having to go through the cumbersome faculty appointment process, soliciting recommendations and submitting the case to the Science Council. Moreover, the number of faculty slots was fixed by MIT, whereas the number of instructors was limited only by the money available.

The instructors were financed from sabbatical and other leaves taken by the faculty. Professors on sabbatical leaves were paid by MIT from a special sabbatical leave fund. Those on *non*-sabbatical leaves were paid by an outside source—NSF contract, Fulbright, the school they were visiting, etc. What happened to the money in the math budget designated for their salaries? Normally, MIT removed half of it in the case of sabbatical leaves, and somewhere between half and all of it for non-sabbatical leaves. But word got out that if you did that to math, they wouldn't have enough teachers to cover the classes.

So math was able to keep most of this money in its budget, available now for paying the two-year instructors. But there were always uncertainties about leaves, and my nails would get bitten way down when spring came and offers had to go out. The budget was not computerized, so the Administrative Officer would keep recalculating it in longhand as more information about leaves, offers, and acceptances came in. Meanwhile, I never did quite understand or even believe what he was doing, so I had also to do it myself.

Salary time was always traumatic. Faculty didn't know what other faculty members were making, but they could find out what their peers at comparable schools were getting: as they got NSF proposals to review, the first page they would turn to would be the last one, which had to list the proposed salaries for the people on the proposal. This led to a lot of difficult visits to my office, but I couldn't blame them.

Of course, the major things were the new faculty appointments and the promotions. As head of the department, I sat in on the committees, pure and applied. Typically the pure and applied chairmen are the primary worriers about the appointments, but sometimes the head has to push a little, or spend his own time, as I did with statistics, for example, in trying to find women and minority candidates for faculty positions or at least visiting positions. These were goals of the administration, but not of primary concern to the two committees, as far as I could tell.

Then you get the promotion and tenure cases. There were easy ones, like Sipser and Leighton. But there were some difficult cases, also. Near the beginning I took an early-promotion-to-tenure case to the Science Council, and it got turned down, probably as much to teach the rookie head a lesson as on the merits.

But even for the "easy" cases, preparing them is a lot of work. You have to supervise the requests for letters the committee sends out, because if they are faulty—in the wording, in the people they are sent to—they can be rejected out of hand by the Council, or even subject MIT to a lawsuit. You also have to write a covering letter. I worked very hard on the covering letters, trying to explain to the mathematical laymen on the Science Council and the Academic Council what the candidate's field was about and what role his work played in it.

Explaining Math to Nonmathematicians: Levinson and the Riemann Hypothesis

Explaining mathematics and the importance of particular fields of research to nonmathematicians sounds like a real challenge.

If you're trying to explain mathematics, you have to decide what the person is likely to know and is likely to find interesting. There's almost always some analogy or something I can find. I remember four of us chair-holders were asked to give short talks to the Corporation, the MIT version of a Board of Trustees; plus the lectures were announced to the faculty as a whole. So it was a good audience, a couple of hundred people. The other three talked about their fields, but I think Levinson had died not very long before, and there had been this dramatic story about his work on the Riemann hypothesis before he died, so I decided to speak about that.

I remember Levinson working on the Riemann hypothesis while he was the department head. He'd be in the head's office, but the desk would be covered with foolscap in his big handwriting. Each sheet would just have a few equations scribbled on it and be tossed aside, and then another one. So when you went to see him, you didn't see him making plans for the department renovations or studying salary figures [laughs]; what you saw was foolscap. He told me what he was working on: I wasn't in number theory, particularly, but I would understand a little bit of it. The previous work had been done by Carl Ludwig Siegel, who some people thought was the world's greatest mathematician. I'd got a chance to see him while I was at Swarthmore; he came to give a lecture, big monster of a man. He had worked on the Riemann hypothesis and gotten a result. Several times I went in to see Levinson. "So, you're solving the Riemann hypothesis?"

"Ah," he said [imitates disgruntled tone], "It's hard to get beyond Siegel. Siegel knew what was going on."

And then he came into my office one day [in a brighter tone]: "Hey!" he said, "I think he missed something. I think Siegel missed something."

Then the next I saw him, a few weeks later, he said, "Yeah, I think it's right." The standard form of the Riemann hypothesis—not the form in which one might see why someone would care, but the standard way of saying it—is that there's a certain function of a complex variable, the Riemann zeta-function, and you have to prove that all the zeros of that function lie on a certain vertical line in the complex plane. And so he came in, he says, "I think I can prove that a third of the zeros lie on the line." What Siegel had proved was that a positive fraction of the zeros lay on the line. For example, if this function has an infinity of zeros, which it presumably has, and you proved that five of them lay on the line, that would not be a positive fraction, because five over infinity is still zero. If there's a positive fraction, that means at least an infinite number of zeros lie on that line. Very small, maybe, but something you can get your hands on. But what Levinson proved raised that positive fraction to one-third. That's a lot of zeros!

Fagi [Levinson] had a theory that it was because of his brain tumor. You know, mathematicians don't usually get new ideas at that age, and Fagi thought that it was the brain tumor that was killing some pathways in his brain and opening up new ones, so the new ones had to take over, and that that somehow was leading to new ideas. I don't know, it struck me as not totally impossible. You know, you think in certain channeled ways, and if you can't think that way any more, then it's possible new pathways open up.

By the end I think he thought he had pushed the fraction of zeros on the line up to something like 99 percent; he sent it in as an unrefereed note to the *Proceedings of the National Academy of Science,* I think, not as a finished proof, but as an indication of a method that was going to work.

Harold Stark, who was a number theorist in the department here and could understand that work, although he didn't work on that particular problem himself, looked at it and said there was one point in the paper, a gap in the argument that he didn't see how to fill. He said he had no clear indication that it was even settleable one way or another. And an analytic number theorist in Michigan, Hugh Montgomery, who was the one closest to the problem, agreed: he did not see any way of filling that gap either.

When I had to explain the Riemann hypothesis to that group of people, who certainly knew nothing about it, the problem was to explain why anybody would care about where some crazy function's zeros were. But there's another way of stating it which *does* make sense to people, and they got excited about it at the lecture. Without that, I would not have dared to use the topic of Levinson's last work.

It's really two problems about prime numbers: first to prove that a certain simple function that Gauss called Li(N) gives a good estimate of how many primes there are less than a given number N. That's called the prime number theorem. Then you want to know how big the error might be in this estimate: the Riemann hypothesis is an estimate for this error. That's the so-called "elementary" form of the problems, the one you can understand. Riemann's genius was to realize that that problem could be restated in a form in which it dealt with the zeros of his complex zeta-function, so that the methods of complex analysis could be marshaled to bear on the problem.

A method or statement in number theory is called "elementary" if it doesn't use complex variables. It's the world's biggest misnomer, because the "elementary" methods are actually by far the most difficult to use and understand. The prime number theorem—estimating the number of primes below N—was finally proved using complex variables 40 years after Riemann's paper, but an elementary proof wasn't discovered until 50 years after that, by Selberg and Erdős, and it was hailed as a great achievement. People who search for elementary proofs in analytic number theory are like the algebraists I once heard Herman Weyl talk about at the Institute for Advanced Study. He was lecturing on Lie algebras or something like that. "Of course," he said, "this was the problem before the algebraists got hold of it. You know what an algebraist is? An algebraist is a person who comes to a river and says, 'I wonder if I can jump across that river!' And then he takes a big leap and he jumps across that river. And then he swims back and says, 'NOW, I wonder if I can jump across that river with my hands tied behind my back!'"

Levinson, an expert in complex variable theory, published toward the end of his life a long paper, for which he won a prize for exposition, explaining the "elementary" proof of Selberg and Erdős, trying to give motivation for each step.[35] Stark read it. "Well," he said, "Norman tried, but the thing is as mysterious as ever."

So Riemann didn't solve the two problems, but he transformed them by turning them into questions about where the zeros of his zeta-function were. To prove the prime number theorem, all you had to do was prove that the zeros lay in the infinite vertical strip in the complex plane lying over the real numbers between 0 and 1. That took forty years. Then he adds that it appears very probable that all the zeros lie not just in the strip but actually on the vertical line running down its middle, "but in spite of considerable effort, I have not been able to prove this."

Fortunately all this is connected with cryptography, the secure methods for encrypting information. The Rivest-Shamir-Adleman cryptosystem depends on finding large prime numbers and multiplying them together. Very large: say, two hundred digits long. Because ultimately, what makes the encryption work—usable but unbreakable—is that if you take two very large prime numbers and multiply them together, that's easy to do, but if you give somebody the result of the multiplication, they will not be able to factor that number. That's why they cannot decipher your encryption.

So you have to be able to find two-hundred-digit prime numbers. In other words, if I give you a big number N, what's the likelihood that there will be primes near N? This is answered by an equivalent form of the prime number theorem, which says if you mark out an interval of length $\ln(N)$—the natural logarithm of N—centered at N, then on the

---

35 Levinson, N. (1969). A motivated account of an elementary proof of the prime number theorem. *American Mathematical Monthly, 76,* 225–245. The paper won the MAA's Chauvenet Prize in 1971.

average there will be one prime within that interval. And anybody could understand that. At the lecture, to illustrate, I picked $N = 1000$, so the interval around it will have length ln(1000), which is about 7. So is there a prime between 997 and 1003? Well, 1001 is divisible by 11. 1002 is an even number. 1003 is not prime. Let's go down instead: 999 is divisible by 3... but there is one, and it's 997—the last one you look at.

But that's what I mean by "anybody can understand that." The prime number theorem then becomes a very reasonable thing that one could want to know, because of cryptography: if I pick a two-hundred-digit number, how far away from it will the computer have to search before it finds a probable prime? That's what makes the cryptography system feasible, that probable primes can be found; you won't search in vain.

So I talked for 20 minutes about what the Riemann hypothesis was, its history, and about Levinson's getting interested in it and working on it, and getting a partial result—which is still the best in that direction, as far as I know—and then thinking he had essentially solved it completely, just before he died, and sending in the paper, but not turning out to be correct. And people were *excited.* At that point, since his failed paper was relatively recent, all I could say was "Well, it has a gap in it and nobody knows if that can be filled." But people were still asking me a couple of years later—friends on the faculty, but mathematical laymen, remembering it well enough to ask me, "Did anybody ever succeed in filling the gap??"

I would say, "Noo, it doesn't seem likely."

[Imitates disappointed tone:] "Well, gosh."

And that's the way it's remained ever since.[36]

Understanding Space

You were very successful as department head in getting the department some more office space, always a challenge in an institution where space is limited.

Yes, that was really my only permanent accomplishment. MIT had this room, 2-390; it was a two-hundred-seat lecture room on the third floor, at the end of a corridor of math offices. It had been originally built in the old style, with steeply sloping seating, so that the ceiling was two stories high. The department was desperate for space. There was no place to expand

[36] The Riemann Hypothesis is one of seven "Millennium Problems" for which the Clay Mathematics Institute of Cambridge, Massachusetts, has offered a prize of $1 million each.

to; graduate students were crammed into multiple offices. So Mike Sipser and Tom Leighton saw me and said, "What about 2-390? Why can't we get that space?" Now, to get a classroom is [laughs]—you don't *get* classrooms. Classrooms are for teaching, you know. Every department wants to take over adjacent classrooms for offices, but the administration and registrar just laugh. So I wasn't about to pour too much energy into it, but they came back three months later: "Well?! What about 2-390?" and I agreed to try.

I knew a certain amount about how the MIT space committee worked and what it took to get space. You had to have something they wanted. And fortunately we *did* have something to trade—without that we wouldn't have gotten it. So I first did a little study proving how relatively little 2-390 was in use compared with other lecture rooms. It was off in one corner of the main building, students had to climb up narrow stairs, it didn't meet code requirements any more in case of fire—all sorts of things wrong with it. As a result, the registrar didn't schedule it for very heavy use, and I determined that all the classes in it were small enough to go somewhere else without any real problem.

When I first became undergraduate chairman, a lab was empty just across the hallway from the new math undergraduate office, and it was *huge*, it was three or four bays—the large MIT windows. We were given a choice, either two bays, and they would fix it up and furnish it any way we wanted; or four bays of unrenovated bare space. I said, "We'll take it bare." You can live in a bare space, but you can't live in a small one, no matter how beautiful it is.

So we had this big room for fifteen years, but we underutilized it. We used part of it for the self-paced exam tutoring and the rest of it for posting solutions to problem sets in the large courses. Handing out solutions was frowned on by the textbook publishers, but it was OK for the students to sit at tables and copy the posted solutions.

That was the space we could trade for 2-390. Margaret MacVicar, the Undergraduate Academic Dean, was searching for good new space that could be used for modern small lecture rooms. Our space was on the first floor, easily accessible, and a decent size. So she consented to the trade, the registrar was willing, the space committee and the provost went along, and it came to pass.

We finally got the space in my last year of being department head. We got the permission in very early '89, and it was planned for and renovated that summer. When they looked above the two-story ceiling, they discovered there was a floor above it, a sort of attic with a skylight. So out of what had seemed to be one large lecture room, we got three floors of offices. We tripled the square footage!

It's very nicely done. The first floor is the fancy offices. There weren't enough windows, so we designed it so that each adjacent pair of private offices could stay private but still share a window onto the river or Killian Court. The second and third floors are graduate student and visitor offices, built around a two-story atrium, illuminated by the skylight. On the atrium floor are several tables and armchairs, so they have their own little enclave. It's very pretty and worth the three- or four-flight climb up from the courtyard.

~ June 14 and 27, 2005

Hartley Rogers

Trinity

J.S.: *When did you first feel attracted to mathematics?*

H.R.: It was really in high school that I began to be seriously interested in mathematics. I went to an independent day school in New York City called Trinity School. It's on the Upper West Side—91st Street. When I went there, it was a very solid kind of place. Tuition wasn't very high. It was children of doctors and lawyers, or clergymen, or businessmen, who lived in the area. It was founded in 1709 and is the oldest independent school in the United States in continuous operation. There's a school that originated with the Dutch in New York that is somewhat older, but it suspended operation during the Revolution. Since I was there, Trinity has become very fashionable. It has a soccer field on the roof of one of its buildings. It's much bigger than it was then; it's now a school "to kill for" among the affluent New Yorkers who don't have enough schools to satisfy them.

I did very well at school. I and one other boy went through as sort of heads of the class. He became a lawyer and editor of the *Harvard Law Review* who died young in an auto accident. I had a very inspiring mathematics teacher: his name was Hugh Riddleberger. I think he later became headmaster of the school. He taught with a great deal of authority but was also very appealing. He was the coach of the football team, among other things. At Trinity, interestingly, there was a boy two years ahead of me, Gordon Raisbeck—Toby, he was called—who came to MIT and studied in mathematics and ended up marrying one of Norbert Wiener's daughters. I remember writing to him about some mathematical problem, and he wrote me back, and oh, that was very exciting. I also was interested in science: when I graduated, I wanted to go to college and study chemistry.

Yale and U-boats

Where did you go to college?

My first choice would have been MIT, but my father wanted me to get a general liberal education, hoping that I would choose to major in economics, so my parents sent me to Yale. I graduated from high school in 1943, at which point I was almost 17, and I then went through Yale on an accelerated program, because it was during the war.

Being in college during the war was a strange experience. I didn't go into service because I'm very nearsighted, and in those days that was more of a handicap for service than it is now. But one was directly conscious of the war. For example, New Haven was in a partial blackout, because the German submarines had had a very successful year—in 1942 and going over into 1943—of attacking coastwise shipping in the U.S. No one ever heard about it at the time, because it was not allowed to be in the news. But a city the size of New Haven would set up a glow in the sky, and the idea was the German U-boat captains could use that glow to silhouette ships that they wanted to torpedo. That period was known among the German U-boats as "the happy time"—a total turkey shoot. I started school in the summer, and they'd take us down in the cellar of the library when it got dark, because of the blackout. The lights were on there, and I could study.

There were lots and lots of military service people there. I lived in one of the upper-class dorms because the so-called Old Campus, the block of the university where the freshmen usually lived, had been totally taken over as an Air Force ground school. It was not part of the university at all. They just used the dorms, and they were in uniform. There was also a Navy program, called the V-12 program, where you were taken in out of high school, and you *were* a student at the university. They were in uniform and lived under a sort of mild form of military discipline and went through college that way. But the Air Force people—that was really serious. They lived under military discipline. They weren't being trained to be officers and gentlemen; they were being trained to *fly*. There was a huge lot of them, and they'd have Sunday drills down on the New Haven Green, which was—oh, must be four or five acres of grass, in the middle of the city. And they sang everywhere they went, marching songs. The Air Force anthem, "Off We Go into the Wild Blue Yonder," had just come into use, so they sang that a lot.

My father was a pilot in the Army Air Corps in World War I. He did his ground school at MIT; his dorm was in Building 1. I didn't know that until

Hartley Rogers punting near Henley-on-Thames, England, with his daughter Caroline, 1974.

he came up to visit not long before he died. We were walking around MIT, and he said, "Oh, that's where I lived."

I majored in English at Yale, and I liked that. One of my teachers wanted me to come to graduate school in English. But I had fallen in love with physics. In the fall of 1945, we dropped the bombs on Japan, and everyone was very, very conscious of physics for the first time. Then, when I graduated in 1946—this was three years after I entered, because of the acceleration at Yale—I was given a fellowship called a Henry Fellowship to Oxford or Cambridge: I had a choice. I talked to some physicists on the Yale faculty, and I said, I've got this, I can study anything I want, and I'm interested in physics. They said, fine; but you need to take more mathematics. So I went to Cambridge, England, ostensibly to study mathematics.

An American among Brits

I *loved* Cambridge. It was just such a beautiful place. There were maybe eight or nine other Americans in Cambridge then. The usual social experience of an American at Oxford or Cambridge is that he's twenty-one or twenty-two, and he's had four years of college. So he's maybe three years older than the starting English undergraduates, because they've come out of school at the age of eighteen. I had just turned twenty, but the British students were all twenty-four, twenty-five, twenty-six, because they had been in service—and *real* service. A much higher percentage of them had been fighting. One guy who lived in the same house that I lived in, who had been a sort of English OSS-type person, had been parachuted into Yugoslavia when the Yugoslavian partisans were trying to get the Germans out. Virtually every British person I knew had been in service. I had some close friends who were twenty-seven years old. Cambridge is in East Anglia, where there were some huge American Air Force bases during the war. And the bases still existed. They had suffered a huge number of losses carrying out their bombing raids. The crews were just cannon fodder; the losses were terrific. I don't think there was any kind of service that was more dangerous to be in. And this meant that the surrounding British population at that time was very grateful for this, very emotionally involved in all these young American guys dying.

That extended to you as well, as an American?

That's right. And not having been in service myself, and always having wanted to be in service—I mean, it was a generational experience that I missed. That was a powerful draw. I wanted to be part of that. So it was a wonderful experience in certain social ways.

And then halfway through the year, I had some health problems. I had been smoking too much, actually. I haven't smoked since then. And I decided I needed some exercise, so I took up rowing, which is a very popular intramural sport at Cambridge and Oxford. I loved the rowing and was pretty good at it. (In the late eighties, I won the Head of the Charles twice in my age class, rowing a single.) I spent more time rowing than I did working on mathematics. I went to lectures, but I didn't study math very hard. I had a supervisor and a tutor in Trinity College, where I was, and I studied with them, but I didn't really get involved very much. I just showed up. The venerable Professor G. H. Hardy was in the same college as I was, but I didn't know him. I was just in Cambridge for a year. I wish it had been longer.

Princeton

Then I came back to Yale and began the study of graduate-school physics and discovered that Yale's program in physics was not very good. There was a lot of emphasis on memorization rather than problem solving. I learned basic physics, but for any kind of future research career it wasn't very good. So I just sort of wasted two years there in graduate school; but again, I had a lot of friends. I had graduate school roommates that I cared a great deal about, and they liked me, and we shared a lot of interests. I had a very good friend who became rather notable later in his academic life as a philosopher: Alan Anderson. He was from Little Rock, Arkansas. He was an undergraduate at that time, at Yale. He would have graduated from high school the same year I did, but he had had an extraordinary experience. There was a program in 1943: the government was looking for truly brilliant teenage students to go to work for what is now the National Security Agency, breaking codes. They were pulled out of high school for that purpose, because the government figured they were bright enough and had a certain flexibility of mind. So he was taken to Washington and put to work with the NSA, learning Japanese and working on code problems. It was kind of an amazing experience. Three or four of them came to Yale, including Burke Marshall, who later became Kennedy's civil rights champion. Alan had come to Yale three years after I did. I entered in '43, and he'd come back in '46.

Alan Anderson was a spectacularly brilliant person, a wonderful person. I got to know him and spent a lot of time with him. He was interested in mathematical logic, symbolic logic, but Yale didn't have much of that. His involvement was through a professor of philosophy at Yale, a rather eccentric character named Fred Fitch. So I got interested in that, and that led me into more mathematics. Then, finally, in my third year, I took a modern mathematics graduate course—a really central, hard course in analysis, an extension of calculus—that was very, very good.

I also started participating in a graduate seminar in mathematics. I had just totally given up on physics. At the end of that seminar, several Yale faculty members in mathematics, of whom Einar Hille was one, came to me and said, "We think you should continue graduate work in mathematics, but we think Yale is not the place for you. We are arranging for you to go to Princeton" [laughs]. And so I did. Princeton at that point was the absolute summit of a place to study mathematics. I was at Princeton for two years, '50 to '52, and I wrote a dissertation in logic and got a PhD.

You did your thesis with Alonzo Church, right? He was a giant in the field, but not the easiest guy to learn from.

That's all true, yes. You didn't relate to him in any strong personal way. He was very eccentric. But he liked me; that was nice, and I liked him. He had suggested a problem for me to work on for my thesis, and I just went and did it. I didn't have daily or even weekly meetings with him. That's my recollection anyway. And he'd have me to his house for dinner once or twice. He had a very outgoing wife. But he looked kind of like a panda bear, this great bulking person.

He was slow. He spoke slowly. He was *meticulous* about preparing the blackboard. Someone else would have been in the classroom ahead of us, and Church would erase it; if there was a little white dot that he couldn't get off the board, he'd just work and work and work to get it erased. Then he gave his lectures, which were very cut and dried. He just laid out the formal results. He had some very good students over the years. It was just *une faute de mieux*[1]: they wanted to study logic, and there wasn't anybody else teaching it properly. But Church had Steve Kleene; he had Alan Turing and [J.] Barkley Rosser, and then Leon Henkin and Martin Davis. These were his early students. I was sort of alone for the two years I was there, more or less. The next few years, he had some really first-rate students.

The graduate students were terrific there. Marvin Minsky was one of the graduate students that entered with me, and I was quite close to him. He became a great guru or patron saint of artificial intelligence at MIT, and I contributed to persuading him to come here. Marvin influenced me greatly because he had a student seminar at Princeton—of which there were quite a few, where students organized by themselves to discuss and learn something that they thought was important, not necessarily with a teacher. Marvin organized a seminar on Turing machines and the work of Alan Turing, and that got me interested in things that have been my bread and butter ever since. One of my acquaintances was another person who had just finished his graduate degree, John Nash, whom I subsequently got to know rather better when we were both at MIT.

Princeton was in certain respects kind of a wonderful place for math graduate students because it was so exalted in the eyes of the profession, and also there was a spirit of irreverence and independence among the graduate students. You didn't have to go to class. There were some incredible practical jokes and making fun of teachers and so on. It wasn't just a bunch of geeks, as it were. These were really strong personalities [laughs].

---

1 "For lack of anything better."

Harvard

In the summer of '51, I had been a year at Princeton, and I just was tired of academic things. So I applied for instructorships at MIT and Harvard. At the same time, I applied for Harvard Law School. And I was working. I had hitchhiked down to New Orleans—this was a sort of summer-long *Wanderjahr*—and got a job working on an oil rig as what was officially known as a roughneck.[2]

You really jumped around!

I jumped around. And I did that for—oh, two and a half months. It was a very interesting time. But while I was there, I had to take the law school aptitude tests. So, I took time off from my work, got on a bus, and went up to Baton Rouge to take the aptitude test. Later I was accepted for a math instructorship at Harvard. I just couldn't make up my mind. I went up from Princeton to be interviewed for the law school. The law school admissions dean said, "I'm going to tell you right now—put it on the table: you have the highest score on this aptitude test that we've ever seen. So we'll accept you as a part-time student." So I went to a certain number of law school lectures. I had some great teachers, but obviously, if I wanted to do it right, I'd have to do it full time.

So to make a long story short, I went to Harvard in '52 and got into teaching mathematics, which I liked very much. I got a position as an instructor, salary $4,000, and was at Harvard for three years. I had a very substantial teaching load, compared to what the situation is these days when, on such instructorships, they give you more time for your own research. I had also applied for an instructorship at MIT; I can't remember if they offered it to me or not. But anyway, I wanted to go to Harvard. I had a girlfriend who was a graduate student in Renaissance studies at Harvard, and that probably was the decisive thing. I got to know her at Bryn Mawr when I was at Princeton.

Then, at the end of my first year at Harvard, I met a woman who had just finished her junior year at Radcliffe. She was twenty. I was, at this point, twenty-seven. And we decided to get married. I still had one year to go, and she wanted to go to medical school. She was from Nebraska. I think we announced our engagement around Christmastime, and I went out to Nebraska to her home town of Columbus, population in those days maybe

---

[2] A roughneck is responsible for the operation of machines and equipment required by the driller.

fifteen thousand. The *Omaha World Herald* announced our engagement, and they had a headline on their society page. It said ADRIANNE ELLEFSON TO WED EASTERNER [laughs].

So by my third year at Harvard we were living as man and wife up on Oxford Street. The first two of those years, I resided in one of the Harvard dormitory houses, a place called Eliot House: it's one of the river houses. There were more tutors living in, in those days, because now any tutor that lives there free—and a few still do—is costing the university the price of two or three students' tuition and room and board fees. We had a very interesting senior common room. Jim Schlesinger was a central figure. He was a graduate student in economics and later head of the Atomic Energy Commission, head of the CIA, Secretary of Defense, general "wise man" in Washington. His wife was a good friend of my wife; they were both Radcliffe undergraduates at that time. There were other very interesting members of the senior common room, both older and younger, in a variety of disciplines, and it was a very special experience for me.

As a resident tutor, I felt obliged to do some teaching in the House in addition to my regular Harvard teaching. And so for a year or two, before I got married, I gave a seminar course, I think it was complex variables. I had three or four students, of whom the most notable by far, as a seminar student, was Richard Friedberg. Another Harvard student, David Mumford, who lived in Kirkland House, was almost equally precocious, but not quite as precocious as Richard. In Richard, you just had the feeling of unlimited raw intellectual power.

Richard Friedberg had gone to a school in New York called Fieldston.[3] I liked him very much. One of the things that attracted one powerfully to him was that he was very sweet, and understanding, and nice, and no great ego at all. He didn't need to have a big ego [laughs]! What he had was so impressive. I was sort of viewed as a kind of mentor to him, but it would be like saying that I was trying to teach Michelangelo sculpture. The pupil was *so* immensely talented.

Another thing that charmed me about Richard was his interest in music, and I remember that in 1955 my wife and I went to a performance of [Gilbert and Sullivan's] *The Yeomen of the Guard* in the basement of the First Congregational Church up on the Cambridge Common. The only instrumental music was Richard at the piano, and he had memorized the entire score. Just playing away! So we were dazzled by that.

---

[3] The Ethical Culture Fieldston School.

Anyway, Richard got interested—partly through me, but partly because it was in the air at that time—in a problem called Post's problem. Post's problem had been around for about ten years, and it had been posed by a rather eccentric—what's the word?—*Einselgänger.* Someone who is very independent and a little bit restless and goes his own way. Emil Post was at City College of New York, which was a better school then than it became later. He developed a whole area of mathematics, which became my area when I went to MIT, and which we called, in those days, recursive function theory. There were other people that developed it too, like Steve Kleene out at Wisconsin, another Princeton graduate, who had developed it a decade earlier. But there was a lot of formalism that had to be learned, with the view that you had to make the subject rigorous, and you had to, in effect, dot every *i* and cross every *t* in your definitions and development.

Post developed it in a kind of intuitive form. It was as if he said, this is like teaching a kid arithmetic. You get the idea of addition and multiplication, and then you can just go on and use them, without fussing over formal, rigorous development. In other words, he did not demand, when you talked about an algorithm, that you have some very precise formulation of "algorithm." Post was very influential among everyone that knew anything about logic. For years I was able to teach people that you don't have to worry about the formality. You *know* and recognize what an algorithm is: it's just a deterministic, step-by-step recipe.

Post's problem related to two kinds of sets of non-negative integers: recursive sets and recursively enumerable sets. A *recursive set* was a set where you had an algorithm for deciding whether any given integer was in the set or not. So, for example, the even numbers are a recursive set because you just take your given number and see if it's divisible by two. A *recursively enumerable set* is a set for which you have an algorithm that you can use to *list* the members of that set in some order—not necessarily in increasing order.

So saying that a set is recursively enumerable is saying *less* about it than saying that it's recursive. Recursive, you have a test for membership. Recursively enumerable, you may not have a test for membership in that set. But at least you know how to get a listing of the set such that each member will occur in the list eventually.

Most examples of recursively enumerable sets will actually be recursive sets. But the existence of recursively enumerable sets that are *not* recursive has been known since the middle 1930s. One person who had a very quick and simple proof of that result was Alan Turing. Post comes along and sees that you can have recursively enumerable sets that are not recursive, and so he raised the question: given a recursively enumerable set, must it either be recursive, or—if it's nonrecursive—must it be of a special kind, called *complete*? A recursively enumerable set C is said to be complete if it has the following special feature: *if* you could have an "oracle" that would simply give instantaneous correct answers to individual questions of membership in C, *then*, given any recursively enumerable set A, you could use this oracle for C to answer individual membership questions about the set A.

Since all previously discovered nonrecursive, recursively enumerable sets were easily shown to be complete, Post's problem was: are there recursively enumerable sets that are not recursive, but are also not complete—that is, that don't have this elite status of empowering you to answer individual membership questions about *any* given recursively enumerable set? It's a tantalizing question, such a simple and natural problem, and it really became known as Post's problem because Post was able to formulate it in such simple terms.

Richard got interested in this, and we talked about it quite a bit. There were earlier papers—by Kleene, for example—that related to this. In fact, there were papers jointly written by Kleene and Post, where Kleene was supplying the detailed definitions and Post was the idea man: that's the way I always looked at them. So Richard was working with these papers and trying to prove things.

Of course, once you run into a piece of pathological behavior like the existence of nonrecursive, recursively enumerable sets—your instinct, your aesthetic feeling, is that you'd like to be able to say that they're all of the same kind. The world is going to be simpler if every recursively enumerable set is either recursive or complete. Unfortunately, the world didn't turn out that way [laughs]. There are, in fact, just a *blizzard* of possibilities in between. The structure of the sets that are recursively enumerable but not recursive, but also not complete, is very complex.

So we come to the last year of my Harvard instructorship, '54–'55. This was the first year of my marriage, and my wife's first year at Harvard Medical School. We lived in a five-room, fourth-floor walk-up on the northern edge of the Harvard campus, and we were very happy with it. The following year I joined the MIT mathematics department, but we lived in the apartment for another four years. This helped us stay in touch with friends at Harvard. 1955–56 was also Richard's senior year at Harvard. Shortly after graduating, Richard went into a trance and saw how to solve

Post's problem, much to the dismay of several hundred other people that were working on it very hard. He wasn't debating with other people; he was just *thinking,* as hard as he conceivably could, of the possible ways to prove one side or the other. After making the discovery, he came right up and showed me on a big blackboard I had installed in my study. His idea involved a kind of circular conveyor belt with buckets on it, and you put numbered marbles in the buckets, and then later on you took some out; you put some in, took some out. Each marble had a certain function, which was to show that a particular algorithm was not going to work. But later on, you might discover that that algorithm, in fact, did work, and so you'd have to go and take that marble out—it hadn't served its purpose—and put it in another bucket. This was a kind of mechanical picture of his method. I was very impressed.

Interestingly, he submitted it as a paper to the *Proceedings of the National Academy of Science,* and that was published,[4] and it turned out that within weeks, a simultaneous solution by Mučnik, perfectly genuine, obviously not the same, appeared in a Russian publication![5] They both hit on the same idea, the same method, which is really a great leap. It's a very simple method called the priority method.

Another thing that happened that year was that Richard had this post-graduation dilemma about what he should do. Should he follow the mathematical guiding light, or should he go to medical school, which is what he felt his father wanted him to do? His father was one of the most distinguished cardiologists in the country at that point and had written what was then the classic text in cardiology.[6] Dick admired his father very much. And he had also gotten engaged to a very bright young woman who was the daughter of a distinguished physician.

He consulted with me because my wife was at Harvard Medical School, and he thought I might understand his dilemma, because I was having this second-hand medical experience.

I wasn't especially helpful to him in this. I was open; I wasn't just saying, "Don't be an idiot." Because he was so friendly in this dilemma of "Should I do this, or should I do that?" I thought that anything he chose to do, he

---

4 Friedberg, R. (1957). Two recursively enumerable sets of incomparable degrees of unsolvability. *Proceedings of the National Academy of Science, 43*(2), 236–238.

5 Mučnik, A. A. (1956). Negative answer to the problem of reducibility of the theory of algorithms (in Russian). *Dokl. Akad. Nauk S.S.S.R. (n.s.) 108*, 194–197.

6 Friedberg, C. K. (1949). *Diseases of the heart.* Philadelphia and London: WB Saunders.

could do very well; and I thought the life of a distinguished physician didn't look all that bad, intellectually or in any other way. But he did not enter medical school that fall, 1956. He decided to take a year off and then decide what to do. I think he entered Harvard Medical School a year later. He did keep an interest in recursive function theory, and we would talk from time to time. We actually wrote a joint paper together: I remember going to see him at the medical school about it.[7]

The first thing that I noticed was that he was getting ready to take an anatomy exam, and he was preparing the neuroanatomy part of it, and if you look in the appropriate anatomy textbook, you'll find that a given nerve can just keep ramifying as it leaves the central nervous system. And these little branches all have names: you can look in the anatomy book, and there they are. Well, the teacher, in those days, wouldn't tell you what you had to know—just to learn as much as you can. But no ordinary mortal would try to memorize this exponentially growing set of names [laughs]. But Richard did. He told me this, proudly. He learned all these names on this ramifying tree, where they were, and so on. I thought, "Oh boy."

Then, toward the end of his second medical school year, he had a kind of breakdown. Whether it was regret at the choice he'd made, or what, I don't know—I hadn't consulted with him. But he just opted out of everything, and he became unengaged to this rather lovely woman, and then he seemed to disappear for a while. But he had this magical reputation—not just with me but with everybody that knew him—of being an intellectual giant.[8]

The next thing I learned was that he had entered Columbia as a graduate student in physics. I remember he came up to a mathematical conference at MIT, and during the lunch hour we went for a walk along the river. He said that he wanted to completely relearn physics, by which he meant that he wanted to start with the observed phenomena but reject all the physical categories that are used in physics and just try and develop

7 Friedberg, R. & Rogers, H. (1957). Reducibilities and completeness for sets of integers. *Zeitschrift für Mathematische Logik und Grundlagen der Mathematik, 5,* 117–125.

8 Friedberg's decision caused regret throughout the logic community. Kurt Gödel, in a letter to John von Neumann dated 20 March, 1956, wrote, "I do not know if you have heard that 'Post's problem'... has been solved in the positive sense by a very young man by the name of Richard Friedberg. The solution is very elegant. Unfortunately, Friedberg does not intend to study mathematics, but rather medicine (apparently under the influence of his father)."

it all anew. Which is actually a very nice attitude to have in research: you go into something and people are sort of stuck where they are, and you just back off and start over again. He was working with a physicist at Columbia named [Chen Ning] Yang[9]; he got his degree and got a position at Barnard, with a research affiliation at the Columbia Physics Department.

In April, 2003, there was a kind of retirement festival for him: he clearly had a close circle of admirers in the Manhattan mathematics/physics scene, and they wanted to have something in his honor. Richard asked me to come and speak, and I did. And he just put on the most beautiful piano concert for everybody while he was there [laughs]. Otherwise he just sat quietly, as people talked either on recent developments in areas of interest to him, or just in frank admiration and praise. He's a very worthy human being.

In retrospect, I wish I had pushed him more in the mathematics direction. I think I have a much better understanding now than I did then of the kind of commitment it takes to do mathematical research. I didn't fully appreciate the importance—for a future mathematician anyway—of his taking advantage [of his success], right then and there. There was a tide in the affairs of this man which could have been taken at the flood,[10] but it wasn't.

Was he aware of that, do you think?

I think so, yes; at some level, he felt that he had had this touch of the divine, and he hadn't followed it.

MIT

Going back to our original timeline—how did you get to MIT?

After we got married, at the end of my second year at Harvard, I went on teaching for my final year, and my wife was at Harvard Medical School. But then at the end of that year, she was doing very well in medical school—she eventually graduated as one of the top five or six people in her class—and I had to find a job, because Harvard promoted instructors only rarely, but I was not "rarely." But MIT said, we'll offer you a visiting job. So I moved to

9 Chen-Ning Yang shared the 1957 Nobel Prize in physics with Tsung-Dao Lee.

10 "There is a tide in the affairs of men / Which, taken at the flood, leads on to fortune." Shakespeare, *The Tragedy of Julius Caesar,* Act IV, Scene 3, lines 216–217.

Claude Shannon.

MIT in '55, and my wife began her second year at Harvard. I was a visiting lecturer for my first year at MIT, and then they made me an assistant professor. And the die was cast.

Were you related to the founder of MIT?

William Barton Rogers. No, no.

Whom did you share an office with when you got there? Whom were you close to?

It was a small office on the third floor, at the top of the corner stairway, and I shared it with two other people, one of whom is still at MIT—Siggy Helgason. The other moved down to Stony Brook: Bill Fox. I didn't know MIT at all when I came, although I had seen it once before when I visited a close school friend studying chemistry there. I think the most visible person, to me, was the chairman, Ted Martin. Second was Norman Levinson. Third was Norbert Wiener. To most outsiders, Wiener was the department, but I'm just telling you my personal involvement. Another person on the applied side then was Eric Reissner,[11] and he and Martin had written an elementary differential equations book together.[12] I liked him. He had a very handsome house out in Weston—sort of separate and in the woods. He went out eventually to a position in California and then died. I think C. C. Lin had come then, and he was a great acquisition. He had unusual international stature as an applied mathematician. Warren Ambrose was, I think, an important figure in the department. I liked him personally.

I had one very good friend who helped me in various ways, and that was Claude Shannon. This would be at the end of my first or second year. I had a graduate student that was working with me: his name was Don Kreider.[13] He was an extraordinary teacher—he won the Goodwin Medal for teaching at MIT—and ended up going up to Dartmouth. And I had gotten interested in a certain area of logic, which came to be called recursive function theory. Eventually I wrote a really major book on the subject, but I had gotten interested in it not on the basis of anything I'd ever learned before: I just began working on it as an independent topic that interested me, and that I felt that I had a better approach to than some other people. So I had a seminar going for a handful of graduate students. But then Claude came to this and got interested, and we became a kind of mutual admiration club.

He had just come to MIT; this was not long after his great theorems in information theory. His theorems are just tremendous: the idea of measuring information the way he does, that was just a wonderful insight. And also his so-called coding theorems—he was the *ultimate* creative engineer. He wasn't centered in mathematics, but the line between engineering and technology and mathematics was completely blurred with him. It was just all a part of his incredible ingenuity. He was such a brilliant guy and very much his own guy. So I think at a personal level, more than a mathematical level, we got on very well. He certainly was a good advertiser for my work and, I suspect, said good things about me when my promotions came up. I recall

---

11 M. Eric (Max) Reissner received his PhD at MIT in 1938. His advisor was Dirk Struik. He was named a full professor in 1949 and worked at MIT until 1969. His work on the Reissner shear-deformation plate theory allowed engineers to model the forces on surfaces such as floors or airplane wings and led to significant advances in civil and aeronautical engineering. For more information about Reissner's career, see the obituary at http://www-tech.mit.edu/V116/N58/reissner.58n.html.

12 Martin, T. William & Reissner, E. (1961). *Elementary differential equations*. Reading, MA: Addison-Wesley.

13 Donald Kreider received his PhD at MIT in 1959.

that one evening we had Claude and Noam Chomsky, another friend of mine whom I much admired, to dinner at our apartment. Noam had audited a graduate probability course I gave at Harvard. The dinner, as a social occasion, turned out to be one of memorable immiscibility.

We live here on Upper Mystic Lake because Claude had a beautiful house on this lake—from our fore bay, you go through the straits over there and down into the main, bigger part of the lake. My wife and I had the apartment in Cambridge, and we started looking for a house. We finally just left it with the real estate agents that if a house came available on this lake—we knew about it from being at Claude's house—we'd be interested in it. When we got our house, in 1959, Claude helped us move; he helped me carry our furniture down three flights and put it in a rented trailer, and he helped me get it into our new house.

The Shannons had an extraordinary house, large, with perhaps twenty rooms, situated way up high, a really dramatic place. Much of the house was crammed with electronic and mechanical items of war surplus: for example, the exquisitely machined, mechanical analogue computers for central fire control on a battleship. The house had a drop of perhaps fifty feet down to the edge of the lake through a long sloping lawn. And Claude, by himself, made and installed a chair lift, where you sat in a chair and pressed a button and you went down, or the reverse, you went back up. It was just for fun, for the hell of it, and to keep his children amused so that they could ride back

Hartley Rogers competing in a Masters National race.

and forth to the edge of the lake. You'd be working down by the lake, and his wife Betty would have lunch ready; she'd call, and you could take the chair lift.

I did a lot of rowing at that time: I'd row over to his house on the lake, and we'd talk one thing or another. And he became interested in the mechanics of the boat. He had already been very interested in bicycles. He was a real Midwesterner in that regard: he was from Michigan, and the Wright brothers were not too far away. One of the things that interested him about riding a bicycle was, what keeps you erect? I said I thought there was some sort of gyroscopic effect from the turning of the wheels. His first response was, have you ever tried to ride your bicycle backwards? I said no. He said don't try it; you'll fall. Because what keeps the bicycle erect is the infinitesimal corrections in steering that you make as you ride along.

So we talked about my rowing boat, and he wondered whether there would be some way to incorporate this into steering my boat. This was a sculling boat, eleven inches wide: normally, when one uses the oars right, they play the role of outriggers and keep you erect. He asked me to see if I could buy him a second-hand boat, which I did. What he wanted to do was to turn the boat into a bicycle—to have a bicycle frame with pedals on top of the boat, and the pedals would drive a propeller, and the steering wheel would be attached to a rudder down under the boat. So I got this beat-up old racing single, and we made this! And we had a great day of trying it out, floating in the lake. He got into the boat, sitting up on this bicycle frame, and I started pushing him, getting up as much speed as we could, bringing this uprightness theory into play. And he just immediately fell over with a giant splash! It was like watching a Buster Keaton movie, these crackpots trying to do something. It was just a total failure [laughs].

We had some great times. He had so many projects. He was interested in tightrope walking. You know, it's hard to walk a tightrope, and it's even harder to walk a slack rope. My own father was a very good slack-wire walker, but Claude had it drawn quite tight. He had it close to the ground in his backyard, because you could fall off and just step down [laughs]. And then he became interested in juggling. And he actually made little model people who juggled.

Like little robots?

Yeah! He just had this dominant desire to *play,* and his children were a perfect audience for him. They became very good jugglers. Curiously, by the way, Richard Friedberg later had a daughter who became a national champion juggler.

Claude loved working with his hands and working in the practical world. He wanted to make a camping bus. Of course the VW camper came along not too long after that, and so did the Winnebago. But he got an old school bus, and for a period of some four or five months, that was all he did: install a bathroom, install beds, install a stove—take the family camping!

Claude and I didn't remain that close. My wife and I were very active in a move to ban motorboats on the lake here, which was unquestionably a good thing to do. But he had a motorboat, and I think he felt that I'd let him down a little bit. It never managed to manifest itself in an explicit way, but I think he was disappointed. He died just a couple of years ago of Alzheimer's, but he'd really been—it started about twenty years ago. His wife lived there until this year.

You mentioned that you'd known Marvin Minsky from your Princeton days as well.

That's right. Marvin Minsky was one of the graduate students that entered Princeton with me, and I became quite close to him. He was just an incredible person, full of ideas. Somewhere between half and two-thirds of his ideas would end up being kind of nonsense. It was just immediate reaction to any ideas that had come to the surface in the conversation, and he'd be off, giving out these suggestions. But having a third of your ideas be brilliant ... [laughs]. With him it was just like watching someone releasing colored balloons in the air! He was very impressive. At one point, he and John McCarthy, who had been at Princeton with us, convened a "what next?" meeting of young faculty from all over MIT who shared the belief that the further development of electronic computers was going to turn the world upside down. I was there; there were about fifteen of us, including Fernando Corbató, who later developed time-sharing, and we all got to know each other. I'm not sure of the date: I think it was around 1960.

Marvin had gone to the Bronx High School of Science, had some military service, and then came to Harvard as an undergraduate.[14] He had a PhD in mathematics from Princeton, which he got by having so many wild ideas that were actually good. He was made a Junior Fellow in the Society of Fellows at Harvard. I was at Harvard at that time as an instructor, so we saw

[14] Minsky attended Phillips Academy in addition to the Bronx High School of Science, then served in the U.S. Navy from 1944–45. He attained his BA in mathematics from Harvard University in 1950 and his PhD from Princeton in mathematics in 1954.

a good bit of each other.[15] Then I persuaded Marvin to come to MIT and to come to the math department, which he did. I think he thought, "Well, there's part of me that's a pure mathematician, and if I'm in the math department, I'll be able to develop that part a little bit."

But Marvin was a kind of mixture, like Claude Shannon, of conceptualizing and conceptual thinking together with practical experimentation and development of practical ideas. One of the things that he was interested in was perfecting what was called a scanning microscope. An interesting piece of practicality: the idea was that by the oscillatory scanning aspect of the thing, you could neutralize diffraction effects and get a higher resolution of the image in the microscope. It made sense, and he said, "Well, I'll build one." He thought it would work, so why not make it and be sure it worked! And Claude Shannon was very similar in that respect, half mathematician and half engineer.

Marvin was one of several friends I had—Richard Friedberg also—who were interested in the same basic idea about the brain, now largely discarded, which is that the brain is no more than a collection of neurons hooked up in a kind of random way that, as the organism developed and interacted with its environment, would organize itself to act and respond in practical and beneficial ways. I can remember we had a program for the better part of a year, which he invited me to, in which he and his students were going to make a robot that played ping-pong and could be a worthy opponent to anybody. It never really got off the ground, because it was just too complex a problem, but Marvin sort of narrowed it down to making an arm which would hold the ping-pong paddle, and then you'd have various other sensors and so on. I remember we got a professor of anatomy from Harvard Medical School to come over and talk about the anatomy of the arm, and the evolutionary aspects of the human arm [laughs]. Marvin really saw artificial intelligence as doing things that it still hasn't achieved.

Bill Gosper, the hacker hero who went to MIT, said "[T]he mathematics department was radically anticomputer. They would literally preach against it. They had essentially driven Marvin Minsky out of the department, into the electrical engineering department." [16]

---

[15] Minsky was at Harvard as a Junior Fellow from 1954–57; Rogers was a Benjamin Peirce Instructor at Harvard from 1952–55.

[16] Albers, D. J., Alexanderson, G. L. & C. Reid. (Eds.). (1990). *More mathematical people* (pp. 103–104). New York: Harcourt Brace Jovanovich.

That is just nonsense. They weren't radically anticomputer at all. Marvin's transfer into the electrical engineering department occurred for one and only one reason. It occurred because Marvin wanted to have one term out of two in each year free for his own work. The math department had the rule that you had to teach in both terms, which was understandable from the math department's point of view, because that meant you had to be in town [laughs], part of the mathematical community. You couldn't just go off and live in a cabin in the woods and think mathematical thoughts. So the mathematics department felt that it couldn't change its policy on that. But it was a very reasonable request from Marvin's point of view, and so he went to electrical engineering, which was prepared to be more flexible for him. The attraction of trying to make his own microscope or his own robot ping-pong player was too powerful.

You also mentioned meeting John Nash at Princeton and that you got to know him better here. What were your impressions of him?

Yes, I knew him when I was a graduate student. He came up here, got married, and lived in Medford. That was when we still lived in Cambridge, and we traded back and forth, one couple inviting the other to dinner and that sort of thing, which people did in those days.

Nash was very highly competitive about his mathematics, and he had a right to be, because until he got sick, he was on the way to being one of the best American mathematicians of the twentieth century. No fooling. Every profession has its eccentrics, and they're colorful, and they make good copy, and you can have movies about them, and you wouldn't think that this was really serious stuff, but with Nash it was. Not his stuff on game theory—the Nobel Prize was for work that Nash himself viewed as rather trivial. He did it because he was getting financial support from the chairman of the Princeton department, and he'd been put on a project connected with game theory. The Nash equilibrium theorem was a big theorem for the game theorists, but it was not central mathematics. It didn't line him up with the really great five or six mathematicians of history. But then what he did start doing, in his serious mathematics—analysis; and very hard analysis—was very, very good and clearly looked as if it was going to bear fruit in an extremely impressive way.

I remember one occasion at our daily graduate-student tea at Princeton when Nash was expounding a mathematical idea to a court of admirers. One of the admirers asked a series of questions, probing for and getting more

Ted Martin.

details. Eventually he said to Nash, "When you publish this, be sure to give me credit." To which Nash replied, "Sure, I'll put in a footnote saying that you were in the room when I had this idea."

The Martin/Levinson Era

Let's go back to Ted Martin and Norman Levinson, who were really responsible for shaping the department in the wake of Norbert Wiener's influence. What do you remember about them?

Ted Martin was a wonderful and dedicated man. I admired him very much. So was Norman Levinson. Norman Levinson had initially been an undergraduate at MIT and majored in electrical engineering. But he had been discovered by Norbert Wiener and worked on problems that Wiener suggested and sort of oversaw. Wiener had written a book with someone named Paley, and growing out of that book were problems that Norman Levinson worked on. There's some very impressive work by Levinson. Wiener had been at MIT since the early twenties, as had also Dirk Struik. I suspect that Wiener was a main figure in getting Martin back[17] and in hiring Levinson, because he was very close to them both. Norman Levinson and Ted Martin, when they were together at MIT in the thirties or early forties, had worked together quite closely. I can remember them telling me that they would get together to discuss their work on Saturday or Sunday mornings. Then, after they worked for a good solid morning, they'd go together and visit Wiener and talk to him, I guess about what they were doing.

Norman Levinson was the better mathematician, but Ted Martin was a very interesting person. They really were very dependent on each other, in some ways, in administering the department. Their main accomplishment, which I didn't realize at that time, was turning the MIT math department into a quality department. And we're there! I mean, we're one of the top five departments in the country, anyway. If you're an undergraduate in mathematics, and you want to go to a really good graduate school, you think of Princeton; you think of Harvard; you think of MIT; you think of possibly Stanford, possibly Berkeley or Chicago.

Ted had been at MIT as an instructor. And then, very young, he was recruited to be the head of the math department at the University of Syracuse. Then, in 1947, MIT brought him back to be head of its math department. So Ted came back from Syracuse and began hiring people that were really good. It was the right time to do that, of course. Up until then there had been a few exceptional people who did exceptional things in certain rather narrow areas of American mathematics. But what really made the difference, what brought American mathematics into the twentieth century, was the refugees from Europe. There were key people, who came among the predominately Jewish refugees, who just raised the standards in American mathematics. And there were other people that came, Japanese and Scandinavians who came to the United States after the war. I mean, America was the place to be! And Ted took advantage of this and got some really good people.

Hurewicz was very good. He climbed one of those very steep Mayan temples and fell off and killed himself: clearly an accident. And we had

---

[17] William T. Martin (1911–2004) joined the Mathematics Department faculty in 1938. He briefly served as the department head of Syracuse University's Department of Mathematics from 1943 to 1946, then returned to MIT, where he served as the department head (1947–1968) and chair of the MIT faculty (1969–1971). Martin retired from MIT in 1976. See Martin's obituary in William T. Martin. (2004, June 11). *The Tech, 1*, 11.

Ken Iwasawa, a Japanese algebraist who was very good.[18] These are just names that occur to me. But it was under Ted, and with the help of Norman Levinson. The mastermind of getting things done was Ted, and the mastermind of identifying quality with confidence—being confident that you're making a good quality judgment—was Norman Levinson.

A second thing that Ted did, and very effectively, through personal determination, was to expand MIT's financial support of mathematics. His underlying argument—which he backed up with detailed figures—was that the math department did an enormous portion of the teaching of undergraduates. I mean, *everybody* took some mathematics. This was a time when MIT was getting a lot of money from Washington for scientific research and a little for mathematical research. And somehow he managed to make the case that MIT's support of the department should be more proportional, fairer, than it had been, and he used this money to get distinguished faculty members. And that in turn led us to a research department.

He was kind of a great man, in many ways, in his ability to run the department. The department was very carefully, parsimoniously run, and Ted was right in his element doing that. He knew where every nickel was being spent and was very careful to squelch any initiatives that could be considered a luxury. As a child of the Great Depression, I felt very close to him.

Another theme, which was not talked about very much but was a significant theme in the department, was the Communist dimension. Ted and Levinson had been members of the Communist Party. Historically, this was perhaps part of the glue that kept them together. They had been members, and Fagi Levinson had been a member, during the Great Depression. You know Fagi. She's a delightful person. I think the issue was much in the air—not only for younger faculty but for other faculty around the Institute. With guys like Ted and Norman, especially, it had been a very idealistic thing. Here you were in the U.S., with all the riches and resources that were evident in the nation, and people were starving! People couldn't get jobs. The whole economy seemed to be somehow stuck on dead center. It's just so easy to see some people—especially mathematicians, who

---

18 Ted Martin asked Kenkichi Iwasawa to come to MIT as an assistant professor in the spring of 1959. Originally intending to stay only a year or two, Iwasawa worked at MIT until 1967. His research concentrated on group theory in his early years, but a two-year stay (1950–1952) at Princeton with Emil Artin convinced him to shift to algebraic number theory. At the 1956 meeting of the American Mathematical Society in Seattle, Washington, Iwasawa first introduced his general method in arithmetic algebraic geometry, later known as Iwasawa theory.

can be a little bit naïve sometimes—saying, "This is the answer. Marxism has got it right." And so they were talked into becoming part of "the Revolution."

Ted and Norman left the Party. I don't know exactly when, but their big disillusionment was the Hitler/Stalin pact. Norman and Ted were pretty intelligent people and pretty much in tune with their environment and with American society. Norman was not from Europe; he had grown up in Revere. Ted had grown up in Arkansas. So they got out; they were completely disillusioned. Then there was a third figure—Dirk Struik—and Dirk did not get out. That's a whole other interesting story, and I don't understand it completely. One of MIT's distinguished physicists, with whom I later worked for two or three years almost full time, was Jerrold Zacharias. And I remember his saying that he thought the Institute had done the right thing in the case of Struik: that is, that they'd maintained his salary and his position but didn't use him as a teacher. And somehow, in the spirit of that time, this worked out.

What was the social scene like in the department in those days?

Ted Martin had parties at his house. They were really quite often; it just seemed to me that it was very frequent that we'd show up there [laughs]. They had a house on Moon Hill in Lexington that they had built. There was a lot of suburban building that occurred in the late forties and early fifties. There's a big project in Concord, and another one I don't remember, and this Moon Hill project in Lexington where Ted lived. You'd go out to his house, and you just sat around and talked. That's where I got to know Is Singer. Lots of the faculty would come, with their wives. Of course the department was smaller in those days than it is now. The cast of characters varied a little bit; not everybody went every time. Not uncommonly there would be a big roast of beef, but it would only be served about ten o'clock at night. You'd be pretty hungry at that point!

Mathematics for Africa

Ted Martin got you involved in the African math writing project that Jerrold Zacharias initiated, is that correct?

Yes. I was there for—oh, four summers. Arthur [Mattuck] came out for one summer. Ted Martin was the head of the department, and he got a large grant through EDC, the Education Development Center.[19] EDC may still exist. It had offices in Newton, and it actually began as a place where they made short films of certain classic physics experiments to use in physics lectures. They started off making these movies in an old movie theatre up on Galen Street in Watertown, very close to where the trolleys came in and out. They made these films available all over the country. The new goal was to develop a set of K–12 mathematics textbooks, for Africa, but written in English, and to use experienced post-doctoral mathematicians to write the texts, because the existing texts, some of them, were clearly deplorable. It was also hoped that when the new texts were finished, they could then be translated into local languages. And somehow Jerrold Zacharias persuaded Ted to run the program, and EDC backed him up and paid the salaries of staff on the basis of a USAID[20] grant for foreign aid. They started it in 1963 or '64 with a summer in which Ted and a few other people met in Entebbe, Uganda, to go over things and see if the whole thing made sense.

The next year, when they began, Ted went around and recruited people and asked me if I could begin with texts for the first two years of high school. The project was called the Entebbe Mathematics Series because they had their first planning meeting at Entebbe. But after that the writers met in Mombasa, which is in Kenya, on the coast. That was very interesting for a number of reasons. Ted tried to make the thing very attractive for young mathematicians to come, but I really had mixed feelings about doing this, because I was on the threshold of building a research reputation, and summers were valuable for research. If I had it to do it again, I probably wouldn't go, but it was a wonderful experience to have. Ted built the whole thing up. There were excursions to the game parks; you could bring your wife, and this *magnificent* old colonial hotel! It was a beautiful old structure, not like Miami or Virgin Islands hotels today, but the location! The views! It was right on a beach, and there was a community of fisherman that still used

---

19 "Jerrold Zacharias, an American physicist . . . initiated an international conference on African education which became known as the 'African Summer Study: The Endicott House Conference.' A major consequence of the conference was the establishment of the African Education Programme in Mathematics and Science funded by the USAID and the Ford Foundation (EDC 1967). The African Mathematics Programme (AMP) introduced the then 'modern mathematics' to Africa under the label 'Entebbe Mathematics.'"

To read more about this and similar initiatives in Africa, see http://fafunwafoundation.tripod.com/fafunwafoundation/id6.html.

20 United States Agency for International Development.

that beach for fishing, casting nets and hauling them in. They also had fish traps, which made a long pier out, maybe fifty yards, and the catching part of it was out at the end, where there was something like a large lobster trap, and the fish would swim in but not be able to get out. Nyali Beach was a wonderful place to be.

The hotel, where we met, was called the Nyali Beach Hotel. It had been "Whites Only," Europeans only, until that year. Ted had recruited some PhD mathematicians who were African, mainly West Africans, who had gone to Europe to get their PhD degrees. They were a very impressive group, from Sierra Leone, from Ghana, from Nigeria, and then we had maybe one guy from East Africa. The African members of our group were the first Africans to stay at this hotel.

And that had a number of interesting consequences. Kenya had just become independent twelve months previously, so there were all kinds of disjunctions in behavior and attitude. The younger, politically conscious Kenyans were one point of view; those who had more traditional expectations and attitudes were another point of view. There also was an expatriate community—mostly British, but from other countries too—some of whom were very welcoming to us, and some of whom were rather grumpy over independence. Ted had a Brit as his chief staff person, who served as a typical, careful, effective, midlevel British bureaucrat.

The ethnic experience was amusing in some ways, sad in others. One summer there, we had a guy named David Blackwell; he's a very distinguished professor of statistics at Berkeley, and he's black. He was just enthralled with the idea of going back to his roots, and this would be a profound experience—using his mathematical knowledge to do a great deal for Africa. And I remember going to visit a local school with this mixed group of textbook writers—West Africans, Americans, and Europeans—and the guide said, "All right, now, I'm going to ask all you Africans to stand over there, and all you Americans and Europeans to stand over here." So they divided, and David Blackwell went with the Africans. And the guide went over to David and said, "I'm sorry, you belong over *here,*" and moved him. David was [laughs, imitating thunderstruck expression]—tragic!

It really was very funny, because the Africans are truly color blind. I mean, part of the education for the rest of us was meeting Africans who did not carry the complex weight, at that time, of being a black person in the U.S. I'd meet these aristocrats from Nigeria, and if anybody felt superior to anybody, they felt superior to me. *They* are color blind, and it helped *me* to become color blind, as an American. Similarly, the

Ghanaians or the Sierra Leonians didn't give a damn. It was possible to be friendly with these Africans because they were so secure in their own status. I had very close friends from Ghana and Sierra Leone in a way that would have been rare here in the U.S. at that time, forty years ago. You just met them as people.

These benefits aside, why not just translate American textbooks into African languages?

Well, the theory was, that wouldn't take full, adequate account of African cultural background and influences. We needed to see how the children were being taught at that time, and that's why West Africans played such a prominent role in the project. And that was very helpful.

And was the result different, indeed, than an American textbook would have been?

Yes, I think so. In some of these countries, such as Kenya—you went to these schools, and they were just total rote learning. It was sort of tragic. We certainly did a lot better than that, but the results of our work were *far* from ideal. I felt at the time that the way the materials were produced and reviewed was flawed. Part of the problem was just the inconsistencies, the stylistic variation that was allowed to occur between the various individual writers' efforts. The overview was through Ted himself, and I think he simply was not interested in riding herd on this group and molding them. We did meet together and talk about things, but he wanted to be sure that everybody was having a good experience. I can just imagine him being guided by some of the MIT people, mainly in the physics department, that were into educational reform. I can hear them saying, in my imagination: "Get the really top people and *leave them alone*." Well, he didn't have really top people: some were very good, some not, and what happened in the summer, the initial writing, should have been followed with an editing process, a serious review. Or possibly there should have been a review and editing process halfway through the summer. I lived for parts of four summers in Mombasa, usually six or eight weeks, and in three of those years my family was with me.

Other Mathematics Department Heads

The department heads that followed Ted included Norman Levinson, Ken Hoffman, Danny Kleitman, Arthur Mattuck, Dave Benney, David Vogan, and now Mike Sipser, and I think that they, in their various ways, were quite different. First of all, there was Ted, who had built the department with his own hands, as it were—who identified people, persuaded them to come. There had been other members of the department who wanted to have people come and who steered the department in certain directions, but it was Ted who really got it done. And who also ran the department in this very parsimonious way [laughs]. But I think the world kind of passed him by a little bit in this regard, which became clear when Ken Hoffman took over. Hoffman realized that parsimony was no longer necessary and that other departments were providing services to their faculty that we ought to be providing. I mean, there were just one or two secretaries, and this was a time when word processing took a lot more human hours than it does now. But mainly, the thing that sticks in my mind is photocopying [laughs]. Ted would *never* have accepted the idea of copying pages out of a book or preparing class notes without arranging ahead of time to make sure you, as the copier, were prepared to pay for it.

And Hoffman just said—

Unlimited! Come on in! It's all free. Ken Hoffman was very ambitious, very energetic. He was unusual in the math department in his semi-aggressive readiness to make changes.

We've had some really good department heads. They saw themselves as servants of the department. Their leadership was "Here's this wonderful, distinguished department. I'm here to help protect them, preserve their ability to do their work." I think that the job of the mathematics department, as Ted saw very clearly, was to get good people doing good mathematics and let the intellectual future of the department evolve within that group. Hoffman was a little more inclined to see the way that the evolution would proceed as something he would guide and suggest and support. He was very ambitious and didn't try to conceal it. He was for good things, but he also wanted to have them seen as part of a new era.

The Pure/Applied Split

What do you remember about the pure/applied split in the department?

Harvey Greenspan was another strong-minded person, more so than Dave Benney. But working together, Dave and Harvey brought about what, I think, proved to be a good thing for MIT to have: a separately labeled group of mathematicians who are working in areas that are more closely related to other things that go on at MIT.

You said that now you see this split as being a good thing. What were your feelings about it back then?

Harvey viewed me, I'm sure, as an opponent. Because I just liked the idea of mathematics as a subject in its own right. That's what I'd come to appreciate at Princeton. But the applied people wanted to be called "professors of applied mathematics."

And you opposed that?

I did, yes. But I feel mellower on that subject now than I was then. It's not my impression that there's that big a split now. And there are some obvious anomalies. When they split, Gian-Carlo Rota, who was a sort of quintessential pure mathematician, ended up on the applied side. And today we have one of his students, Richard Stanley, who is a terrific mathematician, and who is the moving spirit of what must be one of the larger programs in the mathematics department right now: combinatorics. Officially it's in applied mathematics, but I would assert that it is very much of a pure field, or pure *side* of a field, anyway. On the other hand, you have someone like Gil Strang, who could be easily construed as an applied mathematician, but who is nominally a pure mathematician.

We've got a great group in combinatorics, and that happened because we had Gianco Rota and then Richard Stanley. Gianco was an undergraduate at Princeton: I was aware of him when I was there, because he'd go to some of the graduate courses. He always dressed formally—dark suit and tie; I remember that. We had some common interests at that time. Then he went off to Yale and became very interested in functional analysis. But he also developed an interest in combinatorics and came here around 1960. Combinatorics began with the classical problems of counting: if you draw five cards from a deck of cards at random, what are the chances of getting a four of a kind, or a full house, and so on. There was a guy named MacMahon, who wrote a book on combinatorics in the early 1900s, and that was all there was to it.[21]

But Rota somehow established the view that there were some really interesting problems here that went a lot deeper. He eventually won the [1996–1997] Killian Prize, the prize that Is Singer got last year. I was very interested in that because I was the principal moving force in getting the

21 MacMahon, P. A. (1915). *Combinatory analysis.* (Vols. 1–2). Cambridge, UK: Cambridge University Press.

Killian Award for Rota. I just felt that it was shameful that there hadn't been a Killian Award for Mathematics,[22] and that Gianco was just the right person to get it, and so I organized letters and so forth.

Logic and Recursion Theory

In the preface to your book you wrote:

> The decade of the 50s was a time of excitement, enthusiasm and promise for students of recursion theory. The month-long Summer Institute for Symbolic Logic at Cornell in 1957 marked an epoch for the theory (as well as for other areas of logic) at a moment when a significant number of independent and largely self-taught researchers came together for the first time to share approaches and ideas.

What happened at that meeting? If people were largely self-taught, it must have been a really fledgling field.

That was a great meeting. We met at Ithaca for a month, and it was *very* exciting. We met essentially all day. It was a real honor roll of senior logicians and aspiring ones and other kinds of people. My wife and I went up there and lived there for a while. That was after her third year at medical school, so she was able to get away. The next year she would have been interning, and she couldn't have done that.

That's where all the people came together who had independently been working on some of these questions. It wasn't really mathematical logic: the field I'm talking about was more what today would be called algorithms, studies of procedures that could be implemented on a digital computer. We didn't think of it that way then. There was a lot of straight mathematical logic there, too. It was both logic and the kind of recursion theory that I was interested in.

The first work that really tied this whole business of algorithms with computation was Alan Turing's work on his PhD thesis with Church,[23] where Turing devised the notion of what's now called a Turing Machine as a model of effective computability—that is, as a model of a com-

---

[22] Applied mathematician C. C. Lin had in fact received the James R. Killian, Jr., Faculty Achievement Award in 1981–1982.

[23] Turing received his PhD from Princeton in 1938 for a thesis entitled "Systems of Logic Based on Ordinals." His advisor was Alonzo Church.

putational algorithm. And that's what I was expanding on in the late fifties and early sixties. Later on, the big change in this area occurred in the early to middle seventies with the idea of fast algorithms versus slow algorithms. Some of the algorithms in logic are incredibly slow, and some of them are not so slow. Is there some mathematical way of distinguishing between them? The distinction, which was significant, was first made, to my knowledge, by a graduate student named Alan Cobham. I was teaching a logic course in the early sixties, and he was in it and came to me and described the idea now known as "computability in polynomial time." And that was the answer! But I didn't recognize it was the answer [laughs]. I just sort of laughed and said, "Yes, I've heard a lot of ideas like that," and ignored it. But someone else picked up on it—Stephen Cook up in Canada did—and it has been a valuable and useful idea.[24] It doesn't sound like a useful idea, but it turns out, the way it works, that it is.

All this relates to cryptography. You can always find an algorithm that will decode messages if you try enough possibilities for a long enough time. But is there a fast algorithm that will do this? That's still an open question. It's related to the P versus NP question, a big, unsolved problem to which I return every year or so.

One of the seminal moments in the field of recursion theory, it seems, was the list of questions you posed at a logic meeting in Leicester, 1965.[25] *You wrote, "These problems are easily stated and appear to be of central significance in the foundations of* Recursive Function

---

[24] Stephen Cook confirms this: "I was definitely influenced by Cobham. As a graduate student at Harvard I read his 1964 paper 'The Intrinsic Computational Difficulty of Functions,' and I talked with him in person a number of times.

"In my 1971 paper introducing NP-completeness my notation for what is now called P was based on Cobham's notation (script L) for the class of polytime functions. I think it is fair to say that my interest in the class P was inspired by Cobham's paper.

"My 1975 STOC paper 'Feasibly Constructive Proofs and the Propositional Calculus' introduced the equational theory PV (which is now standard in the bounded arithmetic literature), and PV is based on Cobham's recursion-theoretic characterization of the polynomial time computable functions, from his 1964 paper." (Stephen Cook, personal communication.)

[25] Rogers, Jr., H. (1967). Some problems of definability in recursive function theory. In J. N. Crossley (Ed.). *Sets, models and recursion theory* (pp. 183–201). Proceedings of the Summer School in Mathematical Logic and Tenth Logic Colloquium, Leicester, August–September, 1965, North Holland, Amsterdam.

Theory." S. Barry Cooper has written since then that "These easily stated questions have increasingly set the main agenda for basic research in computability theory over the subsequent 33 years and that the broad concerns underlying them promise to mold the direction of such research well into the next century." [26]

I was not aware of that!

You were the first logician to be hired at MIT. Did a whole logic group eventually congregate here?

Yes, though it did not have the strength at the senior level. The book I wrote on recursive function theory was well received and very good, and it had terrific citation statistics.[27] I certainly brought post-doctoral people, and I had a lot of graduate students. Some of my graduate students were quite good, but if I had that to do over again, I would have had them working harder on more difficult stuff. But I was not, nor did I want to be, an academic star of a magnitude that would attract other people here. I did help to get Gerald Sacks to come, who was and is a star. And we had a fellow in the philosophy department at MIT named Hilary Putnam who was really quite good. The philosophy department started a graduate program, centered in logic, but then Putnam left and became a professor at Harvard.

The mathematics department keeps trying to hire more logicians, but somehow it hasn't managed to get a critical mass. They think in terms of getting a single person—that's the way you start in any field of mathematics—and if that person is good enough, he or she will attract graduate students, and you'll get a kind of research community going. But there are a couple of fields where that has not worked. One of those fields is statistics, which is *sui generis*—it just has its own problems, and I could go on at some length on that. And logic is another one. We had a student named Sy Friedman, who now runs the Kurt Gödel Research Center for Mathematical Logic at the University of Vienna—very good group, a lot of good stuff going on there. We had a very good person who was going to be the answer to all our needs, Ehud Hrushovski. We appointed him and everything was

---

26 Cooper, S. B. (2000). Hartley Rogers' 1965 agenda. In S. R. Buss, P. Hájek, P. Pudlák (Eds.). *Logic colloquium '98: Proceedings of the Annual European Summer Meeting of the Association for Symbolic Logic* (p. 154). Natick, MA: A K Peters.

27 Rogers, H. (1967). *Theory of recursive functions and effective computability.* New York: (1987) McGraw-Hill; Cambridge, MA: MIT Press.

fine, but he likes living in Israel and he's got a wife that likes living in Israel [laughs].[28]

I don't think there's been any impediment to hiring more logicians, but mathematical logic has been a kind of orphan in mathematics. There are people who think it's not real mathematics, and there are people who think that it is. There's some gut feeling, on the part of some people, that logic is a little bit inferior. I have doubts myself, as to whether or not it's all that it should be, compared with some of the great central areas of pure mathematics. And I think there are parts of logic where the work done is being elaborated into an analogue of counting angels on the head of a pin. It's just too refined. A portion of mathematics in this country has this property, at the present time—chewing over some very specialized problems, within very specialized areas, within further specialized things, that do not relate very well to central issues of mathematics or of applied mathematics.

Educational Policy

You got heavily involved in administration relatively early on. Can you talk about your experiences as an administrator?

Beginning in 1961—I had been here for six years—I got very interested in broader educational policy at MIT. I was appointed to something called the Committee on Curriculum Content Planning. The initials were the CCCP—this was a case of the acronym governing the name [laughs].[29] Jerrold Zacharias was the chairman of the committee. It was a high-level, Institute-wide committee to consider the MIT's central core educational program. MIT, as an Institute of Technology, has to have a certain amount of math, a certain amount of physics, a certain amount of chemistry. These are the "core" of the "general institute requirements." The whole educational program had been studied in 1949 by a commission chaired by the head of the chemical engineering department, whose name was Lewis, Doc Lewis.[30] He was a very strong-minded person. The

28 Hrushovski is now at the Institute of Mathematics, Hebrew University, Jerusalem.

29 CCCP was the Russian acronym for the USSR.

30 The 1949 Committee on Educational Survey, known as the Lewis Commission, examined MIT's education in light of the changes taking place in the aftermath of World War II. Its recommendations, among others, included strengthening the role of science as a basis

commission included Jerry Wiesner, later a president of MIT, who did some of the writing of the report. The president of the Institute at this point was James Killian, who would become Eisenhower's science advisor in Washington.[31] The commission had concluded that MIT was on the threshold of becoming a truly national institution and a real university, not just a technical school, and certainly not a trade school. This had been gradually happening, and how should they state and present this to the world and to our students?

The Lewis Commission was strongly influenced by the fashion of *engineering science*. That is, that for MIT to be what it should be, it should be teaching more science, less specific technology. If a student came to MIT, he wouldn't start off fiddling with a steam engine; he would start off learning the mathematics that one uses to describe gases and heat and so forth. Engineering science was a doctrine that came out of engineers at MIT. They felt that MIT was not going to have forward-looking, research-oriented engineering students, competent in a variety of industrial situations, unless it got away from merely teaching them how to run a strength-testing machine. In engineering, you'd have these great machines, for testing strength of materials and so on. Well, that's fine, we should have that, but for that to be the central focus of what we were doing—I mean, we were just in the wrong century. We had to get away from that and emphasize basic physical principles. Hence, "engineering science."

So in this spirit, the commission dropped mechanical drawing, for instance, as a requirement for all students. Not everybody thought that was a good move! Some of the crusty old guys didn't like that shift. On the other hand, there were a number of basic science courses being required by individual engineering departments. The commission therefore recommended having a larger *common* required core for all students: two years of physics,

---

for engineering education and establishing the School of Humanities and Social Science. Its chairman, Warren K. ("Doc") Lewis, served as the first head of the chemical engineering department from 1920 to 1929. After this, he devoted himself to teaching, research, and consulting and remained an influential member of the department until his death in 1975 at the age of 92.

31 Among other advisory positions, James Killian served as Special Assistant to the President for Science and Technology in the Eisenhower administration and was involved in the CIA's program to develop the U-2 reconnaissance aircraft. See for example Rudolph, J. L. (2002). *Science in the classroom: The Cold War reconstruction of American education.* New York: Macmillan and Lecuyer, C. (1992). The making of a science based technological university: Karl Compton, James Killian, and the Reform of MIT, 1930–1957. (*Historical Studies in the Physical and Biological Sciences, 23,* 153-180.

two years of mathematics, and one year of chemistry. This did not solve all problems. For example, we still taught matrix algebra in special-skills courses for certain engineering departments. We teach it now, fifty years later, as a central topic in the purest of pure mathematics. But fifty years ago, the only place you'd find it was in one of our elective Methods of Applied Mathematics courses.

Then came Sputnik in '57 and with it a desire to have an educational program that would be competitive with the Russians; they had gotten Sputnik up first. At the same time, MIT began to become a national university in the sense that students from all over were coming here, and the *better* students were coming here. The nature of the undergraduate body abruptly changed around 1959. The numbers and quality of students applying to universities sharply increased. By contrast, in 1951, Harvard had failed to fill its freshman class!

The first calculus I ever taught was during my first year of Princeton, in 1950–1951. This was a big thing for me. It was in the second term, not the regular term, so it was a sort of mixed bag of students. Some were upper-class students, and some were freshmen who decided to learn some calculus a term later than they should have. One of the students, by the way, was a freshman, with sort of long hair. His hair still is a little long. He's listed in my class book as D. H. Rumsfeld. Anyway, I gave—I looked this up the other day in my class book—four As out of twenty students, which was normal in those days. One of the four was Donald Rumsfeld. Today, with the increased flow of better students, it would be half As. When I came to MIT in '55, I taught freshman, sophomore, and junior sections in successive years, just to become familiar with the program. I taught elementary sections of calculus and so on, all the way through. And suddenly, instead of giving three or four As in a freshman section of twenty-five students, I was giving ten or twelve As.

That's fascinating. How do you explain that?

Well, a lot more consciousness of science, a growth of the popular interest with the Sputnik event. Sputnik was the occasion of considerable educational self-examination in the U.S., both in high schools and in colleges. It was a discrete event, but it seemed to trigger a kind of demographic, sociological sea change. It included significant improvements in high school education. And a lot more people were thinking about leaving Idaho or Utah or wherever they grew up and going to college back East or on the West Coast. I could see it in my wife's community. For a while, when she was a girl—well, you went to high school, and then you went to "the U," the University of Nebraska.

So the 1961 committee that you served on was convened to take another look at the conclusions of the Lewis Committee. What was the impetus to get this committee together, and how did it go?

Well, the president at that point was Jay Stratton. And I think the impetus was that there wasn't enough flexibility. If you looked at the MIT catalogue, it looked too much like a trade school catalogue. I'm not saying this is how people felt then: I'm using "trade school" just to communicate in a few words the feeling of inflexibility, the rigidity of programs. You see, the existing system was, you'd divide the freshman class into "cohorts" of twenty-five to thirty students each. And each cohort, until the students separated to go to their professional departments, would take everything together. So I'd have a section teaching them mathematics. Exactly the same students would move to a chemistry section, to a physics section, and so on. But you couldn't do that anymore with the variety of preparation that we were facing. It was clear that the students were coming with a greater variety of talents. We had more students who had already learned some mathematics, so we had to be able to accommodate them.

Another thing that was happening was that you could accelerate the learning for our better prepared students, which made it easier to cope with the variations in student preparation. In the first term of mathematics, we were able to teach, in a single term, what previously had required the entire first year, because the students were better. There was a change in style, less drill work. The students were smarter. You could get them to the same place while exercising their higher creative abilities. And there was also a certain "Let a hundred flowers bloom" aspect to this. If you want to get students to exercise their higher creative abilities, you needed to try various things, to see what inspired approaches we could take.

So we recommended cutting the basic calculus in half, only requiring two terms instead of four terms, because many of the students were ready to go ahead into more specialized major courses. And we recommended similar 50 percent cuts in physics and chemistry as well. They could find electives to take that would be better for them, and better for their development, than simply going on for the previously required years in the basic sciences. We suggested that reduction, and MIT adopted it.

Our final report and recommendations were written by me. The introduction took the form of a hypothetical dialogue in the style of Galileo, who was suggesting a major change in the way we looked at the universe. We were suggesting a major change in the way MIT was looking at its undergraduate education! What we did in the early sixties has served us pretty well. It has lasted to today. Bob Silbey is now chairing a committee that will be producing the next installment.[32]

Then, in 1966, Howard Johnson became the president of MIT, and the administration and faculty decided to have another committee, the MIT Commission on Education, which, rather than just being concerned with the basic requirements, would look at the whole educational process. This was a big committee, and it was hoped that it would deliver profound wisdom on the basis of which further radical changes in the curriculum, or perhaps in the style of MIT as a whole, as an educational institution, could occur. The chairman was Ken Hoffman: this was before he became head of the department. But the Commission had a tough time and eventually ran completely aground. It made a report, but it was aground on Vietnam, student rebellion, and so forth. Then they appointed another committee to look at the Commission report and kind of sharpen it, determine what practical consequences might emerge from it. This new committee, the Special Task Force on Education, included Frank Press, who was later head of the National Academy of Sciences;[33] Sheila Widnall, who was later Secretary of

---

[32] From "MIT reviews what, how it teaches" (*Boston Globe*, 17 March 2004): "The Massachusetts Institute of Technology . . . is launching a two-year review of its undergraduate education, examining its required courses and other student experiences in light of new developments in science, as well as the changing interests of students.

"The review . . . will take a sweeping look at MIT's demanding roster of required courses, developed more than a half-century ago when MIT was an engineering school populated chiefly by men.

"MIT needs to study how teaching methods and the frontiers of science have been transformed in recent years and how those changes should be reflected in the curriculum, officials said.

"'There's much more biology in much of the research that we do,' said Dean of Science Robert Silbey, who will lead the review committee. 'There's much more interest in economics and social science among our students than there was 20 years ago. It is important for any great university to ask: What are its students like? And how should we go about preparing them for life?'

"The review begins as Charles M. Vest's 13-year presidency draws to a close. . . . The review committee, made up of faculty and students, will spend the next two years collecting data, conducting surveys, and consulting with educators and employers outside MIT. Recommendations will ultimately be voted upon by the faculty."

[33] Frank Press later served President of the U.S. National Academy of Sciences and Chairman of the National Research Council (1981–1993) and Science Advisor to the President of the

the Air Force,[34] Benson Snyder of the MIT Medical Department, who had a long-standing interest in curricular changes at MIT,[35] and myself as chair. It was a pretty good committee. We worked for four or five months.[36]

Did your work on these committees influence your own math teaching or shed light on the teaching in the math department?

Well, in reviewing the basic curriculum, I became more aware of what Arthur Mattuck was doing in his courses and was deeply influenced by it. In his calculus courses, he had little books which were called "Supplementary Problems." These successive books had problems in them which weren't routine, crank-turning recipes for how to solve certain kinds of specific problems—"how do you integrate this" and so forth. They touched on whether you really *understand* this. It forced me to face up to the possibility that there were more mature mathematical themes that could be brought into an elementary course. Today, this is just taken for granted in the math department, but Arthur was responsible for it.

Arthur has had other interests in our undergraduate program. He has, for years, taken charge, in effect, of the teaching skills of our teaching assistants, instructors, and younger faculty. If you come to MIT now, as a post-doctoral instructor, the department wants to be sure that you're good at teaching! You'll have had recommendations and so on, but one of the

---

United States and Director of the Office of Science and Technology Policy (1977–1980). Before that he was a professor of geophysics at MIT and chairman of the Department of Earth and Planetary Sciences.

[34] Sheila Widnall was Secretary of the Air Force from 1993 to 1997, the only woman to head a military service. She is a professor of aeronautics and astronautics at MIT, where she served as associate provost (1992–1993). She was appointed Institute Professor in 1998.

[35] Benson Snyder, a psychiatrist, was the author of a book called *The Hidden Curriculum.*

[36] The Special Task Force on Education reported on December 8, 1971. According to the Friday, December 10, 1971, issue of *The Tech* (Vol. 91 No. 54), The Task Force proposed:

> 1) the expansion of current experimental undergraduate research programs to form an extensive part of most students' regular curriculum;
>
> 2) administrative reorganization to create a Dean for the Academic Program responsible for providing an 'educational-focus for the faculty's exercise of its undergraduate educational responsibility';
>
> 3) a more-formal commitment of MIT resources to programs in educational research, including specifically the establishment of a separate 'Educational Division' to provide the needed organization, coordination, and support.

things that happens is that early on, you are videotaped teaching a class, a real class. Arthur takes the videotape and writes a very good—spirited, but kindly—but firm—report. Elementary things, such as, if you've written something on the blackboard, don't stand in front of it so other people can't see it [laughs]. But also questions of organization and exposition, and more subtle questions of affect: are you encouraging questions; are you friendly, are you hostile, and so on. It's made a big difference, because everybody in the department is conscious of Arthur's doing this, and one is subject to his gaze, no matter *who* you are. It's been highly effective: what was once a trade-school experience for our undergraduate mathematics students has become, if you like, an intellectual voyage.

The Student Rebellions

You mentioned Vietnam and the student rebellions. How did those events affect life at MIT, in your experience?

In 1969 I was appointed the chair of the Special Panel on the November Events. Jerry Wiesner was the provost at this point, and his office was occupied by students. The student rebels had gotten themselves organized and were sitting in, disturbing classes, blocking classrooms, things like that. I can remember seeing trails of tear gas across the street from the main entrance of MIT, wafting over the lawns, and I can remember a big march the students had on the Instrumentation Lab and their being met by the Cambridge police in riot gear, with masks, batons, and shields, and the whole futuristic business. The student revolts were very worrisome to the faculty, and I had been a visible figure with the faculty and with the students. I took the events very seriously and worried a lot about it.

MIT appointed two *ad hoc* committees. One was a committee on outside agitators, and one was a committee on our own students: How should this disturbance be handled? What disciplinary steps could be taken? The administration wanted to expel some students, but they felt that would be seen as a conflict of interest, so they wanted to have a kind of independent grand jury, as it were. The first committee, on the outside agitators, was chaired by Mert Flemings, who later became head of the Materials Science Department. This committee became bogged down in controversial issues and was unable to finish its work. I was asked to chair the second committee, which would consider the role of our own students in these disturbances. This looked as if it might be a fiasco, because of its diverse membership;

they had assigned six faculty members, two graduate students, two members of the administrative staff, and two undergraduates to this. The six faculty were full professors, associate professors, and so on. I had the guy who founded the cancer center [at MIT], Salvador Luria. He was a Nobel Prize winner.[37] Then I had quite a conservative faculty member. One of the graduate student members was identified with the rebels. It just looked on paper as if everybody was going to disagree with everybody. But it met very intensively for about a month. And I was just lucky: I was able to find a common feeling of common goals. It was, in many people's eyes, a kind of miracle. I have the report in my office. We took a somewhat legal approach, where we talked about rights and responsibilities and so on. But on the basis of our report, the standing faculty committee on discipline was able to meet and expel the more extreme agitators among the students.

The two principal agitators—there was a guy named Mike Albert, who was a real creature of the times. I remember, in the fall of '68 or '69, President Howard Johnson addressed the freshman class in Kresge Auditorium. Mike had gotten himself elected as the head of the student association. So Howard Johnson got up and welcomed them and presented his good wishes, hopes, ideas and so on. And then he said, "And now we'll hear from the head of our student association, Mike Albert." So Mike Albert kind of swaggered up to the podium in blue jeans and began, "Well, now that you've heard all that shit—"

Mike was an undergraduate. There was another guy; his name was Jonathan Kabat. He was a graduate student in biology and quite smart. These people kept popping up in one's life in those days: I was introduced years later to a staff member at the University of Massachusetts Medical School in Worcester, and his name was Jonathan Kabat-Zinn. He was very much into meditation when I last talked to him.[38] Mike Albert was working in a book shop the last I heard.[39] They had the administration *paralyzed* with fear. The administration just didn't know how to handle it.

One of the issues was the Instrumentation Lab, which produced inertial guidance mechanisms for missiles—that MIT didn't have any business

37 Biology professor Salvador E. Luria shared the 1969 Nobel Prize in Physiology or Medicine with Max Delbrück and Alfred Hershey "for their discoveries concerning the replication mechanism and the genetic structure of viruses."

38 Jon Kabat-Zinn founded the Stress Reduction Clinic at the University of Massachusetts Memorial Medical Center in Worcester, Massachusetts, and is the author of several books, one coauthored with his wife, Myla Kabat-Zinn.

39 Mike Albert founded South End Press and wrote for *Z Magazine*.

becoming a production facility for armaments. I think a lot of faculty felt that way, too. Another issue was ROTC [Reserve Officer Training Corps]. I was present when they blockaded the ROTC offices in the old Building 20. That was the only time that a shot was fired in all this time of student unrest. I remember the guy who was on the "campus patrol"—then the name of the campus police—had a pistol, and he thought he was going to scare the rebellious students. It was in a hallway at the foot of some stairs, and he pulled out the pistol and shot it upward, and the bullet entered a beam over a door. You could go see it, if you wanted, at least for a few days.

The MIT faculty was more conservative than faculties at other places. Harvard dropped ROTC. They regret it now, and their president talks about getting it back. But we kept ROTC.[40] BU and Harvard and Northeastern all ended up sending their ROTC scholarship students to MIT for their ROTC classes.

The students made up some issues that they felt were important and that the administration was being obstructive about. One of Albert's early issues, for example, was to have unisex bathrooms [laughs].

Other Chairmanship Positions

You continued on in administration after that, right?

Yes. In 1971, I was elected chair of the overall MIT faculty for a two-year term. It's sort of a figurehead position, not a position with any administrative power, but one that has seats on various administrative committees as a kind of voice of the faculty. As chairman of the faculty, I met with the Academic Council, which is the council of all the deans for administrative and academic purposes. And then I also chaired a faculty-student committee called the Committee on Educational Policy for those two years. It was a very, very

---

[40] From "A Brief History of the ROTC Programs at MIT," part of the March 20, 1996, Final Report of the MIT ROTC Task Force:

> The growing national unrest over the Vietnam War brought the ROTC programs into the debate of the [MIT] faculty once again in the Spring of 1969. During the same general time period, ROTC programs were abolished at Tufts, Boston University, and Harvard. At a faculty meeting, held on May 14, 1969, in Kresge Auditorium, with an attendance of four hundred fifty seven faculty members, after lengthy debate, a motion to abolish ROTC was defeated by a large margin.

A longer excerpt from the report is available at http://web.mit.edu/committees/rotc/rotchist.html.

good committee, the central committee of the faculty. I realize now that it served as a kind of decompression chamber for MIT, helping things settle down after the strains and divisions of '69 and '70. We met every Thursday for two hours, with lunch. I had a wonderful time. I had really good people: Mike Dertouzos, now dead, who was the big computer visionary for the Institute.[41] A guy in the humanities department named Travis Merritt, also recently deceased.[42] Several members of the administration, including the provost, also attended regularly. At the end of this committee, Travis wrote a verse play in iambic quadrameter couplets about our committee and its life, and they put on this show in partial costume—Dertouzos wore a hard hat—at our final meeting at the end of the year. The title of the play was "Hartley Ever."

I still was teaching in those days. I was very close to Arthur Mattuck, who had been a graduate student with me at Princeton, and he kept me teaching, even though I was doing all these other things. I would always have a lecture each term, usually a big lecture three times a week.

I had another very close friend named Constantine Simonides, who was a vice president of the Institute and very influential person with the senior administration.[43] Very lively Greek guy. He invited me to a lunch over at the Harvard Club, in the Back Bay. It was in a private room. I went in with him, and the table was set not just for two people but for three other people as well. I didn't know what this meant. And suddenly into the room came Jerry Wiesner and Walter Rosenblith and, as I recall, Paul Gray. They had come to offer me a position in Wiesner's administration as associate provost, a new position. The administration was set up with Jerry Wiesner as president, Rosenblith as provost, and Paul Gray as chancellor. I was to be number four as the associate provost and would take a general view of everything. The titles "chancellor" and "associate provost" implied somewhat different administrative responsibilities then than they do now: I didn't have specific duties but would serve as a kind of deputy for the provost. We four would form the Senior Executive Group and meet periodically.

---

41 Michael L. Dertouzos served as director of the Laboratory for Computer Science and was the author of numerous books. *Business Week* referred to him as "the guru of user friendliness."

42 Travis R. Merritt was a professor of literature, director of the Humanities Undergraduate Office, and dean for undergraduate academic affairs.

43 Constantine B. Simonides served as secretary and vice president of the MIT Corporation until his sudden death at age 59, apparently of a heart attack, while playing tennis.

TO

Professor HARTLEY ROGERS, Jr.

Chairman of the Faculty

After the Tenure Review

SPRING 1973

The Members of AASG

"Certificate" of thanks presented to Hartley Rogers after the spring tenure review by the members of the review board. In the bags are "prime grade shredded tenure cases, no additives."

And that was a mistake. I had been offered something on the basis of my personality and not on the basis of my experience, but I wasn't able to have that experience because I had no operational responsibilities. So that cost me in various ways. But it was an interesting time. I did eventually have one operational responsibility: to set up and oversee a new Office of Minority Education with a variety of educational and student-related responsibilities in accord with its name. I led a search for the first director of OME, and we appointed Wesley L. Harris, a junior faculty member in the Department of Aeronautics and Astronautics. He and I enjoyed working together and have remained close friends. He is now the head of the Department of Aeronautics and Astronautics.

Early in '79, I was asked to chair a group of six faculty, including Richard B. Adler from electrical engineering, to prepare a presentation for an institute-wide accreditation review by the New England Association of Schools and Colleges. I wrote a historical essay, of which I was quite proud, on the growth and changing emphasis of MIT's academic program through the years.[44] I continue to believe that it would be useful reading for any group preparing to consider such matters.

Then from 1974 to 1981 you served as chairman of the MIT Press Editorial Board. How did you come to that job?

Constantine Simonides got me into that as well. Constantine took over responsibility for the Press, and there was a significant belt-tightening move in the administration: "What the hell are we doing with this? It's just money down a hole. Let's get rid of it." Constantine fought that successfully, so that was the background.

Our job was to look at prospective works that the Press acquisitions editors would propose that we accept or pursue. We weren't always all that malleable, but we were sympathetic. I wanted very much to do it, and I think I was very good at it. I did it for a long time—seven or eight years. It was really fun.

So your job was to help the acquisitions editors have a better sense of whether this book was really valuable work in the field?

That's right, yes. Did it meet a certain scholarly standard? Was it a contribution? Was it consistent with MIT's educational mission? We had a wonderful time; we could pick the people that we wanted on the editorial board. Gee, they were good. Harry Hanham, the dean of humanities, was an excellent person. Bill Porter was the dean of the School of Architecture and Planning; he was excellent. Just the other day I saw Carl Kaysen: he's a distinguished, Washington-connected, guru type. I had Ernie Moniz, who was the Undersecretary of Energy for a while, and who's also been head of the physics department at MIT. He's of Portuguese ancestry, and he's from the New Bedford/Fall River area, a very interesting person, very bright, very intelligent, and an experimental physicist, still at MIT. We had all these super people, whom I just loved to work with.

Teaching Again

You left the MIT Press position in 1981 and went back to teaching.

Yes. The period from '80 to '81 was primarily a sabbatical year. Starting in the fall of 1981, I began teaching a calculus course for first-term freshmen

---

44 "Basic Educational Requirements at the Massachusetts Institute of Technology: A Report Prepared for the New England Association of Schools and Colleges, Inc., in Connection with the Accreditation and Re-evaluation of M.I.T., April 8–11, 1979."

who had advanced standing. It took the number and name of an existing course, but I made it harder. I also had slightly over six hundred students in a single lecture room, 26-100. The course was too hard for them: the bottom half of the class was kind of lost. Though they learned what they needed to know; I never had more than five or six failures. They were just letting off steam, so all kinds of things happened. One day some guys came in and fought a fencing duel up and down one long aisle and around and up the other.[45]

And is it true that you used to give out cookies in class?

The Newton-Leibniz Awards. It was for the best performance on the hour-tests, and it's genuine. If you got 100 on all four hour-tests—nobody ever did, but you could come close to that—then you'd get a Newton Award, which was a box of Fig Newtons. And the Leibniz Awards—Leibniz cookies, they're made in Europe—were a little bit lower. I was very careful about choosing the people. I'd get them to the front of the room, and there would be wild cheers, and that's still true! I still do that, and they really enjoy it. And they value those cookies! I announce the names, and if someone isn't there, he shows up later to get his prize.

Every year one of the fraternities would give me a six-pack of beer during the last lecture. One year, when I knew that was coming, I took a bottle out and started to drink it, and then I went into this sort of Jekyll and Hyde routine, where I was sort of struggling with my throat and slowly sank from view [laughs]. I had a wig with me, a sort of rock-star wig that belonged to one of my sons. It's just this long, black hair, it sort of hangs down and around, and it's *very* deceptive. And while I was out of view, I put on the wig, and then I slowly emerged with this huge crop of hair! And I thought there'd be great laughter—there was just *stunned* silence. It was such a successful thing [laughs]. Another year I sort of set things up by saying that I had a graduate student who really wanted to give part of the next lecture, and so he would be coming. So for the next lecture, I came in, you know, in blue jeans, sloppy clothes, and with this wig on, but I couldn't say anything because they would have known right away.

---

45 For a fuller account of this student hack, see http://www.everything2.com/index.pl?node_id=1078225.

Hartley Rogers.

You are now teaching courses in logic again. How is it different now than when you started?

It's better. These are ideas that I've thought about, and I'm writing a book, not research but exposition, the title of which is *Logic as Mathematics.* I'm thinking of new and simplified ways of communicating the existing field and am discovering a few new things as I go. And I have a coauthor: I've done almost all the work on it so far, but he's got a lot of special articles that he wants to put in it. It's a guy named Henry Cohn. He went to MIT as an undergraduate; he went to Harvard as a graduate student, where he had a very distinguished thesis, and he's now at Microsoft, in their Mathematical Research Group. I'm working on the first half of the book, and I use that material in my course.

What's the course called?

Mathematical Logic. It's the first-year graduate course for students interested in this.

So it must be very different than in 1955, when you first started teaching.

Oh, it is. And I think it's very different from what it is at a lot of other places right now. I think the title of the book says part of it: *Logic* as

Mathematics. The teaching of logic gets kind of complicated because logic was set out, in part, to justify theological arguments, starting in the Middle Ages—how one defined rational argument, and what does it mean to say that if A is true, then B must be true. Syllogism was a kind of trivial part of medieval logic. And then you go on into the nineteenth and twentieth centuries: logic goes on to try and be mathematical. Mathematical logic developed—and this is really part of the definition of logic—as a codification of valid methods of argument in mathematics: what makes a proof correct or incorrect, and so on. But now, in mathematical logic, we're trying to go back and look at fairly delicate mathematical questions, and there can be a confusion between the justification aspect and the research aspect.

Part of the problem, you might say, is ontological: what really exists in mathematics? I'm stating this as a philosopher now, and it's not something I'm very interested in, but what objects exist? We talk about an infinite set of numbers, but we don't experience it in our physical universe. Maybe the universe is finite, in fact. Subjectively, mathematics is very Platonistic. That is, you assume that these are very precise objects; they exist in some realm of ideas, or in some heaven somewhere; you know exactly what the integers are and what the geometrical properties are, and so on. But *subjectively,* most mathematicians tend to think of these infinite objects as really existing. So when I talk about "Logic as Mathematics," I take this more Platonistic route. The only way to look at the philosophical questions properly is not to try to begin with them, when your students are naïve, but teach them some mathematics.

The unusual thing is that mathematicians *know*—good mathematicians have a very precise concept of what constitutes a correct proof, even though they usually haven't analyzed it very well. And why is that? It's because there is this sort of God-given sense of what the subject matter of mathematics is and how we can work with it. And we're better off doing that than we are trying to go back and teach in a way that reexamines various blind alleys.

I approach logic just as a part of mathematics. We observe what the classical definition of a correct proof is, and it has many interesting and mathematical properties. And then we turn around, in the course, and try to develop new research on the questions that have been raised by logic, and so forth. I'm working very hard, and it's good. I'm realizing things about myself and about math that I wasn't aware of before.

Putnam Team Coach

The MAA, the college mathematics teachers' organization, holds the William Lowell Putnam competition every year for college students, both teams and individuals. MIT's had a very good record with its Putnam team recently, under your coaching. Can you say something about the exam and its level of difficulty?

It's a very tough exam that's set every year, and it occurs on the first Saturday in December. It's been given for sixty-plus years in the U.S. and Canada. Another exam like it is the International Math Olympiad for secondary school students, but the Putnam is much harder than that. The theory is that there will be no questions asked that go beyond the sophomore year of a standard math program. BUT— the huge *but* is that there are no questions asked about the math that you would *normally* study as a junior or a senior. But there's a lot of math included in the Putnam that's not studied in a standard curriculum: combinatorial things, certain algebraic areas, certain geometrical areas. If you've practiced on previous Putnam exams, you know what kind of things there'll be. There are twelve questions. You get ten points a question, and that means the maximum score is 120 points. For the past few years, the number of students taking it has been about three and a half thousand. And the median score, most recently, has been either 0 or 1.

I am the co-coach together with Richard Stanley. We won the Putnam Exam in 2003 and 2004 and were in the top five in 2005; I don't know if we will have won this year [2006] or not.[46]

What has been the score of the winning teams at MIT, roughly?

Usually they're up in the seventies and eighties.

These students are really smart. How do you help students like that?

Well, it's like a football coach: you don't help them if they're good; you just recruit them. And MIT has been very good about admitting students of that kind. Richard Stanley and I have a seminar through the fall. We alternate lectures, where we lecture on certain areas of mathematics that are not in the standard curriculum very much but are somewhat advanced, and then we give them problem sets every week on what we've lectured on.

---

[46] The MIT team placed third in 2006, after Princeton and Harvard. Two of the four Individual Putnam Fellows were MIT students.

Do you ever feel like, "My goodness, this kid's a better mathematician than I am?"

Oh, I think they're all smarter than I am, but it's a special talent, a kind of puzzle-solving talent. That doesn't mean they're going to be great mathematicians, necessarily, in the future. You're under a time pressure, and it's a puzzle, and you know that there has to be an answer, because otherwise the question wouldn't be on the exam. That's a big step forward for anybody doing research, if they know that a question has been properly formulated. That's a big leg up on getting a solution. It's interesting to note that several of our best Putnam takers have been somewhat autistic. But one of the characteristics of autism is that you can get very interested in a particular topic and go more deeply into it, and one such topic is mathematics.

Does the Putnam Exam have real importance in popularizing mathematics?

I think it does, just as the International Math Olympiad does. There is a whole subculture of high school and college students who are interested in this. There's a great sense of community; they all know each other, and the high school students know who the really good ones are, and where they're going to college. It's a lot like having a kid that's interested in playing soccer: a parent will take time to facilitate this. A Romanian told me this. I said, "Why are there so many brilliant students from Romania?"

And he said, "Because the parents feel about doing math competitions the way you might feel about having your children play Little League baseball."

RSI *and* SPUR

You're involved in two other vehicles for encouraging gifted students, the summer programs you run, known as the Research Science Institute (RSI) and Summer Program for Undergrad Research (SPUR). What is the idea behind these, and how did they get started?

RSI is a program that's run by an organization known as the Center for Excellence in Education. It was founded in connection with a special program for high school students that was initially called the Rickover Science Institute—RSI.[47] This was funded by admirers of Hyman Rickover, an admiral who introduced nuclear power into the Navy, most notably for submarines, but also for aircraft carriers. He accomplished what he did by being a kind of testy,

[47] The name was later changed to Research Science Institute.

Hartley Rogers with his sons at the headstones of his great-grandparents, Orlando and Cephronia Rogers, in Albion, New York, c. 1975.

outsider type. He'd gone to Annapolis, but he was very tough, and willing to shake people up, and kind of a disciplinarian with people that worked with him. When they went to him and asked him what would be the best memorial thing they could do, he said, we've got to provide opportunities for really good students to go on in science and learn what science research is like.

The paradigm of the program was that a Washington office, now called the Center for Excellence in Education, would publicize it and get applications for high school students at the end of their junior year to come for a six-week summer program. After the students had applied and been selected, the administrators of the program would contact faculty members or people in industrial labs—industrial or commercial firms of one kind of another—to be mentors for the students that came, in the city where the program was held.

They had a lot of trouble getting math faculty people. A chemistry professor will have a lab; he can take an RSI student in and introduce him to his graduate students, and they can work that way. But that isn't satisfactory in math. And you can't ask a math professor, the main part of whose research time comes in the summer, to suddenly start spending an hour—or if he has two students, two separate hours—every day. RSI wanted to have individual mentors for individual students, but this is a very inefficient way to use math faculty.

So they were racking their brains: how can we get professional mathematicians involved? A colleague of mine at MIT named Hal Abelson[48] asked me. Could I do something about this? So I proposed using MIT mathematics graduate students as the responsible mentors, and that has worked like a charm. We have a *very* impressive graduate-student community. I could never have gotten into MIT in my day if the students then had been as good as ours are now! The graduate student mentors suggest the projects, I give them a little money from the department, and they do the mentoring. Each student has his or her own mentor. For a five-week period, student and mentor meet with each other every day for an hour or so. They talk over what the student has been doing and the problem that the student has been given to work on, and [the mentor] makes suggestions. It's usually a problem for which the answer is not known until the student works on it.

RSI has been in Boston/Cambridge for fifteen years. Most recently, there are about eighty students every year in all areas of science, with about twenty in math. There are some great success stories. The RSI students go back to their schools for the senior year, and they can take the work they've done at RSI, develop it further, present it as a science fair project, and enter it in one of two national competitions. One is what used to be called the Westinghouse National Science Talent Search and is now run by Intel. And the other is called the Siemens Competition in Math, Science & Technology. These competitions are in all areas of science and technology. I'd say that on average at least once in every four years, the winner of the Intel has been one of our RSI mathematics students. At a rather more senior level, I would note that a winner of one of this year's Fields Medals, Terence Tao, was an RSI student in 1989.

SPUR also uses graduate students, but they're mentoring undergraduate students?

Correct. I started SPUR simply as a version of RSI for undergraduates at MIT. The programs are very satisfying for me because the graduate students are so good, and they appreciate the opportunity: they spend all the rest of their time learning, and now they can teach! I've run these summer programs for fifteen years now, and I decided this year that if I could find a good faculty successor, I wouldn't do them any more. I have found the best possible successor in my colleague Professor David Jerison. RSI and SPUR

---

[48] Hal Abelson is Class of 1922 Professor of Computer Science and Engineering in the Department of Electrical Engineering and Computer Science.

have been a very happy thing for me, and now I want to use the time for other projects.

~ December 12, 2005, and March 15, 2006

Gilbert Strang

A History's Beginnings

J.S.: *The idea to conduct a series of interviews with the department's senior mathematicians began with you. How did the idea come up?*

G.S.: My wife, Jill, who is a historian, suggested that a history of the department would be good. If we don't make a record of things, when people know about them and remember them—then we just won't have a record! When the department had its annual social event at Endicott House, the idea was fresh in my mind, and I spoke to Singer and one or two others. The response was positive, so ultimately I spoke to Mike Sipser. I just put the idea forward to see what happened.

It connects with one chief thing that I've done in recent years, having my linear algebra lectures recorded. Soon after that, OpenCourseWare started up, and Arthur Mattuck and others have had their lectures recorded since. My motivation also came from thinking, shortly after Gian-Carlo Rota's death, what a shame it was that we don't have his lectures. I think this department, historically, has been a very important one, and that when something is sort of unique and of wider than [just] internal importance, we should pursue it.

Math Undergraduate at MIT

How did you first become interested in math as something to study?

I think, like probably a lot of others, I knew early that this was the easiest route, and I was interested in it. I remember in school figuring out how you would square 999 in your head: the fact that 999 squared is the same as (998 × 1000) + 1. I always hoped somebody would ask me, but it never happened. And similar little bits of algebra. Probably even earlier, I tried to count to a million and never succeeded, didn't have the patience. But anyway, numbers were fun.

It was certainly easier than doing labs, so I never seriously considered the alternatives. I never considered another major at MIT, even though at that time, the number of mathematics majors was quite small: ten, maybe. It was more of a service department than it is now. There were a few math majors and certainly some strong faculty, but the full scale of the department wasn't realized until later. All the people you have interviewed, the heart of the department, were coming to MIT in those years, the late fifties.

It was a different university then. One side issue was, it was quite normal to flunk out! Now it really doesn't happen so much. The whole selection process was not what it is today: some clicked, others didn't. People I knew would disappear for a year and return. Perhaps they almost all graduated in the end. Our math undergraduates now are the best I have ever seen.

What was your first introduction to higher mathematics?

In those days even a decent school didn't have a calculus course: high school mathematics at that time ended with trigonometry and solid geometry. I did those early and didn't have anything in the senior year, so I read, in a kind of desultory way, Thomas's calculus book.[1] George Thomas was just writing his book; it was in sort of paper notes at that time, but it had all the exercises. I remember that he offered ten cents for an error: a very wise author! It was just trivial errors in the solutions to problems, because for authors, as I now know, the most miserable part is solving the exercises. I would *gladly* give ten cents or much more for corrections of errors. So when I came to MIT, I went to George Thomas with a list of ten errors, and he counted out ten dimes. I never had George in a class, but he was a very friendly, straightforward guy.

Who else do you remember of the MIT math faculty in your undergraduate years?

The great stars in the department—Wiener of course, but I never had a class [with him] and he wasn't a good teacher; Levinson, who was superb in all respects; and then on the applied side would have been C. C. Lin. I guess I did have Ted Martin in a class. It wasn't a topic I focused on heavily. I had George Whitehead for calculus. He was a very reliable, straightforward, nice person. George Whitehead was our topologist, along with Witold Hurewicz,

---

[1] George B. Thomas (1914–2006) served on the MIT mathematics faculty from 1940 to 1978. His *Calculus and Analytical Geometry* was first published by Addison-Wesley in 1952 and is now in its eleventh edition as *Thomas' Calculus,* with added material by other authors.

and then that part of the department grew as Frank Peterson[2] arrived and, more recently, Haynes Miller and Mike Hopkins and strong people in between, too. Singer had been here previously as a Moore Instructor but had left by the time I came here as an undergraduate. He would return a few years later.

I had [Kenkichi] Iwasawa for linear algebra, a member of the faculty until he accepted an appointment at Princeton, where he continued a remarkable career.[3] I remember perfect lectures, from a very quiet, retiring person. He wrote on the board what he was saying, because he wanted to get everything exactly right: there were a lot of words written on the board, and absolutely accurate. So I saw a real mathematician at work—at a very low level of course, but yet, you saw the precision. That kind of experience with Iwasawa led me to think of mathematics at that point as getting theorems in a row, so to speak: mathematics came in units of theorems. I don't think that way anymore. Now I'm more interested in the underlying ideas. So my linear algebra class, 18.06, is a hundred-and-eighty-degree change from what I had with Iwasawa. His class was purely theorem/proof, and that was okay with me, but I don't do it now, because it's not okay with the engineers and scientists that I really want to connect with. In that way, my view of mathematics and certainly of teaching mathematics has evolved from what I saw as a student.

I can't say that I had a mentor within the math department. I did the courses, but I didn't become a mathematician as an undergraduate. I wasn't especially thinking of an academic future, because I felt a little more naturally inclined to get involved with concrete problems. I imagined that would happen in industry rather than in a university. One has a sense of what parts of mathematics click, what results you remember and incorporate in your experience and can use to solve future problems. Linear algebra and related problems were for me the core ideas. I took that linear algebra course, which was on the purer side, but it was really the next year in Oxford, reading a book by Mirsky on linear algebra[4] that was much more concrete: I found that natural to read, and learn from, and get new ideas about. So maybe that contrast of Iwasawa versus Mirsky—both names from the far past—reflects my own tendency towards more concrete problems in mathematics.

---

[2] For an appreciation of the life and work of Franklin P. Peterson (1930–2000), see *Notices of the AMS,* November, 2001, 1161–1168.

[3] See interview with Hartley Rogers, page 127, n18.

[4] Mirsky, J. (1955). *An introduction to linear algebra.* Oxford: Clarendon Press; (1990). New York: Dover.

Rhodes Scholar at Oxford

As an undergraduate you appeared in the January 7, 1955, issue of The Tech. Front page center: "W. Gilbert Strang awarded coveted Rhodes Scholarship." It says, among other things, that "Mr. Strang is completing in three years the usual four year undergraduate course of studies." What led you to cram MIT's undergraduate experience into three years?

Mostly that it *could* be done in the math department at that time. One could take advanced placement exams. I read the differential equations course in the summer, while I had a summer job, and then came back and took the exam, which then officially meant that I understood differential equations. But as a result, I've never taught 18.03 and never will, because I *don't* feel I really understand it from just reading the book and passing an exam.

So yes, I was an undergraduate at MIT from '52 to '55, and then this scholarship when I was a senior took me to Balliol College in Oxford for the next two years. I did a bachelor's degree again, but it was a little bit advanced. So it was appropriate, sort of graduate study.

What was it like, living in a different country for the first time?

The experience of being abroad is eye-opening for everyone. You assume that everything you had grown up with was the way things were, and then

Photograph of Strang that appeared on the front page of *The Tech*, January 7, 1955.

you see that there are different ways that societies can be organized and different educational systems. One experience, certainly, was to meet students who were not engineering or science students but [rather] were involved with politics. Just reading the newspaper: the English newspapers were far better then than they are now. Seeing the international politics, and traveling in Europe, and getting behind the Iron Curtain—it's a much wider world, to do that. One memorable couple of weeks was after we won the English University basketball championship, mainly because we had a lot of Americans at Oxford and then were invited, as the English champions, to Poland and Czechoslovakia. So that meant going as far as Berlin, being prepared, and then going into Poland. They were a little surprised to find this largely American team, but they were much closer to professional teams and beat us easily!

It was interesting to see Warsaw and Gdansk—Danzig—and everything in those years. This was serious Cold War. Even in Berlin, which was the first step, we naturally went into East Berlin. It was safe enough: this was all before the Berlin Wall, of course. You just went in. East Germany was poorer than West Germany, which wasn't all that rich either, at that point. So you saw the gray, kind of monolithic structures and architecture of the Stalinist era even without being in Russia itself at that time.

Later, as a professor at MIT, I had an invitation to Moscow, which was also still in the Cold War period. Every floor of the hotel would have a—concierge, I'll say, but that's not the right word. Her job—and it usually was a woman, I think—was to maintain security, see that nothing happened that wasn't supposed to happen on each floor of the hotel. It was a very rigid society. If you wanted to have a conversation, you had it out walking in a park, rather than inside a room. So I've seen a small, tiny bit of that era before it ended.

Rhodes Scholars always went to Oxford, but Oxford had a less illustrious mathematical history than Cambridge, although G. H. Hardy had been there, and E. A. Milne, and of course Michael Atiyah came later.

Yes, that's right. Maybe less illustrious than Cambridge, but still pretty good. But that wasn't my chief consideration. I knew and respected some men who had been Rhodes Scholars, and so I applied for it. The decision was made early December of '54, before I'd even had to think about applying to graduate schools. So it was the first thing I applied for, and when chosen, it was the natural thing to do.

Whom did you study with, and how did the experience affect you?

I had classes from [Edward Charles] Titchmarsh, who wrote a wonderful book on complex analysis.[5] And [Graham] Higman in algebra. But the central part of the Oxford system was your college tutor, and I was very lucky that the tutor in Balliol College was a friendly South African who had got his PhD in Princeton. Jack de Wet had just a good general knowledge, and his friendship meant more than the precise mathematics that I learned from him. Again, I didn't take that step to be *professional,* if I may say it that way. I chose to continue with mathematics, of course—never considered other alternatives. But I think even those Oxford years were not the years in which I became a mathematician, so to speak. It was fairly straightforward to prepare for the exams that came, all in a rush, at the end of the two years in Oxford. And I suppose all along I would think about mathematical questions that were not necessarily in the course. So I kept doing mathematics, but I wasn't in the world of mathematics.

Oxford is divided up into colleges: these beautiful buildings, centuries old, most within easy walking distance of each other. Why did you choose Balliol? Do the different colleges have different characters?

Balliol is one of Oxford's twenty-five or thirty colleges. Everybody going to Oxford or to Cambridge has to choose a college, and usually it's just that somebody you know has been at such-and-such a college. You know that one name, and therefore you seize on it. I was lucky in the random choice of Balliol—lucky in many random choices in my life. Balliol is an unstuffy, active, lively college. My own tutor, Jack de Wet, was that way. And it remains so among Oxford colleges: it has the reputation of non-stuffiness.

Going back a century even, Balliol people were going to positions in what were then the Colonies, and probably, as a result, Balliol was known. And prime ministers: Ted Heath and Harold Macmillan in our time. I guess Balliol is sorry but proud of the fact that both Tony Blair and Bill Clinton applied to Balliol and were not accepted! They both went to other colleges in Oxford.

UCLA and Post-Doctoral Years

From Balliol you went to UCLA for your doctorate.

Yes. Again, random process. I won a National Science Foundation Fellowship—that wasn't anything out of the ordinary, really—and I wrote

[5] Titchmarsh, E. C. (1932). *The theory of functions.* Oxford: Oxford University Press.

to a mathematician I knew of and outlined my interests in concrete mathematics. He was on the West Coast and recommended UCLA for the kind of mathematics I could imagine doing, which was applied but not physical mathematics, more interested in matrices and numerical solutions. So I went. I'm very happy to see in yesterday's AMS Notices that UCLA has won the AMS Award for this year as the outstanding department in all it does.[6] So it really has grown in strength, and it had good people in areas that I was interested in. Magnus Hestenes was a superb mathematician from Chicago.[7] So my experience there was a good one. Essentially, I found a problem to work on, worked on it, wrote the thesis, without knowing exactly what the next step was going to be.

Do you remember how you found the problem you worked on?

Yes. Peter Henrici, who turned out to be my advisor, had a seminar where each of us took a recently published paper and spoke to the class about it. I happened to be assigned a really seminal paper by Peter Lax, who became my hero, and Bob Richtmyer about stability of difference equations.[8] Von Neumann had ideas about that, when computational science was just being born. Lax—his papers have ideas, and yet they don't close out the subject completely. I think many of us found something useful to do as a result of seeing ideas that Peter had opened up but not closed. So I saw some further problems in a natural development from the Lax-Richtmyer paper, and that was my thesis.

I can remember Peter coming to UCLA to give a seminar. I was just a member of the audience, and he was a young whiz: it's amazing to think how young he was at that point, probably not thirty.[9] He was recognized as a star from the very beginning, and when he came to New York from Hungary as

---

[6] The UCLA Mathematics Department received the American Mathematical Society's 2007 Award for an Exemplary Program or Achievement in a Mathematics Department.

[7] Magnus Hestenes (1906–1991) earned his PhD from the University of Chicago in 1932. He moved to UCLA in 1947, where he served as department chair from 1950 to 1958.

[8] Lax, P. D. & Richtmyer, R. D. (1956). Survey of the stability of linear finite difference equations. *Communications on Pure and Applied Mathematics, 9,* 267–293.

[9] Lax was born May 1, 1926, in Budapest, so he would have been slightly over thirty at the time. He subsequently received many awards, including the Chauvenet Prize (1974), the Norbert Wiener Prize (1975), the National Medal of Science (1986), the Wolf Prize (1987), the Steele Prize (1992), and the Abel Prize (2005).

a teenager, he met Courant, so he was part of the Courant Institute. When he had to serve in wartime, he had the year in Los Alamos coming to know and work with von Neumann, which changed his life. Anyway, Peter became for me, and for many others of my generation, a real guide and inspiration in directions to take, in applying real mathematics—functional analysis and other tools of mathematics—to solving problems. So maybe that clicked with my natural tendency or my natural direction: to use mathematics, but to use it in solving concrete problems. He just came through, gave a one-hour talk, but because of the experience of working with his paper, I went to the talk, and I remember it better than most of the thousands of talks that I've been to since.

Did you have any more personal contact with him over the years?

Subsequently, yes. Certainly not then. When I came to MIT as a Moore Instructor, '59 to '61, the general description of the area was stability for difference equations. By using the Lax–von Neumann ideas, that reduced to problems in complex analysis: whether some function stays less than or equal to one, is a typical question. You're multiplying at every step by some growth factor: if that growth factor is greater than one, it's unstable. So you're always trying to prove that the growth factor is less or equal to one. Somehow, I proved stability for a family of difference equations and boldly sent a copy of the theorem—not email of course, in those days—to Peter Lax. And he replied! He was very pleased by that theorem; it was sort of a natural question, and the answer was good. And then, he even connected me to a *great* man, a mentor of his, Gábor Szegő at Stanford, who knew everything about polynomials and complex analysis and sent a *neat* proof of a key step in that theorem, which was better than my proof. So that was fantastic, to have a link with Lax that continued and even that link to Szegő.

Back to MIT and Back Again

You said you came back to MIT as a Moore Instructor. Who brought you back here?

That would have been a department decision. My guess is that Ted Martin would have remembered me as an undergraduate. He was the kind of guy who would keep track of people, so he would have remembered that I went to Oxford and maybe learned that I went to UCLA.

When the opportunity came to return to MIT, that was the clear choice, even though it wasn't what I had been preparing for. I certainly never regretted it. It's a wonderful part of MIT, the large number of instructors and post-docs who were influenced by those two or three years at MIT and then have their careers all over the country—all over the world. There was a bunch of Moore Instructors at that time, all very interesting people.

One exciting, entertaining part of the department's history happened when Ed Thorp, who was also a Moore Instructor at that time, suddenly announced a winning strategy for Blackjack—Twenty-One. Ed had also come from UCLA and went back to UC Irvine ultimately, so that's where his career has been. Anyway, he had thought there could be a winning strategy against the fixed rules that the house must play: if the right set of cards were still in the deck, the odds changed in favor of the player. The gambler makes his decision first, and then the house will follow a fixed strategy when its turn comes. So if the cards still in the deck are right, Thorp discovered a slight edge for the player. The whole strategy is defeated by reshuffling the deck after each deal, but that's not what the house does. I think he published it in the *Proceedings of the National Academy of Sciences*, so it became national news.[10] A gambler was seen to drive up in a Cadillac to Cambridge to meet Thorp and offer him his money to go and gamble with.

Did he?

Well, I think he did. Yep! And successfully, to the point where he was recognized by Las Vegas casinos and excluded. I thought, "Okay, that's probably the end of it." I assumed that the house would change its rules slightly, enough to defeat the strategy. But not at all. So Ed wrote a book called *Beat the Dealer*[11] and thought about other winning strategies, and then later thought about winning strategies in finance and became very rich, a little bit anticipating what mathematical finance has gone on to do.

There's a recent popular book, *Bringing Down the House,* by Ben Mezrich[12]—again, MIT students who, far from doing their math homework, went out in a team to Las Vegas and to casinos all over the country and made serious money.

---

[10] Thorp, E. (1961). A favorable strategy for Twenty-One. *Proceedings of the National Academy of Sciences, 47*(1), 110–112.

[11] Blaisdell (1962); Random House (1966); Vintage (1966).

[12] Free Press (2002). The movie *21* (2008) is loosely based on Mezrich's book. Another book that discusses Thorp's findings, as well as those of others, such as Claude Shannon, father of information theory, and John Kelly of the "Kelly criterion," is *Fortune's Formula,* by William Poundstone (Hill and Wang, 2006). See also the Wikipedia entry "MIT Blackjack Team."

The Oxford experience must have been an important one, since you continued to travel back and forth to England.

Yes. I met my wife in Oxford, when we were students. We were married during the California time, then we were in Boston at MIT with the Moore Instructorship. That was a two-year appointment, '59 to '61. As it came to an end, because my wife was English, we wanted to have a chance for her family to see our children, their grandchildren. A NATO Fellowship opened up, so I went back to Oxford. It was just a one-year fellowship; then MIT wrote to offer an assistant professorship. Maybe UCLA did too, and others. The effect of Sputnik was terrific growth in mathematics departments, so it was a time when people had a lot of possibilities. But this one, to return to MIT, was the one we chose.

I don't know how it happened, but I think it probably was [Gian-Carlo] Rota, in the year that I was away, my post-doc year, who maybe suggested to Ted Martin that my interests could connect with MIT's department, and that led to the assistant professorship. Certainly applied mathematics was growing and I was somewhat applied, although the work was chiefly in fluid dynamics, and that wasn't a subject that I really knew about. My connections were still with core mathematics. I was interested in applications, but I didn't work in *those* applications.

Back in England for a year, and then back to MIT, so MIT three times. Then I didn't leave for what's now getting toward fifty years. But some years later, an American who had been at Balliol created an exchange between MIT and Balliol for faculty. So I did that for a couple of weeks, twice, and my wife and I found we enjoyed being in Oxford again, for a few weeks or a month. Now we do that each year, and Balliol elected me an Honorary Fellow.

What was Gian-Carlo Rota like?

Oh, he was a most remarkable person. I was just a few years younger. He was in this group who came to MIT in those years we're speaking of—late fifties, early sixties—and really, they were the ones who made the department. He was unusual in being seriously interested in difficult parts of philosophy at the same time as doing mathematics.[13] He came to MIT as a student of Jack Schwartz in functional analysis, but his life's work became the creation of the field of combinatorics, to put combinatorics on

---

[13] "Rota was given the title Professor of Applied Mathematics at MIT but in 1972 his title was changed to Professor of Applied Mathematics and Philosophy. He is the only professor at MIT ever to have such a title."—MacTutor biography of Gian-Carlo Rota. For more on Rota's life and career, see http://www-gap.dcs.stand.ac.uk/~history/Biographies/Rota.html.

an algebraic and deeply mathematical foundation. After Rota came Richard Stanley, who's now the star, internationally recognized as a leader in combinatorics on the algebraic side. With Danny Kleitman also, we became the best department in the world in combinatorics, in this field that Rota gave shape to.

Did you work with Rota directly?

We wrote one joint paper, a very, very minor one, about something called the joint spectral radius, that took a very short time. No, I think we were just friends. He was friendly with a lot of people and had an interesting, remarkable life, and died suddenly and sadly.[14] He was a very important member of the department, an outstanding lecturer and exceptional writer. He had his own style, and he edited a lot and was very active in the publication of mathematics, editing journals and writing books. And he wrote about people in a very penetrating and unusual way.

One figure who is rarely mentioned today is F. B. Hildebrand, the numerical analyst, who was a member of the faculty for forty-four years [15] *and central to the department's teaching mission.*

Well, perhaps I could, more than anybody else you could speak to, say something about Francis Hildebrand. He was a professor when I arrived as a student, so I guess I would have heard him lecture a few times, and he was always friendly to me. His strength was his course on mathematical methods for engineers, so thousands of engineers remember the MIT math department as Fran Hildebrand. But he was a very quiet, retiring person. He was an author and teacher when I knew him, not really a researcher. I'm sure he

---

[14] Rota died in April, 1999, at the age of 66. He was multilingual, hugely prolific, and was described by his thesis advisor as a "gourmet of mathematics."

[15] Francis Hildebrand, according to his MIT obituary, got his PhD at MIT in 1940, joined the faculty that year, and served for 44 years until retiring in 1984. "When Professor Hildebrand joined the faculty, engineering and computational fields were increasingly in need of a standardized mathematics curriculum tailored to their studies, so he developed his famous courses on advanced calculus for engineers. The courses resulted in a textbook, *Advanced Calculus for Engineers*, which became the standard reference for engineering mathematics.

"In 1952, Professor Hildebrand wrote *Methods in Applied Mathematics*, another influential teaching and reference text. In 1956 he published *Introduction to Numerical Analysis*, which contained techniques he himself developed. The book played a role in positioning numerical analysis as a major influence during an early stage of computer design."

published papers, but probably they're not read today, and I didn't have any occasion to read them.[16]

In a way, I suppose I've filled his shoes in the connection of the math department to the engineering departments. He had created this course, Mathematical Methods for Engineers, actually two courses and two or three important books. But time passed, and computations became important, and that was not in Hildebrand's books. He did engineering mathematics, but not scientific computing or computational science. I'm happy that Hildebrand's courses still exist, using special functions and complex analysis and infinite series, as 18.075 and -6. The problems that could be solved by analysis and special functions and analytical techniques were very limited, but from the ones you can do, you still get great understanding about more general problems. Now the world is entirely different. Everybody can solve much more general classes of problems, if you define *solve* to mean compute an answer; and for many purposes that *is* to solve the problem. A big activity of mine, for twenty years now or more, has been teaching a course I created, 18.085 and -6, as a sort of modern Hildebrand. The new courses are about how you would really compute an answer using the tools of finite differences and finite elements and so on that we now have. I'm writing a new and more computational textbook for that course every day of this year, except for this hour with you as a break. This is probably the most comprehensive book I've ever tried to write.

About a decade after joining the MIT faculty, you became involved in department administration as well. Frank Peterson became chairman of the pure committee and you later replaced him as graduate chairman.

That's right. Frank Peterson served a number of years as chairman of the pure group. He had been the graduate chair, and I took on that role of making admissions decisions for those applying to graduate school on the pure side: part of the pure/applied division is that the graduate admissions are done separately. But I didn't have any department-wide discussions about it; I just read the applications and asked advice and made choices. And tried to persuade them to come: that's what the department graduate person does.

Later I became chair of the pure group, but again, I felt my part was to help move toward decisions of faculty hiring and so forth, very much with

---

16 Hildebrand's MIT obituary indeed focuses on his contributions as an author and educator but adds that "As a researcher, Professor Hildebrand contributed to the studies of numerical solution of integral equations and the theory of elasticity. During World War II, he worked for two years in the Radiation Laboratory."

advice from those who were really genuine—who were pure mathematicians and knew the stars, the right people to approach. You certainly learn who's who nationally, among young mathematicians, because MIT and Harvard and Princeton and Berkeley are competing for the same group of young, talented mathematicians. That was an interesting experience and quite fun. I put a lot of time into the department's work in those years.

Did you have a definite agenda in chairing the pure group?

No, I don't think I did. That's why I didn't continue with it. I was happy to help people who knew where different areas—say, group representations or topology or geometry—were moving. But my life's work was not in the department administration but in moving linear algebra and computational science forward by teaching and lecturing and textbooks. That's the main thing. My direction has been, "Where should our teaching go?" and backing that up with writing books, more than the upper-level picture of department issues. Teaching at MIT is an absolutely wonderful job, the best one could imagine.

Writing and Publishing

Your interests have generally led to books on the topics you teach, books that are regarded as unusually lucid in their writing and relevant in their real-world approach. Can you explain how you took on the task of book-length writing, first about the finite element method, then about linear algebra?

Well, a quick history. I was in the math department doing sort of concrete problems but still writing in the mode of theorem-and-proof until about 1971 or '2, so that's ten years after coming here. Up until then, my Peter Lax connection was to finite differences, but then something called the finite element method was in the air. Engineers had created it, were using it, and it was a big success. So I thought, "Well, I'll find out what it is," because it wasn't in math departments at that point.

It happened there was a younger guy at Harvard, George Fix, who had written a thesis about the finite element method. So I asked him: "What is this?"—intending to prove either that it was only finite differences in disguise or that it was not really all that the engineers said it was. But my experience was entirely opposite and very important for me, to find that the engineers had created an alternative to finite differences that was more appropriate for a lot of problems: more flexible geometry, and with a good mathematical foundation. Its foundation was in the Galerkin, Rayleigh-Ritz world. Courant actually had written a paper that suggested the finite element idea, but of course he hadn't followed through the way a hundred engineers had followed through, because they really had problems to solve! This business of meeting people who had real problems to solve, who needed a method that worked, and then exploring the mathematical foundation of the finite element method, was quite exciting. Because people cared! Engineers cared, and others cared, and it was just the right moment.

Then I thought, "All right, I understand this subject. I'll write a book about it." My first effort in the book was limited to equally spaced, regular meshes. In other words, it was entirely mathematical, but it didn't capture what had excited the engineers about having unstructured, irregular geometries. So after writing most of that book and at the same time talking to engineers, I abandoned that book and wrote another one, which was my first published book, in 1973: *An Analysis of the Finite Element Method,* which still exists.[17] George Fix was a coauthor, and he contributed ideas for early chapters and contributed all of the part that had been his thesis topic. We worked together, but in the end I was the one who enjoyed the writing and really turned the effort toward a book.

Most important for me was to realize how much I enjoyed writing a book and organizing a subject. That turned out to be a life-changing experience. Soon after that I was put in charge of the linear algebra course, which was a basic course. But at that time, it was what I had seen years earlier as a student: it was for math majors. And my experience of going to applied math meetings was that linear algebra was the core knowledge that people needed. The purer, abstract linear algebra course remains, still taught as 18.700, but I created an applied option that engineers very quickly picked up on. The population of the linear algebra course grew quickly, and I could see that a textbook was needed for this approach to teaching linear algebra.

The first sentence of my textbook is "I believe that the teaching of linear algebra has become too abstract." That's probably what I would be best known for—the effort in textbook writing and teaching and lecturing to move linear algebra into the center where it belongs, at the center of applications, and taught in a way that helps engineers see its value. Linear algebra is more important to more people than calculus. On my Web site

[17] Strang, G. & Fix, G. J. (1973). *An analysis of the finite element method.* Englewood Cliffs, NJ: Prentice Hall.

there's an essay, one page, "Too Much Calculus," because I do think that the American undergraduate education in mathematics often begins and ends with calculus. Calculus is a three-semester course in most universities, and then it would be followed by differential equations by those who make it that far. So that's now four semesters, and that's just about it for almost all students except math majors. Which means that engineers and economists and many, many other people don't learn linear algebra, which is actually more important to a larger number of people than calculus. To have calculus and differential equations four semesters and linear algebra zero semesters—four to zero—is way, way off balance. So the teaching and pedagogical parts of my career at the undergraduate level have been to push applied linear algebra, to make its importance better recognized and provide textbooks that people with a different training could use in teaching. It was the time to move; it was the right direction to move, for a very large number of students, and now linear algebra is normally taught that way. When engineers take linear algebra now, except in a very classical or older department, they would expect to see it in this more applied, concrete form.

The first of now four editions of that linear algebra book[18] was in 1976, and there have been three editions of another linear algebra book.[19] So that's thirty years ago, and it completely changed my work. My teaching became focused on these courses, the basic linear algebra course and the Hildebrand course that developed into computational science and engineering. Those are large classes. Writing books became a major activity, which is something I had never thought about.

Your emphasis on linear algebra was one way in which the books you've written have been somewhat revolutionary.

That's the fundamental difference, yes.

But your books are different stylistically as well. Reviewers, professional or otherwise, consistently use the word conversational to talk about your writing. One Amazon reviewer wrote, "Reading Strang and Fix reminds me of reading Lanczos: the book is dense with ideas which are both mathematically and practically interesting, but the presentation is so

---

18 Strang, G. (2006). *Linear algebra and its applications* (4th ed.). Pacific Grove, CA: Brooks/Cole.

19 Strang, G. (2003). *Introduction to linear algebra* (3rd ed.). Wellesley, MA: Wellesley-Cambridge Press.

smooth that one can almost read it as one reads a novel." That's not something one commonly hears about math textbooks.

That's a style that has developed. And it's still there: the new book will be quite personal. I'm sure that many readers don't approve of a conversational style, but others say to me, "I can hear you speaking as I read your book." I guess I regard the books quite personally. They're like children that are part of you: you take a few years to give birth to them, and they're pretty special. So I don't see writing about mathematics as just putting things on the record. I regard textbooks as speaking to the student and persuading him or her that this is worth learning and making it clearer and clearer and clearer, as far as possible. I don't think you can take the attention of students for granted. Maybe you could when I was a student, but certainly not now.

What made you decide to start self-publishing your books as Wellesley-Cambridge Press, rather than relying on academic publishers?

That happened ten years later. The first self-published book was 1986 on applied mathematics.[20] Essentially, I wanted the adventure. And you work so hard on the book, it's like your child, and I didn't want to put it out for adoption. Not that I had such terrible experiences, though editors and publishing houses come and go with lightning speed, as I now know. That one linear algebra book has been through multiple publishers. But I thought I'd have the adventure of being in contact with people who use the book, which is still a pleasure. And also of being able to change it, being able to write conversationally if I wanted to, though I think that would have been acceptable. And just staying with the book rather than writing it and saying goodbye to it.

So you supervise the typesetting, cover design, interior design?

I do. Not that I know so much, but I care so much, and maybe that makes up for not knowing what you're doing. I'm still discussing the cover for the new book. I have views about it and the interior design, and you just have freedom to try things and innovate. It's interesting to me.

And certainly I didn't do everything right. The least successful has been my calculus book.[21] This was after *Linear Algebra and Its Applications,* when there

---

20 Strang, G. (1986). *Introduction to applied mathematics.* Wellesley, MA: Wellesley-Cambridge Press.

21 Strang, G. (1991). *Calculus.* Wellesley, MA: Wellesley-Cambridge Press.

was a reform movement in calculus. It was a need nationally recognized. I remember going to a conference in Washington, sitting way at the back of three hundred people, and the issue was, how can we change calculus? I thought, "Okay, one possibility is write a different book"—because everybody was complaining about the books. So I tackled the calculus book, trying to do it conversationally. But I learned that calculus is a different world from linear algebra. The big publishers really take calculus seriously. They live or die on that. You've got to visit math departments, and whole committees make these decisions. They think it's safer, and undoubtedly it *is* safer, to have six people making a decision. But you get a conservative committee decision out of that. That's been a wonderful thing about the MIT math department: We don't have a lot of committees. The decision on which calculus book will be used is made by the people directly involved and not by some large committee. The decision on how the linear algebra course should go, or how my 18.085 computational science applied math course should go, is directly one person. And that really encourages you to change a course, because you have the freedom to do it. And then, if you're inclined to writing, to write a book for it.

I hadn't taught calculus year in and year out, so I didn't have the sort of special experience that you can bring to a book when you've taught the course many times. So that was less successful. Now it's available on the MIT OpenCourseWare for everybody. And I think this is coming with books generally: books will soon be on the Web for free. The older system is going to fade.

Wavelets

You wrote a book on wavelets, as well.[22] *How did you get into that topic?*

That's a topic in electrical engineering. Wavelets are functions created in an unusual way: we don't have exact formulas for them, but still we can produce them. We have algorithms rather than closed formulas. The finite element link was with mechanical and civil and aeronautical engineering. Those are the finite element people. Wavelets just sort of happened, luckily. Again, it was a field that was emerging, and I thought I'd try to find out what it was about. And I found what was for me a key link back to finite elements. The question was, what's the accuracy of an approximation by wavelets? That was a natural step after the question of approximation by finite elements.

---

22 Strang, G. and Nguyen, T. (1996). *Wavelets and filter banks.* Wellesley, MA: Wellesley-Cambridge Press.

So there was a mathematical link, and the link there was something that has become called the Strang-Fix conditions. We found the conditions for a given accuracy in finite elements, and they applied to wavelets, too.

So that underlying mathematical link gave me a connection, and then, I just like to learn a new field. Overall, where other mathematicians will go deeply into a subject, I don't go as deeply, unfortunately. But I like to learn about things that are new to me.

Wavelets have been very widely applied, am I right? Earthquakes, image compression, HDTV....

Yes, the problem is how to represent a function, an alternative to Fourier series. Fourier series are never going to be replaced, but wavelets have some advantages. They're more local, so that if you have a function that has sudden jumps or discontinuities, if it's a local feature you're looking for, wavelets give you a more accurate local description. It's all part of harmonic analysis.

Wavelets turned out to be the choice in 2000 to replace the standard JPEG compression systems. So JPEG 2000 is not based on Fourier series, cosine series, as JPEG originally was, but is based on wavelet series as a way to compress files. Think of images. Modern high-definition TV can't send all the pixels that it has, so it has to compress the image. Many, many applications now: we're just flooded by data; it's the century of data. And to make any sense out of it, or to transmit it or to visualize it, you have to compress it. So you could take a few terms in a Fourier series, but better to take a few terms in a wavelet series, because it gets the local features more clearly.

One application you wrote about was law enforcement's compression of the tens of millions of fingerprints they need to store.

Fingerprints, right. That was sort of an important moment—and I had nothing to do with this, so I'm just reporting what I learned. The natural image compression was JPEG and Fourier cosine series. But when the FBI experimented with wavelets as an alternative, they were more efficient. They gave better compression.

How do wavelets relate to earthquake research?

Well, earthquakes would be another signal that is local. What's important in an earthquake is happening on a short time scale, and that's where wavelets have a chance. For a symphony, Fourier's going to win. But for an earthquake or a sudden shock, wavelets are better.

Gilbert Strang lecturing.

President of SIAM

Despite being at least officially on the pure side of the aisle at MIT, you served as president of the Society for Industrial and Applied Mathematics from 1999 to 2000. How did you come to that position?

Well, SIAM was the natural organization for me. It was going to SIAM meetings that really led me to see how important linear algebra was and how to approach it. So that was my world—still is, I guess. They were looking for a Vice President for Education, so somehow because of *Linear Algebra* and because I was giving lectures, they picked me for that. I took that seriously, and then quite a bit later on they needed a couple of candidates for president, so I was one.

What was your main goal as President of SIAM?

I suppose what I wanted to do was communicate well, to present SIAM and its activities well. Like every organization, it's a bunch of people doing their thing. There were good things that SIAM was doing, so those were quite active years. As president I took that seriously, also testifying to Congress and in Washington. [Kenneth] Hoffman had done early work in Washington, starting from zero and getting some activity, recognizing that other subjects—chemistry, physics, others—had homes in Washington and were lobbying and were politically active.[23] That was the time that all the societies were appointing people to represent them in Washington. And math wasn't. But that was the beginning.

How did that go?

I enjoyed that part, the personal testimony to Congress, a hearing before a subcommittee that was responsible for the NSF budget. Funding is normally the biggest issue. SIAM, or applied math, has big activities in the national labs, where the Department of Energy is the important agency. But even within SIAM, I'm on the academic side, so I'm in the NSF world.

Funding went up, but not because of me! First of all, it was the right time, and then came support from the NSF director, Rita Colwell. Also Philippe Tondeur, a very active and energetic and really charming pure mathematician from Switzerland and Illinois, was appointed as director of the Math Sciences division at the NSF. He moved the math sciences budget way, way up.[24] Relatively speaking: of course, it's infinitely smaller than the National Institutes of Health. But I just happened to be the SIAM president in those years, so it was a very nice experience working with Philippe and with the other societies.

In a 2003 interview about SIAM you said that "the job of the President and the Council of the society is to see what new actions and what new work the society should be doing. That depends on where applied mathematics is moving."[25] You started two new activity groups within SIAM, which indeed seem a clear indication of how applied math is developing: one group in life sciences and the other in imaging.

Yes. SIAM has an annual meeting of the whole society in the summer. But the active meetings are these groups that meet every two

---

23 See the interview with Kenneth Hoffman in this volume.

24 "Thanks to [Tondeur's] work and power of vision, the annual NSF budget for mathematics has approximately doubled in four years, whereas over the preceding ten-year period it had increased at an annual rate of 1.5%." *EMS (Newsletter of the European Mathematical Society),* September, 2003, p. 21.

25 *Newsletter of Institute for Mathematical Sciences, NUS* [National University of Singapore], Issue 2, 2003, p. 9. For further information, see http://www.ims.nus.edu.sg/imprints/interviews/GilbertStrang.pdf.

years. There would be one on dynamical systems, for example, or one on applied linear algebra. It was clear that life sciences was growing, and that imaging was growing: that came from the wavelets people and the imaging people that I met. NIH has a billion-dollar division that does medical imaging, MRI and CAT scans and so on. There are Nobel prizes that are basically awarded for computational science, and CAT scans would be one.

So those were two groups that I established. And computational science was coming into focus, so I was an early part of helping that activity group. Everybody thought the reason we don't have an activity group in computational sciences is—that's what SIAM does! But still it turned out a very good thing, because then there's a group that has newsletters and has a meeting every two years, and things happen. Otherwise, it's just people at the annual meeting, but now SIAM does it in a more organized way.

Mathematics Ambassador

You were the first recipient of the International Council for Industrial and Applied Mathematics (ICIAM) Su Buchin Prize, awarded this year, created on the initiative of the Chinese Society for Industrial and Applied Mathematics (CSIAM).

Yes. Several prizes have come in the last year. This international prize for encouraging and teaching and writing mathematics. One from the local Math Association of America, the New England Section, and then from the MAA, the Haimo Prize for university teaching. The Henrici Prize is to be awarded in a few weeks at the International Applied Math Congress. It's named after my advisor, so I'm especially happy to have been chosen for that—and happy, too, that it's not for teaching: it's for analysis, mathematical work.

Perhaps more than any other faculty member here, you have represented mathematics internationally. In 2003–2004 you served as chair of the U.S. National Committee on Mathematics, which "represents U.S. interests in the International Mathematical Union (IMU) and promotes the advancement of the mathematical sciences in the United States and throughout the world," according to the department's annual report to the president of MIT.[26] You have traveled a great deal: your citation for the 2007 Su Buchin prize says, "He has visited China eight times"—discussing math and teaching with Chinese

[26] This report is available for viewing at http://web.mit.edu/annualreports/pres03/11.05.html.

SIAM President-elect Gil Strang with Lin Qun (on Gil's left) and others in Baoding, China, in a photo from *SIAM News*, Jan/Feb 1998.

students, researchers and teachers—and "He has visited many other developing countries, including Vietnam, Malaysia, Singapore (5 trips), Brazil, Mexico (4 trips), Tunisia, South Africa, Egypt, India, Korea and Cyprus etc." [27] *Is that part of your conscious agenda, trying to lift up math in the developing world?*

Not especially for the developing world. There was a period in my life where I felt anxious or didn't feel comfortable traveling, and then that passed, fortunately. Now, if you invite me to Egypt, I go! Invitations came to visit China at an early point, when China was opening in the early 1980s, so that led to friendships in China. And MIT has a close link now with the National University of Singapore, so I play a small part in that.

It's not a mission, and I don't have any sense of giving something. It's more I'm being *given* the chance to travel and to meet people. It's fun to meet people. I try to be encouraging and helpful. But I certainly don't feel I deserve any special credit for being a missionary. I just do what comes naturally, which is to try to explain things clearly.

We're living in the virtual age, and your lectures, through MIT's OpenCourseWare, have reached far more people than you could hope to know personally. A grad student from Singapore, for example, calls you "the god of linear algebra." [28] *The MAA Focus announcement of the 2007 Haimo Awards* [29] *reports that your courses in linear algebra,*

[27] To learn more about this award, see http://www.iciam.org/Prizes/PrizeNominations2007.pdf.

[28] See http://www.ntu.edu.sg/home5/OMID0001/favorites.html.

[29] Adams, C. (2006). 2007 Haimo Award winners to speak at the January joint meetings. *Focus, 26*(9), 4–5.

"Mathematical Methods for Engineers" and "Wavelets and Filter Banks," are the first, third, and fourth most popular math courses on OCW. Your linear algebra course was one of the earliest lectures to become available.

The OpenCourseWare is pretty significant for a lot of people far, far from MIT. Seventeen hundred courses are available in some form freely on the Web, and a few of those courses have video lectures. I mentioned that Rota's lectures were never recorded. It would be wonderful to see Norman Levinson again, and Ted Martin; even Norbert Wiener. This department history should be electronic, too!

The chance for 18.06 came somehow—oh, I remember: the physics department had a grant to videotape Professor Lewin in the basic physics course, 8.01. My lectures in linear algebra were in the same room the following hour, so I said, "Maybe you could just keep the cameramen there for another hour and let them shoot it live." So mine is very different from Lewin. He goes back; he fixes any mistake. Mine are just as the class ran that particular semester. Then half a year later this OpenCourseWare idea emerged at MIT. So it was natural to link the videotapes that we had just made to OCW. Having the videotapes on the Web is more alive than reading a syllabus, so people look at the video lectures very widely. I get a lot of email, students from all over the world saying "Thank you."

~ June 18, 2007

THE WORLD OF
MATHEMATICS

Kenneth M. Hoffman

Department History: The Early Years

J.S.: *Every department like this one exists within the context of a larger institution. Can you trace some of the history of the mathematics department's relationship to MIT?*

K.M.H.: Yes. In modern times, the math department has enjoyed, I think, very strong support of the central administration of MIT. You've got to think of the people who were in the math department. Norbert Wiener, of course. Claude Shannon, Mr. Information Theory. Marvin Minsky and Seymour Papert, the stewards of artificial intelligence at MIT: they started in the math department.

But mathematics has been built into the fiber of MIT from the beginning. One of the small group of founders of MIT was a mathematician, John Runkle. He was the first head of the math department and became the second president of MIT, after the founder William Barton Rogers, and was president for about a decade. He then rejoined the math faculty, stayed there for another twenty-five years or so.[1] Runkle gave mathematics some real roots in the institution, and that stream of the centrality of mathematics has continued to the present time.

Runkle was very well known to the students all across the institution. He was called Uncle Johnny by all the students, for *many* years. He taught one of the very first two classes ever taught at the Institute. The Institute was formed in 1861, but, of course, getting it really going had to wait until the Civil War was over. So it was 1865 when the first classes were held in the

[1] John Daniel Runkle (1822–1902) was a student of Benjamin Peirce at Harvard, then "the preeminent mathematician in the United States" (Moore, C. C. (2007). *Mathematics at Berkeley*. Wellesley, MA: A K Peters). Runkle was later a member of the 1860 committee that drew up the "Objects and Plan for an Institute of Technology." He joined the faculty as professor of mathematics in 1865, served as president pro-tem 1868–1870 and as MIT's second president (1870–1878), then remained as professor of mathematics until 1902.

new "School of Industrial Technology." There were fifteen students when they opened.

See, the Massachusetts Institute of Technology was a broader conception, which included the school plus a museum, plus several other things that I can't remember.[2] Most of these things never came to pass. MIT was originally located downtown in Boston on Boylston Street, near the Public Gardens. That's where they built the Rogers building; it opened in 1866, and the Institute was there until it moved to Cambridge in 1916. It was known as Boston Tech.

Starting in the late nineteenth century, there was a huge Boston redevelopment effort, one of the biggest urban redevelopment projects ever undertaken in the world. It was on an unbelievable scale. Boston was originally an island in a semi-swamp. It is still an island. But what is now the Charles River was tidal. All the rivers around there were tidal. So when the tide went out, the water level went down practically to zero, and all this yucky, gucky, stinky mud was exposed. The stench around was just terrible. And yet many aspects of the place were beautiful.

So Mayor Quincy, after whom the town of Quincy is named, got this idea that they were going to just fill in the backwaters—the Back Bay, as they called it—on both sides of the Charles River. The City of Boston had seven hills; they cut down six of them to get the dirt to use as part of the backfill and left only Beacon Hill. The project took decades. One of the great things to see is to go to the MIT Museum and look at the photograph of the very first building at MIT going up after the Back Bay had been filled in. It's like an isolated building out in the middle of flatland—surrounded by this *huge* area, which was perfectly flat because the engineers had just finished filling all that in. It's a wonderful thing to see that photograph, a very striking scene.

So Runkle became president, and then afterwards continued on in the math department. But during all that period, the math department was a service department. It wasn't there for the purpose of doing research in mathematics; it was there to teach the engineering students what they needed to know about mathematics. One of the founding principles with MIT was that the engineering students should always learn mathematics from mathematicians and physics from physicists. That has always been the case at MIT. So it was a sort of schizophrenic relationship in some ways because of the

2 The 1861 charter of the Massachusetts Institute of Technology calls for "a society of arts, a museum of arts, and a school of industrial science."

important *service* role that mathematics courses played for the education of engineers, and now for scientists as well.

The legacy of the role of the service department continued, as testified to by the fact that the courses in mathematics—subjects, as MIT correctly calls them—didn't have a number prefix indicating the department when I first arrived here; they just had the service letter "M." All the subjects were numbered; freshman calculus must have been called M1 or something. It was about 1960 that they started using 18. I think the math department had a number, Course XVIII in MIT talk, going quite far back historically—back to the point when the eighteenth department would have come up, right? But it wasn't used for subject numbers. And people didn't talk about Course XVIII the way they talked about Course VI.

When the math department got rid of its M, it was a big day. Dan Kan, the topologist, made up this cockamamie numbering scheme for the subjects, which is still used, I gather. The approximate idea was just the division of mathematics into algebra, geometry, analysis, and applied math; it wasn't much more finely tuned than that. It was so complicated, I never could understand it [laughs]. But it didn't matter, because it logically made sense.

When all that M stuff had gotten started is unclear to me, because Wiener was at MIT pretty early on, right after World War I, and *he* was focused on research, let me tell you. And yet there was no strong focus on mathematical research at that time. It wasn't as if they didn't know about mathematical research and didn't have any concern about it, but it just wasn't really a big focus.

After all, Norbert Wiener was the towering figure at MIT in the 1930s or 1940s. The place ran on his brainpower. That's another reason the tradition of strong support of mathematics stuck in people's minds at MIT. He was a great man, the mathematical giant at MIT, like G. D. Birkhoff at Harvard.

Wiener was still here when you arrived. Did you have any direct contact with him?

My interaction with Wiener was limited, because we only overlapped by a few years, and I was just a young upstart. He talked like a fire hydrant. Everybody said that about Wiener. He couldn't keep a secret: Levinson told me that he was used by the Intelligence people to spread disinformation.

You didn't have a discussion with him; you just listened. I didn't talk with him a great deal. He did corner me a number of times, trying to talk about extensions of his own work in harmonic analysis and generalizing it. I listened for a long time and finally realized that what he was

Julius Adams Stratton (President of MIT, 1959–1966), Norbert Wiener, and Claude Shannon, c. 1960.

doing was trying to invent what Gelfand had done in 1937, when he invented something about commutative Banach algebras. But you didn't tell Wiener things like that: he wouldn't have heard you, anyway! Those were practically the only mathematical discussions I had with Wiener.

Of course, the annals of MIT are full of Wiener stories, some of which are probably true. But nobody knows which ones, I think. My favorite one was where he met the graduate student on the stairs there in Building 2, on the flight of stairs, and he stopped midway to talk to the graduate student for a minute. When he got through, he said to the kid, "Which way was I going when you met me?"

"You were going up."

"Oh, good. Then I've had lunch."

Wiener seems to have been pretty much a force unto himself. And having one very, very bright guy there is not the same as having a really organized research effort.

And that's the lesson of it. But they did bring some other people. Philip Franklin, the number theorist, came [in 1924]. I think he was a known research person. Not a Wiener, but he was strong. And somewhere in that period, probably in the 1920s, was when C. L. E. Moore must have made his imprint. He was a faculty member who pushed hard for beefing up the research strength and the faculty. That's why they named the Moore Instructorships after him, because he was the big research-pusher.[3]

Another *very* important figure in the math department, and in the history of MIT, for that matter, was H. B. Phillips. I heard a lot of the history from Julius Stratton, the former president of MIT, in the late seventies. We used to talk by the hour about the old days and about people like Phillips. The old-timers used to say that Phillips was very influential in the education of students all around MIT, because he was a young man, and he knew modern science. Henry B. Phillips got his PhD at Johns Hopkins in Baltimore, then joined the faculty at MIT as a young man, and he was a very important force during the twenties and continuing on. He became head of the department and then retired after World War II,[4] and Ted Martin took over the department.

Now, MIT does not like to have it known very widely, but the truth is that after the move in 1916, as they went into the 1920s, the quality level at MIT deteriorated rather badly: they degenerated practically into a trade school. Why the deterioration took place, I don't know. Who can say, in institutions? Just lack of focus, lack of attention, lack of leadership. No one ever says, because they don't want this period talked about. But it was not a great period. I think it probably was around that time, in the twenties, when Harvard tried to buy MIT. Why they would have wanted it in the shape it was in, I don't know.

And that's part of what Henry B. Phillips was helping to combat. He was one of the leaders who pushed, pushed, pushed to get the young kids to learn about modern science. I was told by Jay Stratton that Henry Phillips used to run around the institution "like an errant ion," grabbing students, and he'd drag them into classrooms and talk with them about modern science and mathematics, because he was so concerned that the students weren't learning anything about modern science in the classes. Phillips not only taught all kinds of mathematics, he taught the first course in relativity ever taught at MIT. He taught the first course in quantum mechanics taught at MIT. And he pushed with another group of faculty, at the time that the presidency was

---

3 Clarence Lemuel Elisha Moore (1876–1931) is commonly said to have awarded the first mathematics PhD at MIT. It was awarded to William Fitch Cheney, Jr., for a thesis entitled "Infinitesimal deformation of surfaces in Riemannian space," in 1927. See, for example, Kleber, M. (2002). The best card trick. *Mathematical Intelligencer, 24*(1), 9–11.

4 Henry Bayard Phillips was head of the department from 1934–1947.

about to change, in 1929 and '30, to get some kind of leading scientific figure brought in to rebuild the institution. The Corporation listened to that, and that's how they got Karl Compton.[5]

Compton was the greatest president of MIT, I think everyone would agree: just a phenomenal person. He rebuilt the engineering school and also built basic science. And this is what really put MIT on the map, because when World War II came, *all* that was brought to bear on the war effort. The whole institution was practically a branch of the military in World War II. The events of World War II lifted the institution from where it was in the thirties up to a whole other planetary level.

But all that was because of some influential, far-sighted people like H. B. Phillips. He's one of the big names anybody should know if they want to know the history of the MIT math department.

Who were some other important figures from the early years?

Well, George Wadsworth was a figure after World War II. He and a graduate student of his named Enders Robinson took part of Wiener's work and applied it to seismic oil exploration, how you could bounce the signals off the earth and read the echoes back.[6] It was this work of theirs that led to the equipment that Cecil Green invested in to found Texas Instruments, which was founded to do seismic oil exploration.[7] You can read about the importance of this work by Wadsworth and Robinson in a history of geology at MIT written by Bob Schrock, an old time geology professor at MIT.[8] In fact,

---

[5] Karl Taylor Compton (1887–1954) was president of MIT (1930–1948), then chairman of the Corporation until his death.

[6] For a brief but excellent account of Robinson's work under Wadsworth, see Robert Dean Clark's biography of Robinson at the Society of Exploration Geophysicists' Virtual Geoscience Center, at http://www.mssu.edu/seg-vm/bio_enders_robinson.html.

[7] Cecil H. Green (1900–2003) attended MIT from 1921–1924 and received a BS and MS in electrical engineering. He worked for and later bought (with three other partners) Geophysical Service, Inc. (GSI), which used refraction and reflection seismology to find petroleum deposits. The company placed increasing emphasis on electronics work, begun in World War II, and in 1951 changed its name to Texas Instruments Incorporated (TI).

[8] Robert Schrock (1904–1993) was chairman of MIT's Department of Geology and Geophysics (1949–1965) and author of *Geology at MIT, 1865–1965* (MIT, 1977). Together with Cecil H. Green, he initiated the GSI Student Cooperative Plan, a summer program, designed to give students practical experience in geophysical exploration, that ran for 17 years.

he used to say to me that the Green building should have been given to the math department [laughs].[9]

The math department paid no attention, by the way. They didn't know that Wadsworth was important. This could have been exploited easily for fundraising for the department, but the mathematicians failed to take advantage of it. To the best of my knowledge, up to the point when Jim Simons gave the money recently, I don't think anyone had ever given any significant money specifically for the math department in the entire history of MIT. This is why this history of the department is terribly important. The department had this big impact on the institution; part of that old MIT tradition is the sense of the primary place of mathematics in the landscape of the institution. And it's important to know the history of that, because the day will come when perhaps it won't be as deeply engrained, and you will need to remind people of it.

Also after World War II, Phillips and Martin brought C. C. Lin and George Whitehead, the topologist, to MIT.[10] So that was a significant step in bringing some more strong research people there.

Lin was a phenomenon in many ways, a brilliant fluid dynamicist who worked in all *kinds* of different fields, depending on what decade you were in. He was the one who was influential in bringing people like Harvey Greenspan in. In that period, they also had Eric Reissner, a leading figure in elasticity.[11]

Norman Levinson and the Buildup of the Department

Levinson was another great research person brought into the faculty early on. What a phenomenon he was. Norman Levinson came out of this little Jewish ghetto in Revere, Massachusetts. His father was a typesetter, and on that typesetter's salary he put Norman through MIT and his sister through Radcliffe. Norman graduated in electrical engineering, as an undergraduate, and one year later, he got his PhD in mathematics. And then, after he got his

---

[9] The Cecil and Ida Green Center for the Earth Sciences.

[10] Chia-Chiao Lin came to MIT in 1947 and retired in 1987; George W. Whitehead came in 1949 and retired in 1985.

[11] Eric Reissner received a PhD in mathematics from MIT in 1938 and remained on the faculty, serving as full professor from 1949–1969.

degree, it was Wiener who intervened to get Levinson put on the faculty.[12] Wiener was, of course, a strange man and eccentric; but *he knew talent* when he saw it. He wanted Levinson there and he pushed for it. You know what the issue was. There were too many Jewish faculty members. And that wasn't fully broken until the day many years later when Jerry Wiesner became the president of MIT. I said to myself, "It's over. It's over."

I am told that Norman was quite a brash young man, when he was young [laughs]. He was smart *way* beyond what most people think of as smart. Somewhere in the archives of the math department I kept the textbooks he used when he was an undergraduate at MIT. These were totally graduate-level things, very advanced books on mathematical analysis and so on.

But Norman became, as he aged and mellowed, what I always called the Godfather of the MIT math department. First of all, during the years of the major development of the department following World War II, it was probably Norman and Is Singer who were the two principal advisors to the department head, Ted Martin, on which people to hire. They both had infallible instincts about talent in young mathematicians. Many great mathematicians don't have that ability to spot the young talent. And the build-up of the department was because of that work behind the scenes, as it were, by Levinson and by Singer.

One thing that is worth noting is that the development of the department at Berkeley and the department at MIT were very similar. They both had some leading mathematicians, even before World War II. But their focus wasn't heavily on the research. If you asked somebody to name the top eight or ten math departments in the United States, you wouldn't have found MIT or Berkeley on those lists before World War II. Well, by 1960-something, you would. They were the two new kids on the block, and they developed by the following process.

During that growth period, there was a period of about three years, '55 to '57, during which each of those departments hired fourteen assistant professors—same number in each school. At MIT, those people were Arthur Mattuck, and Hartley Rogers, and Gian-Carlo Rota, and Dan Ray, the probabilist, who died early, and so on. Singer came in that group, even though he was older. He was really farther along. And many of them stayed and some became permanent members of the faculty and world leaders in the field

[12] Norman Levinson entered MIT in 1929 as an undergraduate, received a BSc and MSc in electrical engineering in 1934 and a PhD in mathematics in 1935; he was appointed an Instructor in 1937. He served as head of the department from 1968–1971 and died in 1975.

they were in. That's what made those departments. I've always thought that was a very interesting period. It put these places on the map, and they've stayed there.

Norman, like Singer and others—C. C. Lin in applied mathematics—were pushers for *constantly* increasing the standards of the department. Always reach higher and higher for people. Always try to hire people who are better than you are. They were just relentless about this. And that gets into the air in a place. That's what made the department grow so well during the sixties, seventies, and so on. C. C. Lin was the third advisor to Ted Martin, specifically on applied mathematics. He was a very quiet man, but he certainly was an outstanding applied mathematician, so his advice and counsel were used a lot in the early years about which people to hire. There was a period of a few years where it was really Levinson, Singer, and Lin who were the—I don't know if it's a kitchen cabinet, but the prime advisors to the department head.

But Levinson was much more than that. He was the one who took care of all the extremely difficult situations that arose in the department. When Witold Hurewicz fell off the pyramid in Mexico and died, there was the monumental problem of getting his body out of Mexico. It was Norman who took care of that. When John Nash finally became critically ill, mentally, it was Norman who dealt with the situation with the psychiatrist. When any critical situation of that type arose, it was always Norman who took charge. He was just a *pillar* of the department.

Another thing Norman did, as the build-up of the department occurred in the late fifties and early sixties: Norman was the one who insisted that we watch out. Because, he said, every great buildup is followed by a letdown; and the day is going to come when the student enrollment figures won't support this big growth. "Don't build up the number of assistant professors any further," he said, "because the day will come when you will rue that." So he argued that we were to freeze the size of the faculty at fifty, and then we had an agreement with the administration that we would use all the money freed up by people going on leave to hire post-doctoral instructors. This way, the institution and the department didn't have a big exposure in case things ever had to be cut back—and some cut-back years did occur, years later. It's like this big accordion, you see? Something that could absorb the blows when money got tight. You expand the instructorships when times are good, and you shrink them when times are bad. And it brought all those young people to MIT, which was valuable scientifically and very valuable for them.

But Norman foresaw, you see. That's the point. He foresaw all these things. And he set up a policy. I don't know what the faculty size is now, but I tell you that size, 50, stuck for many years. When I was department head, we had 50 faculty members and 36 postdoctoral instructors. So instructors were a significant fraction of the teaching staff, but they weren't line items in the budget. In fact, during most of those years, there wasn't any money in the budget for the instructors! It was the job of the department head every year to patch together the money to keep it going.

Ted Martin built this Moore instructor program, the C. L. E. Moore Instructorships, and got grants to support them from the Air Force. Martin used that funding and his relationship with Merle Andrew, who ran the math program in the Air Force Office of Scientific Research, to build up the program. And he'd get nationally prominent, high-quality people.

Another thing that you need to understand about the history of mathematics in the United States, post–World War II period, is the *critical* role played by the Department of Defense. After all, much of MIT grew with support from the Department of Defense. The Research Laboratory of Electronics started with a grant from the Army Signal Corps, I believe. The original grant was six hundred thousand dollars. This is 1946. Can you imagine what six hundred thousand dollars was in 1946? The person who had to spend it was Jerry Wiesner, who later became provost and president of MIT. He told me, "I had no idea how I could spend that much money!"

The primary supporter of the math department at MIT after World War II was the United States Air Force. They just understood that mathematics drives the whole engine of science and technology: you've got to support it. When I was a young faculty member, we used to go on leave every fourth year. A genuine sabbatical! And many times it was paid for—lock, stock, and barrel—by the Air Force. But the grants that we had in those days were very different than what grants are now. They paid for all kinds of things.

Ted Martin was a strong leader. He had his own Levinson-like insights. One of them was that, in the running the department, he realized that the way you get faculty salaries up is to increase the salary of the Moore Instructors. He couldn't do that, you see, without raising the other salaries. He used devices of that kind.

Now, it's quite possible that Ted got this idea from Norman Levinson. But you give a leader credit, right? *He knew which people to listen to.* When he had to decide about talent in people, he knew to listen to Levinson and Singer, and that was responsible for a great deal of success. And that's the same thing I did when I was department head, I can tell you, up until Norman died.

Ted Martin.

The early years of the instructors—it's like a Who's Who of American mathematics. Incredibly powerful people. Years later, they were then able to use that to hire from the outside, as it were: the Stephen Smales of the world, bigger-name people. When I was department head, we had a small group of faculty in geometric topology that consisted of John Milnor, Dennis Sullivan, Bill Thurston, and John Morgan. It was the strongest geometric topology group in the entire world, hands down—and in the span of a year and a half, we lost them all. When I went to Jerry Wiesner, the provost, to cry on his shoulder about this loss, he said, "Look! What a great testimony to your department that you're so strong, you have people like that to lose!"

All the same, that really does sound regrettable, no?

Oh, of course it was regrettable! I really did need a shoulder to cry on. But it's typical of MIT mathematics: what you *don't* do, if that happens, is try to get somebody else in geometric topology. You move on. See, one of the guiding principles that Levinson and Singer also pushed in the hiring of faculty was to get the best people you can get *independent of field.*

The only area in which we did pay attention to field was applied mathematics, because the criteria were of a different character there. You had to use different judgments about people, and if applied math didn't have some autonomy, it would get squashed by the pure mathematicians. So this limited autonomy was created.

It seems, though, that there's a logic to thinking strategically: "We can't be good in every field. So let's build up this or that specific group."

Yes. It would seem to people that that is so. But a person like Norman Levinson would say, "That is rubbish. That is just how you become mediocre. Just don't worry about it." He was a great mathematical analyst, was Norman Levinson. But was he pushing all the time to hire analysts instead of topologists or algebraists? No! He would have argued against that. "Get the best people you can get."

Norman had great practical horse sense: just astonishing. He was *highly* respected all around the institution. People all around him in MIT thought the world and all of Norman Levinson. Everybody knew who he was, and people talked about him with great respect, people from the engineering and science schools, for example, or the engineers and scientists who were in the upper administration.

How did people from other departments know who he was?

Through interactions, I suppose. I remember the year we were about to give a PhD to Harvey Friedman, a young kid who was a logician. He had come to MIT as a freshman. He was bored with the freshman curriculum but very smart, so part-way through his freshman year the math department made him a graduate student. He had been a graduate student in the department for a few years, and we were about to give him a PhD. He already had serious faculty offers from Stanford and Harvard and other places, but there was a big ruckus from the graduate school when it was time for the awarding of degrees, because he hadn't completed the language requirement or something for the PhD. To the old, conservative engineering faculty this was an atrocity of the first magnitude. I mean, you can't do that—give a degree to somebody who didn't meet the requirements! So they tied things up in the graduate school policy committee for weeks, debating it.

It was Norman who convinced them to proceed without the requirements, and he did it with one single line. He said, "I don't think the quality and reputation of MIT will be determined by the people to whom it does *not* give degrees." And that ended the discussion!

Levinson died in 1975 of a brain tumor, a day before his sixty-fourth birthday. You were the head of the department and must remember his final months well.

Yes. I remember vividly how I first knew that there was something deeply wrong with him. In the evenings, I would very commonly be on the phone with him, talking about whom to hire, what to do; and these conversations would go on for a long time. Suddenly, he just disappeared: the conversation was cut off. Then a minute later, Norman would resume, picking up right where he was in mid-sentence, with no awareness of the fact that he had been gone for those sixty seconds.

You know, if that happens once, you don't think anything about it, but after the second or third evening that happened, I called [his wife] Fagi the next day. She had noticed that he was having trouble buttoning his shirts and things.

She said that he had been having severe headaches also.

I didn't know about that, you see. I just stumbled into this accidentally. But clearly she was aware of some things. And that was a sad, sad period, I'll tell you. Because the cancer in his brain was just—they did operate finally. But his prognosis was never good.

Norman was just a genius. Not only a mathematical genius, he was a practical one. He was not an administrative person, and he was not an academic politician. He wasn't noisy; much of what he did was done behind the scenes, as it were. He just was a pillar. He was like a tower of strength inside the institution, and that was widely recognized. I think surely in the history of the department there has never been another figure like Norman.

From Occidental College to MIT

Let's backtrack a little and talk about your own story. You came from a different undergraduate background than a lot of the other senior faculty, who went to places like Princeton, Harvard, Yale—or MIT itself.

Yes. I'm a community college person. I grew up in Altadena, California, north of Pasadena. At that time, Pasadena's city schools had the 6–4–4 organization plan: you went to six years of elementary school, four years of junior high, and then four years of junior college. So the eleventh and twelfth years of what we would now call high school were the first years of this four-year junior college thing. I attended John Muir Junior College.

Now, this had many advantages to it. One was that you got many more people to go on through the first two years of college. I just drifted into this thirteenth and fourteenth grade because it was in this school system, you see? But it also meant that serious things were going on in eleventh and twelfth grades because you were being taught by the faculty of the junior college. I was very fortunate that there was a man where I went to junior college named L. Clark Lay. He was a gifted teacher, worked at it very hard. He wasn't inspiring; he was just very well organized and knew what success in mathematics required.

Then they told me I should go on to college. I said, "What's that?" The world was so different when I was growing up: I didn't even know there were such things as separate colleges or universities. I'm sure my parents did, but I'd never heard anything about it. And then there was a question of where I should go. Well, I was very interested in track and field, so I went to Occidental as a junior. I was a shot-putter. In those days, Occidental College had a nationally known track and field team. They competed in track and field, head-to-head and nose-to-nose, with USC, Stanford, UCLA. The year I graduated from college, *six* people from our track and field team in that little college were on the U.S. Olympic team.

But I was also always in math, from seventh grade on. I always thought I was going to be a math teacher, I suppose, though my view of what that meant changed every year. I only went to Occidental College for two years: by the time I was in my second year, Mabel Barnes,[13] who was half of the math faculty—they only had two math faculty in that college—told me I should go to graduate school. And I said, "What's that?" I didn't know! She explained that to me. "Well, okay," I said, "I guess I could do that. And where would I go?"

She said, "Why don't you go to UCLA, because my husband's on the faculty there." Her husband was a distinguished professor of electrical engineering at UCLA. He was also Executive Vice President of North American Aviation. And I think that I went to graduate school at UCLA without ever applying [laughs]. It just *happened*. That's the way my life went.

---

[13] Mabel Schmeiser Barnes (1905–1993) began her education in a one-room schoolhouse. In 1933, she and Anne Stafford were the only women in the first group of mathematicians at the Institute for Advanced Study, where Schmeiser was taken aside by the Director of the School of Mathematics and warned that "Princeton was not accustomed to women in the halls of learning and I should make myself as inconspicuous as possible." She later served on the faculty of Occidental College for 21 years. In 1988 she gave an address on women in mathematics to the centennial meeting of the American Mathematical Society.

Do you remember meeting Singer at UCLA?

Oh, vividly, yes. He was wonderful. I met him the very first semester I was there in the fall of 1952, and we've now been close friends and colleagues ever since. He was a young assistant professor there. UCLA was then building up. They had hired about three former Moore Instructors [from MIT], I think, as assistant professors at UCLA, and Singer was one of them. I don't know how we came in contact: it was not through a course he taught or anything. But he got me to reading books. I think Lynn Loomis at Harvard had just written a brand-new book called *Abstract Harmonic Analysis*.[14] Singer gave me a copy and said, "Here, read this," and I would discuss with him things that were in there. That's how we first got acquainted.

Then I wrote a Master's thesis under him. In those days, the Korean War was on. You had to get a Master's degree on your way to a PhD because deferments were only for two years: you had to have an educational objective that wasn't more than two years away.

And I always remember, there was an oral examination for the Master's degree. I think Singer was there and probably the great algebraist Bob Steinberg, who was just a super-outstanding guy at UCLA. Just brilliant. The unbelievable questions they were asking me! My God. See, I was strong enough that I knew the routine things well. So the first forty-five minutes was spent with them firing questions and me firing back the answers as soon as they got the questions out. Then they started getting mean. They asked me something, I forgot what it was, where you had to know the invariance of domain theorem, which I didn't know about. The invariance of domain theorems tells you that if you have two homeomorphic sets in Euclidean space, and if one of them is an open set, then the other one is an open set. That was the issue that stumped me during the exam. Anyway, it worked out well because it got the faculty people into a big debate amongst themselves. That's always the best tactic in an oral exam.

Then Singer went on leave, and he recommended that I carry on under Richard Arens, who was the second winner of the William Lowell Putnam Competition when he was an undergraduate at UCLA. The first winner of the Putnam was Irving Kaplansky, a Canadian. Kap was an incredibly brilliant person. He was later the Director of MSRI, the Mathematical Sciences Research Institute at Berkeley.

---

14 Loomis, L. H. (1953). *An introduction to abstract harmonic analysis*. New York: Van Nostrand.

When and how did you come to MIT?

1956. After I got my degree at UCLA, I was working at the Ramo-Wooldridge Corporation in Los Angeles, doing research on questions which were important to the future of their R&D business. One of these concerned orbits related to going to the moon. They were nice people at Ramo-Wooldridge, but I didn't feel that I was contributing very much there. I wanted to get back into the real mathematics racket, and I was in contact with Singer by phone about possible job openings in math departments. I've forgotten where he was that year. He may have been at the Institute for Advanced Study or Columbia, on leave from UCLA. He said, too bad I hadn't called him a few weeks earlier, but by now all the Moore Instructorships had been decided, so it was too late for that. "But let me see what I can do," he said.

The next day, I got a telegram from Ted Martin, the head of the math department at MIT, offering me what's now called a regular instructorship, a post-doctoral instructorship, at one-third the salary I was earning at Ramo-Wooldridge. So I packed up my family in the summer and went off to MIT.

Nobody believes it, but I really didn't know MIT from a tree stump. I didn't! I thought of it much the same as my landlord did in Westwood, California. He said to me "MIT… MIT… Oh, that's that place in Massachusetts that's like the place in Pasadena"[15] [laughs]. That was the sense I had.

The very first day I arrived, I went over to MIT, and I couldn't find a living soul at the math department. It was August. Then I ran into this one gentleman who later turned out to be Norman Levinson. He was the only person around. He actually knew who I was and that I had been offered this instructorship, so he got me located somehow, in practical terms.

What was the working atmosphere like in the department, when you got here in the fifties?

Oh, the working atmosphere was *very* exciting at that time at MIT. The instructors and the assistant professors all had those little offices there, right along the river in Building 2, and we used to pack ourselves into those things. There was so much mathematics going on, it was bewildering. I recall one day, we were in Is Singer's office, and I looked up to find that there were thirteen people jammed into that little cubicle.

---

[15] Caltech.

Warren Ambrose was around. Alberto Calderón, who was a really outstanding analyst. And then there were scads of young people, instructors and so on. We talked constantly. The graduate students were mixed in: there was no distinction between the graduate students and the instructors and the faculty. We were just worrying about mathematics.

The contrast between MIT and UCLA must have been quite intimidating, like going from the minor leagues to the majors.

Yes. At the time I went there, UCLA was at the margin almost, just starting to really build more seriously. They were so excited at UCLA: they celebrated and celebrated that one of their PhD people was going to MIT. The man who was department head was Magnus Hestenes,[16] a great man. When I left UCLA environs—I was at Ramo-Wooldridge, but I still checked back with the department—he hauled me aside, said, "Listen. Here is what you have to do. You are going to meet more smart people than you thought existed on the planet. And the risk is that when you are around those people, they are doing all of these interesting things, that you will get absorbed in what *they* are doing. *Never let a day go by that you don't work on your own research.*"

He looked at me, and he said, "Never—let—a day—go by!"

And it was great advice. Of course it wasn't a day; it was a night. Because we didn't have time to work on mathematics in the daytime, we worked at night. Singer and Ambrose and those people really worked very late at night. Took their coffee breaks at two in the morning.

Must be hard on your family.

What family? [Laughs.] Yes, it is hard on the family.

See, the great thing that strong departments do for the young people is they cause them to work on more difficult things than they otherwise would have worked on. That's what the atmosphere does to you. You work on the *most* challenging things, you see? Just because it's what's in the air. And when you get to some other place that isn't quite as strong, you don't do that. You work at publishing papers. That is a different line of work.

---

16 Magnus Hestenes (1906–1991), was a professor at UCLA from 1947–1973 and department chair from 1950–1958.

You wrote your Linear Algebra[17] *book with Ray Kunze in those early years, right?*

Yes. In those days, Kunze and I were office mates, and we wrote this linear algebra book, which was certainly started during the period we were Moore Instructors, in '57 or '58. We tested it out in our teaching, and it was first published about '61.

Steven Kleiman, one of the few faculty members here who attended MIT as an undergraduate, remembers you testing out the book. He took both your freshman and higher-level courses.

I remember Kleiman as a student He was brilliant, one of those kids that show up at MIT that seemed to have a brain bigger than the whole world. Some of the students who come there are pretty intimidating characters, because it's clear when you talk with them that their brain works so much faster than yours. It's the young whiz kid that appears and blows everybody away, the kid who walks into your office and says, "Say! Have you ever wondered...."

No one ever struck me as brilliant as Steven Orszag, however. I first met Orszag in the second-year graduate course in mathematical analysis one year. I walked into the opening session, and there was this little kid sitting in the front row! He was sixteen years old. He was far and away the strongest student in that class. Another one of those kids that came out of the Peter Stuyvesant High School in New York City, which has produced more mathematicians than any other school in the country, hands down. Peter D. Lax: good God, there's a really great mathematician. Paul Cohen, who just died; another one of these brilliant young people.

So you got a Moore Instructorship your second year here?

Yes. I came one year as what's now called a regular instructor, a postdoctoral instructor. The following year, I was made a Moore Instructor, and I stayed that for two years. And after that, I was kept on as an assistant professor. Now, the department had a flat policy that they never promoted Moore Instructors to assistant professors: they didn't want the young people competing with one another to see who could stay on the faculty. You get terrible morale problems from such things. But it violated its own policy

---

[17] Hoffman, K. M. & Kunze, R. (1971). *Linear algebra* (2nd ed.). Englewood Cliffs, NJ: Prentice Hall.

several times: Arthur Mattuck was such a person. Sigurdur Helgason was another exception. But the exceptions were made rarely enough so that they could maintain the policy.

Such a concern for morale is surprising, in a way. Academia is so competitive.

It is. It is. But you see, it's Levinson again. You have to always remember, if there's anything about the math department unusual or sensible that they did, you could be almost certain it was Norman Levinson who was behind it. He foresaw the morale problems.

And this is *exactly* what happened at Berkeley, because their department developed in similar ways; they had a large body of young people, too. They were all called assistant professors: they didn't have these instructor positions. And then the problems set in. They got terrible morale problems. Until, finally, years later, they changed the system: they invented lectureships. So what we would call the post-doctoral instructor, they call the lecturer, and only a few people were called assistant professors. And they put in the same kind of guidelines: "This is a two-year instructor position. Don't think you're going to stay around."

I always marvel that Berkeley was able to develop its department as well as it did, because it was very democratically run, and it had endless discussions about everything. That didn't go on at MIT. We always benefited from the history of having started as a school of engineering. Historically, schools of engineering were run by the president and heads of the departments. And there almost wasn't any paperwork. The object is to get things done. So if you got this young kid out there calling Singer from somewhere near UCLA, "Is there any position I could get from you?" and Singer talks to Martin and says, "We really ought to hire this guy"—"Okay." He just does it. Knowing that he'd find the money somewhere.

Did you and Singer actually work together?

Oh, Singer and I wrote several papers together. Singer was involved in many parts of mathematics; he had many interests. So I kept bugging him about the parts I was interested in, what is now called function algebras. At one point, we hit upon a list of unsolved problems that Gelfand had published in *Uspekhi,* the Soviet Russian journal.[18] Singer and I solved most of those problems, so we took the solutions, wrote them up, sent the paper to Gelfand, and

18 *Uspekhi Matematicheskikh Nauk.*

suggested that he might want to have it translated and published in *Uspekhi,* which he did! That was unheard of in those days. It was very much of a Cold War atmosphere already, and Russia didn't publish things in English, but they did translate that into Russian. Gelfand always remembered that. That was the first connection, I guess, that we had with Gelfand. But that went on for many years. Both Singer and I stayed in touch with him. When he was able to visit MIT years later, it was a very emotional event. Talk about brilliant! There really are no words to convey how brilliant Gelfand was.

I understand that the Russian paper later came back to these shores.

It appeared, I guess, in an American Mathematical Society volume of translations of Russian papers. They were looking for a translator [for this paper by] "Goffman and Tsynger," and I think they contacted Singer—a total coincidence, apparently—to see if he could suggest a translator. So he says, "Why don't I just send you the English manuscript?"

What was your main line of work as a mature mathematician?

Well, starting from the late fifties, a batch of us were studying a whole set of interesting questions about bounded analytic functions in the unit disc in the complex plane. Lennart Carleson, who just won the [2006] Abel Prize, solved the most difficult question about them, which was to prove that if you have a finite number *n* of such functions, and if the sum of the absolute values of the functions is bounded away from zero—so there's no sequence along which they all tend towards zero—then the ideal they generate contains everything, that is to say, contains the function one.

Very bizarre stuff, actually. You see, they form a Banach algebra, the ring of bounded analytic functions, which has a maximal ideal space, which is a certain compactification of the open unit disk. And because it is a compactification, which allows every bounded analytic function to extend continuously to it, it has to have an awful lot of glop glued on there to take care of that. Picture this maximal ideal space as the open disc, plus a ring around it of junk added on. When we were at the Institute for Advanced Study in the late fifties, Andy Gleason from Harvard showed us how he could analytically embed a disc in the fringe. I remember the day that he told us about that. He was all excited.

So the main question I labored with is: find *all* of the analytic structures in that space, of any dimension, anywhere. Not a thing of prime interest, but to do that stuff about those analytic structures required really deep and difficult, complex analysis—like what Carleson had been doing, only not as difficult.

Lennart Carleson was my hero at that time—still is. Nobody, for sheer analytical power, is on a par with Lennart. One of the great mathematicians in the world. I mean *unbelievable.* Carleson finally proved the Lusin conjecture, that the Fourier Series of a continuous function or a square-integrable function on an interval actually converges, point-wise, almost everywhere to the function. That had been an open question from the beginning of Fourier series.

The day he spoke about it at Harvard, he talked in the lecture hall in the old building where the math department used to be, in there with foreign languages and such. It was a real old-fashioned lecture hall: tiered, sloping upward to the heavens, with all kinds of moving blackboards and such. Lars Ahlfors, a great Finnish mathematician at Harvard, introduces Carleson. Carleson gets up, goes up to the blackboard, picks up the chalk, and begins the proof. He doesn't say what the theorem is! He *assumed* that everyone who was there knew why they were there. He just began the proof. It was a most incredible, dramatic thing to do.

When he finished this tour de force of hard mathematical analysis, I walked out of the lecture hall and down the sidewalk with Mike Artin, a great mathematician of MIT, son of another great mathematician. We walked along quietly for a few minutes and then, out of the blue, Mike said, "I feel just like a little kid."

Carleson did a whole string of deep and difficult things. He is a great analyst, an amazing man, and part of a line of great Swedish mathematical analysts. He was a student of Arne Beurling[19] for years at the Institute for Advanced Study. I once asked him how the Swedes managed to produce, with their little dinky population there, this unending string of great mathematical analysts: Carleson, Beurling, Torsten Carleman, Lars Hörmander. I said, "It must be that you put all of your eggs in one basket, or you focus all of your efforts on one or two people."

"No, no, no," he said. "No, it's just because we are taught not to sit around and philosophize—just *attack the problem.*"

---

19 Arne Beurling (1905–1986) was already a well-known young mathematician when he cracked the German *Wehrmacht*'s code for the Swedish Defense Staff Headquarters during World War II. In 1954 he was appointed a professor at the Institute of Advanced Study. (See F. L. Bauer's review of Bengt Beckman's *Codebreakers: Arne Beurling and the Swedish Crypto Program during World War II*, in *Notices of the AMS, 50*(8), 904.)

Did you make some good headway on the problem you were working on?

Oh, yeah, I finally did it.[20] I worked on it for five or six years, probably. It's such a *bizarre* space, worthy of any bizarre planet. It everywhere looks like the picture of Dorian Gray [laughs]. You know the story? This is the guy who was so evil and corrupt inside that when they finally unveiled this painting of him, it was the most horrible thing that anybody had ever seen. It could keep you awake at night. When I told my friend John Wermer at Brown about my results, he invited me to come down and speak to the Brown faculty. I said, "On what date do you want me to come?"

He said, "Why don't you come on Halloween?" [Laughs.]

But for me it was a triumph. You can't be Sigmund Freud on yourself, but I suspect the fact that I did that was one of the reasons that freed me up mentally to start doing some administrative things. I had a higher level of confidence than I'd ever had about my own mathematical ability. And I sort of went out at my peak. Probably *was* my peak; probably wouldn't have done anything else, but you never know. It doesn't pay to sit around and try to psychoanalyze.

Levinson and Department Administration

In 1967, the same year you published your results, you became pure chairman and Norman Levinson became department head. What made Ted Martin step down?

He'd been there twenty-one years. He had stepped down two years earlier, but they couldn't get anybody to agree to become the head of the department. In the math department, nobody wants to become the head of the department! That's one of the great things about the place. But somebody had to do all that stuff.

After exploring, the only choice to be the department head after Martin went was Levinson. But Levinson wasn't going to run the math department on a day-to-day basis! [Laughs.] He only agreed to do it on the grounds that everything routine would be run by Harvey Greenspan and me and that he would just handle the really important things. So I did most of the day-to-day administration. Somebody had to do it, and I admit I had some interest in administrative things, unlike most people in the department.

One of the great strengths of our department is that people don't want to do such things. And they also don't want to sit around in a lot of silly

[20] See Hoffman, K. (1967). Bounded analytic functions and Gleason parts. *The Annals of Mathematics, 86*(1), 74–111.

faculty meetings, yakking. I remember when I was the department head, you couldn't *call* a department meeting. You could say we're going to have one, but nobody would show up. They just said, "What do you want to talk about? *You* do it! You worry about it."

They wanted the administrative people to worry about all these things. And they supported the view that when you had a committee in the department—a committee of the calculus course or something like that—that these committees should have only one member, appointed by the head of the department. Then, if things weren't going well, the department head would just change the person [laughs]. Other departments have a lot of baggage, because they have a democracy. They get committees, and they vote, and they waste more time. One of the great things about the MIT department has been its resistance to hot air. It just doesn't want to do all that stuff.

You have to remember: in most university departments, the question of who gets tenure, for example, is voted on by all the tenured members of the faculty. That's not true with the MIT math department: we went to a representative form of government in the sixties somewhere, on Levinson's recommendation, long before Martin was out. What happened was this. We started getting into trouble because when new appointments were proposed, you couldn't get a strong enough positive vote. Negative views started to dominate, so you couldn't move forward. Levinson said, we've got to get rid of this. We're going to go to the following system: we had a group of eight people appointed by the department head, and *they* would decide the tenure cases. This was discussed with the faculty, but it was just *done* by the department head. As long as the upper administration supported it, and Levinson supported it, and Singer, and so on, it was fine. And it helped enormously. Because then the discussions were much smoother.

And then, when he became department head, Levinson put in this two-part structure, a pure math committee and an applied math committee. He says, "Look at the criteria for appointing people in applied math. They're different!" It's not that the quality level is different. It's just different things that they're taking into account. For example, you want to hire a mathematical biologist. The applied people want to know, does this have any significance for biology? That idea is foreign to the pure mathematicians: they're interested in pure mathematics for its own sake. A high percentage of our core group of applied mathematicians was classical fluid dynamicists, and that was not something that was ragingly popular with the pure mathematicians. Not all the pure mathematicians looked with favor on the strong development of the applied group. You wouldn't have thought of anything like an applied group in the thirties: they just existed as an integrated part of the math department. It was only when they got *really* strong, and also because of the efforts of Harvey Greenspan, that the desire to have greater autonomy emerged. When the applied mathematicians wanted to have a separate department, Jerry Wiesner, the provost,[21] said, "Ridiculous!" He started to laugh. He said, "MIT is one great big institute of applied mathematics. That just doesn't make any *sense!*"

Whenever there are separate groups and they want things, there are misunderstandings that arise, and feelings get misplaced. Always very complex. There was always this risk that the applied mathematicians would get treated like second-class citizens, because they were outnumbered by the pure mathematicians, so they couldn't get what they wanted on the basis of a pure vote.

So how could you structure this thing so that it would work? In order to have the appointment and promotion processes run smoothly, you have to give some autonomy to the applied mathematicians. And so they invented this two-part structure. It was Levinson—as usual, see? Do the sensible thing. "We'll have a department head and two chairmen."

Having—and sustaining—an essentially autonomous applied math group within the larger realm of the math department was very unusual.

It was very, *very* unusual, I would say. Very MIT-like; very Levinson-like. Very *sensible.* Levinson made up this model: he said, "This is the way it ought to function." He would have taken this model, and he'd talk to the provost and say, "This is what I want to do."

"Okay. Do it!"

And then he'd talk with people. And, of course, you had to talk about it with the applied people so they would see what this limited autonomy meant: that there was some kind of division of resources and appointments and so on. But we were still one department.

It didn't have to be adopted, just talked about. So this was discussed, but it's not like you had to have a referendum or something. The department head could just announce: that's what we're doing. They don't get to vote on it. They may *think* they want to vote on it, but the department head runs the department: Norman could do whatever he wanted. The department head is in charge of the curriculum; the department head is in charge of

21 Wiesner served as provost from 1966 to 1971, when he succeeded Howard Johnson as president of MIT.

appointments and promotions—on paper. Now, you don't *behave* that way as a department head if you've got any brains, because you'll get killed by the faculty. But at critical moments it becomes very important. That's the way you do things at MIT. That's part of the legacy of an engineering school.

The Student Uprisings

You had been pure chairman for about two years when in 1969 you became chosen as chair of the Institute-wide Commission on MIT Education, formed at least partly in response to the student uprisings of those days. What was the atmosphere at MIT at that time?

Things were brewing; the radical student vibrations were going on. They were in a siege mentality. All of the people in the upper administration carried walkie-talkies with them. The radical student movement at MIT was run by graduate students, not by undergraduates. They were smart as hell and great politicians. They were master organizers: they could run the whole thing with a strong hand and have it appear that it was a total democracy and nobody was running anything. It looked like anarchy. What was that kid's name who really ran—?

Jonathan Kabat?

Kabat, yes. God, he was a strong kid. Very talented. And I used to say to him, "Jon, you've got to look out; you're going to get into trouble now." All radical revolutionary movements of any kind, if you read history, make the same mistake. They start to get up a head of steam and momentum; they're making some progress on the establishment, and they're persuading people, right? And then they get to a plateau point, a critical juncture. They have to decide whether to move closer to the establishment—or to become more radical. And they always make the same mistake. They become more radical, and the establishment destroys them.

See, when we were standing in the president's office trying to protect the president's office from the radical students, I'd look up and I'd be standing right beside Noam Chomsky.

Who is perceived as quite a radical himself.

Yes. But not if you're going to mess around with our institution. And the time that the students interrupted a class? Oh, my God. The faculty came down with a giant iron fist.

It was a wild time. But we got through it; we got through it.

The administration of MIT did such a good job during the radical student days that very few people were aware of the amount of turmoil there was on campus. But it did create a great deal of day-to-day tension. When I was off doing this commission on the future of MIT, my office was in Building 39. It was a big target because it was then the main computer center on campus—those were the days of mainframe computers—and we had bomb threats virtually every day, phone calls to one of the administrative offices. I've forgotten what the maximum was, but I'm sure we had a day where we had more than twenty in one day. And they feared that something could happen at MIT like happened at Madison, Wisconsin, when the radicals loaded up a truck with TNT and drove it up under the archway of one of the main buildings and blew it up. It was a *gigantic* explosion.

So they had to heed these threats. We all had to parade out of the building, and they had to search the building, I guess. Now, I have to say that there were a few times we didn't bother to leave the building [laughs]. The fifteenth one in one day—I mean, it was very hard to work there. And then I was meeting with the upper administration leaders, like, every other day about whatever was going on on the campus. So life was a little chaotic and it was very difficult to think about the future of MIT.

We decided we would set me up a second office out at Endicott House,[22] where every other day or so I could go out there and think quietly about what was going on in the world. MIT was already actively using it, particularly the Sloan School of Management. They were the ones, from the beginning, who made the most extensive use of Endicott House for these multiday seminar/training sessions, so there was always a staff there. So that academic year, 1969–1970, I had this other alternate office, which was rather more luxurious than my office in Building 39. It was my peace and solitude in those days. While I left them struggling with their radical activities, I was out there thinking [laughs]. I did a huge amount of reading in those days—the most intensive reading period in my life, I think.

You later inaugurated the annual department get-togethers at Endicott House, is that right? The department had grown too big, by this point, for the gatherings that Ted Martin

22 MIT has operated the Endicott House in Dedham, Massachusetts, as a conference and training center since 1955.

and others used to host. So your refuge there drew Endicott House into the math department landscape, at least once a year.

Yes. When I became department head I was looking for some way to have a bit of a different event for the faculty, and particularly for the newcomers. So in the early 1970s I decided we would indulge in a little luxury and have an event there at Endicott House in the fall. And I guess they've continued that. I get invited to it, but I live a little far away [in Madison, Maryland] for me to motor out there.

I knew Endicott House pretty well. I met the then-reigning Mr. Endicott several times and talked with him about the house and the history of it. He was in his forties, I would judge, about the same age I was, and he would come there maybe once or twice a year to look the joint over, see how it was going. It was this young man's home: he grew up in it. It was an incredible thing to stand there, in that big mansion, and try to picture any *kid* growing up in such a house. The Endicotts were one of the richest families in New England, as well known as the Cabots, the Lowells, and the Lodges. They had made a fortune in the shoe business.

But they gave that mansion to MIT,[23] and they had the great wisdom to endow the maintenance with it. One reason they did this was because this was and still is a very special piece of property. When the Endicott shoe family built that place, they put trees from all over the world on that property. They bought them all in pairs, and they gave the second one to start the Boston Arboretum. So that estate is packed with an *amazing* array of trees if you walk around it. They didn't want any of them to be lost, and they also knew that a university would have a heck of a time maintaining a big mansion, so they endowed the maintenance.

Why was the Commission created, and how did you come to chair it?

The president of MIT at that time was Howard Johnson—not the man of the restaurant chain but Howard Wesley Johnson, who incidentally turned out to be one of the great presidents of MIT, even though he was only in the job for five years. And in 1969 Howard decided that it was time for MIT to have a look at itself. They had had, immediately following World War II, a commission—the Committee on Educational Survey, it was called—

---

[23] The estate had belonged to H. Wendell Endicott, descendant of the family that ran the Endicott Johnson Shoe Company, who stipulated in his will that the house and 25 surrounding acres "be donated to an education, scientific, or religious organization," according to the Endicott House website at http://www.mitendicotthouse.org/about_history.htm.

which was highly successful. It was chaired by Warren K. "Doc" Lewis from chemical engineering, the founder of chemical engineering—not the chemical engineering department, but chemical engineering. He chaired that commission, and it had a big positive impact on the place.

So the president had the idea we'd appoint this commission, and I got stuck on the planning committee to propose to the president a plan for how this commission should be structured and what its charge should be. It met over one summer. When it came time to write up a report of the planning committee, they had somebody working on it who was doing a not-so-great job. So I just took over the writing and wrote the report.

Well, that was my downfall; because then I was dumbfounded when the president asked me to chair the commission. I was thirty-nine years old, a pure mathematician that nobody in the rest of MIT ever *heard* of. It was a pretty astonishing thing for the president to do, I'll tell you.

The committee included not only faculty, but also students, right?

Yes, and alumni.

Didn't this lead to an age and an attitude gap that was hard to bridge? MIT students were protesting the ROTC at that time; they were picketing companies who came to campus to hire, racial issues were coming to the fore, there was tear gas—even a shot was fired.

Oh, yes. But, my God, the assets we had. The young faculty were just outstanding: four nontenured assistant professors, all aged thirty, thirty-one. The first task I had as the chairman of the commission was to go to the heads of the departments of those four young faculty and say the following: the president wants your absolute guarantee that this person will get tenure. Now, this wasn't told to the young faculty members, but it's absolutely gospel, and it's typical of MIT. The president had to be told that; because if he was going to ask them to spend this kind of time on something, you had be sure that it wasn't going to screw up their chances for tenure. And there was only one way to do it.

So did they choose outstanding people who would have gotten tenure anyway?

Probably yes. They were outstanding and were also interested in teaching and a wide variety of things. They weren't just geeks or nerds. One was Sam Bodman, who was then an assistant or associate professor in chemical engineering, later ran Fidelity Investments, and is now the United States

Secretary of Energy and former head of the Cabot Corporation. He and I wrote the report on the finances of MIT.

As the title of the Commission report, "Creative Renewal in a Time of Crisis," suggested, MIT did indeed seem to be renewing itself. Margarent MacVicar had started the Undergraduate Research Opportunities Program (UROP), for example, and the pass/fail system for freshmen and Independent Activities Period (IAP) were introduced around that time. Did the Commission have anything to do with those changes?

The Commission had a major role in getting the Independent Activities Period created. It did the politics. The IAP was Jerry Wiesner's idea. He floated it with quite a number of faculty people, and he ran into a stone wall. He was trying to shorten the fall semester to eliminate what had always been a sort of stupidly awkward thing, that the fall semester continued after the Christmas break. The students would return after that break and then have to finish the fall semester, which was artificially longer than the spring semester. So Wiesner got the idea: let's cut off that part of the fall semester, stop it when the Christmas break comes, and not begin classes again until the first of February! And this way we would create what he decided we could call Independent Activities Period in the month of January, in which all kinds of interesting, educational, and intellectual activities would go on on the campus.

Well, he just met with more resistance than one could imagine. Faculty in science and in engineering had been in the habit of teaching their fall semester course in a certain way for ten years, all right? They've got a syllabus, and they've got lectures planned, and they've worked out a scheme for how to handle the Christmas holiday break, what to do after it, and so on, and this was just upsetting their apple cart. *And* it was reducing their instruction time, see?

So I was sitting in my office one day; the phone rang, and there was Jerry. He quickly described where he stood on his idea of the Independent Activities Period, and he called to ask if the Commission could help. By "help" he meant persuade more people on the faculty that this was a good idea. I got hold of Sam Bodman, because he was a man of action. I called him in, and I explained to him: "Sam, Jerry's got this Independent Activities Period idea. Here's the idea, in a nutshell, and he really wants to get this done, but he's getting a lot of faculty resistance, most of it coming from the School of Engineering. So what I want you to do, Sam, is take time off from whatever you were doing for this Commission and work full time: you become the floor manager for putting this through the faculty of MIT."

Sam looked at me and he said, "Consider it done."

And he left. He never talked to me further about what he did, but when it came up next in the faculty, it passed easily. So when he became the Secretary of Energy of the United States, I said, "I wonder if Bush knows what he has hold of there." [Laughs.] This guy is a *real* doer!

The Independent Activities Period, contrary to the predictions of these resistant faculty, was a *smashing* success from the day it happened. The worry was that the students wouldn't come back after the break, you see. They'd just stay home. Quite the contrary! We had ninety-some percent of the students back on campus. The activities that went on were just fantastic, from the beginning. One of the most popular ones was how to repair your own automobile, run by some guys in mechanical engineering. Over four hundred people signed up! Things in music, and in art, intellectual topics that were narrower and specialized, that nobody ever had time for in the regular courses. All these kinds of things. It was a smash hit from the first day it started.

People tended to think the Commission was some kind of think tank or study group. It was partly that, but it was a very action-oriented group, in communication with hundreds if not thousands of people on the MIT campus, discussing issues of education, research, governance of the place, and so on. The younger faculty members on the Commission used to hold regular sessions, where they would get chairs and go sit out in the corridors, or in the lobby of Building 7, and put up a sign to get faculty and students to come and talk with them about their ideas about education at MIT and how to improve it. So they did a mass frontal assault, I would call it. I and the other graybeards would have a topic we were working on: we would know what kind of people to approach in different departments to discuss what they thought about various things, and then we would try ideas out on different people.

But beyond just talking with people, we actually had *action* programs, things that we were implementing and doing. So we had one staff member named Stuart Silverstone, who worked full time on improving aesthetic things around MIT. Not writing reports about it, but making changes happen. He was an art person, came out of the school of architecture. He worked on how to improve the look of the main entrance of MIT: the lobby of Building 7, where you enter off Massachusetts Avenue. He decided he didn't like the way that place looked, so he just undertook to change it!

The thing that caused all the phones in the Commission to ring right off the hook was the day he started auctioning off this huge sign thing that had been on the wall there for forty years. It was a map of MIT with lights at different key locations on the place and a little index at the bottom with push buttons. If you wanted to find Building X you pushed that button, and it would light up. It was a famous artifact of MIT. My phone rang one day, and somebody informed me that Stuart was auctioning off this sign, and the provost wants to know: who authorized Stuart to auction off MIT's property? They became hysterical, because everybody believed that that sign was one of Jim Killian's favorite things. Jim Killian was the great figure at MIT at that time; he was chairman of the Corporation and former president. He was a great man.

Well, the answer was that Stuart had authorized himself. We had to explain to him that, you know, there are some procedures you have to go through [laughs]. I talked to Jim Killian later about it, and he told me, "I've always *hated* that sign! I wish he had sold the damn thing." So eventually Stuart did manage to get that down. Then he put carpets in the lobby of Building 7 and had people sitting around on the carpets. These caused a real flap all around MIT. People said, "You can't do that! That's sacred territory."

The last remaining student mural.

People who know MIT now probably would not believe that in 1969, the corridors of MIT were all battleship gray. Stuart says, "We've got to change this." So he went and he talked with the physical plant people. He talked with some authorities, and he got a bunch of students from the architecture school and other departments who were artists, and they agreed that they were going to paint murals in the corridors. So one day at five o clock in the afternoon, the physical plant crew came in, and on all the panels of the walls in the main corridor of MIT, they put up new—I don't know what you call the fabric, but it's like heavy wallpaper that somebody could actually paint on and have it last.

Unfortunately, the kids didn't start [painting] that night. So by early in the morning, the phones started ringing at the Commission: people were shouting at me from every direction. What had happened during the night was, the entire main corridor of MIT, which comes in from 77 Mass Avenue and goes straight across to Building 6, these clean, white, fresh walls had been turned into graffiti. But Jerry never flinched. He still backed us, and we got the physical plant people to go back that very night, and we put new paper up on the walls. And the kids came in immediately, and the following morning, all the main corridor of MIT had gone from battleship gray to these murals. Great art work, about international students around the international students' office, murals going down the walls in the stairwells, and the most famous mural of them all, the dollar bill painted around the bursar's office, which I think is partially still there.[24] Incredible, what these kids did.

But still my phone never stopped. We got more complaints about that than anything else we ever did. So the administration went back and took out all but about eight or ten of these things. But that began the active campaign of redoing the corridors of MIT in a whole new color scheme. This section is red, that one's yellow, and all that. So that was when it all started, and the hero of all that was Stuart Silverstone. I remember the day he came to me and said he was leaving. "Goodbye," he said, "thank you." And he said, "I've just gotten rid of my car, and given up my apartment, and broken up with my girlfriend, and I'm off."

I said, "What are you going to do?"

He said, "I don't know, I've given up planning, too." He joined a commune.

---

[24] The dollar bill mural was recently removed when the area was turned into a lounge. A small reproduction of it remains on display.

You said that Howard Johnson became one of the greatest presidents of MIT. He's written his memoirs,[25] but can you say something about him?

Howard had been the dean of the school of management. And he made president, I'm sure, because of management skills he was going to bring to the job. Well, in any ordinary sense, that wasn't what he needed, it turned out. What he needed was a backbone bigger than anybody else's. MIT had more serious radical student activity than almost any institution in the country. That isn't widely known; and that's a tribute to the administration, and especially Howard Johnson, who was extremely skillful at handling these things. He became a *towering* figure in the minds of those of us who were close to him at that time, just a towering figure.

Howard governed MIT during those radical student years with one simple principle, which he would repeat over and over. "As long as the faculty is together, we're indestructible." And he would make decisions and do things on the basis of that principle. Whenever there were close votes in the faculty, dividing the conservatives from the liberals, so to speak, he would sometimes—though he was quite conservative himself—vote with the liberals, for the simple reason that he knew that the conservatives would still hang with him, right? But if he voted with the conservatives, the liberals would blow a gasket and whine forever [laughs], and he'd have a radical revolution going on. The faculty was meeting in Kresge Auditorium in those days, and tie votes occurred. He had to cast the deciding vote himself, as the president, on major matters. I know twice in which he voted the opposite of what he believed, and he did that to hold the faculty together.

So when he said "as long as the faculty hangs together," he didn't mean necessarily that everybody would agree with each other.

No. But it wouldn't get torn asunder.

The other part of his policy was to protect the students. Because MIT does not allow the city police on its property.[26] And when the students started demonstrating out in the *streets* of Cambridge! Oh, my God. They were terrified, the administration, of what could happen. One day there

---

25 Johnson, H. W. (1999). *Holding the center: Memoirs of a life in higher education.* Cambridge, MA: MIT.

26 Howard Johnson's memoirs imply that the events of 1969 forced the administration to make occasional exceptions to this rule. See Johnson, (1999), 176, 178.

was a big demonstration planned and riot squads were brought in from police in all the metropolitan jurisdictions around Boston—gathered in the Commonwealth Armory near Boston University. And Howard and Jerry Wiesner, his good friend and the provost, were down there at four-thirty in the morning. They went around the armory, and they shook the hand of every policeman, and they talked to them about the students: "These are not criminals," they said.

Jerry Wiesner told me, "If I had been the president during that time, MIT would have blown apart! I couldn't have held it together." But Howard did it. During a time when he was suffering from acute back trouble, he stood on his feet for five straight hours chairing contentious faculty meetings. He was just a great man.

There is a moving story in Howard Johnson's book [27] *about a party that was held for him at Singer's house after all this was over, and you were there. Do you remember that?*

I organized it! The party was held for Howard and Jerry Wiesner. It was a surprise party. Because I knew the strain that those two men were under, and that there were strains within the faculty, in spite of the fact that he was holding them together—and potential bad strains with the administration. Truth is some faculty members had behaved rather badly. They didn't do what they had promised. And I knew that these guys, Johnson and Wiesner, were strained right out to the limit, and that we needed to do something to bolster them up. So we organized this party, Singer and I. We used Singer's house. I got a list. We made sure that some of the people who had behaved badly were at the party; in fact, one of them presented the gifts to Johnson and Wiesner. But we had very conservative faculty, very liberal faculty. We got Mrs. Johnson and Mrs. Wiesner in on it so they could tell them they were going to the movies or something.[28]

I haven't seen much of Howard in recent years, nor have I ever discussed the event with him, but I know from other sources that he never forgot that party. It was deeply moving for him, because he was really stretched to the breaking point. It sort of carried him through, as it were. When he retired as chairman of the Corporation years later, there was a big event at the Faculty Club, where the upper administration people and the corporation members

---

[27] Johnson (1999), 202–203.

[28] Johnson recalls this minor point differently, remembering a request to "come to an emergency session of some faculty members." Johnson (1999), 202.

were there—and I was there. His executive secretary happened to be sitting at the same table I was, and she said to me, "I know how the invitation list for this event was put together. What are *you* doing here?"

"Howard invited me." And that was the signal to me that he remembered.

But he deserved the support. He was just amazing. And it practically killed him. After five years as president he was exhausted.

The commission's report, "Creative Renewal in a Time of Crisis," didn't really fall on fertile ground.

That is an understatement. First of all, when it came to the educational part, I don't think we had anything very creative to say. Because you have to say things that were both creative and implementable in kind of a simple-minded way. Like UROP—this is an innovation; but it's not mathematical research. It has limited scope. Much of what we did have to say about education had to do with the role of the humanities and social sciences in the education of the MIT students. We said that there were critics of science and technology roaring around, and you need to bring them inside your institution. One role of the university is to have people on its faculty who are social critics, right? And that may include criticism of the role of science in this society. We do have some of them: Noam Chomsky; and Joseph Weizenbaum in electrical engineering, who warned everybody about some of the danger of computers, for example.

But these were not popular things. And the main tenor of the report that got it into trouble was that it was asking MIT to stand up and take it a little bit on the chin and listen to what people are saying and try to adjust themselves. But people weren't in the mood to listen to all this stuff. They wanted to hear, "MIT is wonderful!"

Now, part of the motive of the president was to help defuse the radical student situation, to demonstrate that there was some emphasis on looking forward, on positive things. We did do that; but most of the rest of what we had to say wasn't really listened to, other than the part about the financial management of MIT. I can't summarize all of it; but the main thing was, you've got to open the budgetary process. Open it up at least to the faculty and make it much more transparent and visible. To give you a sign that that was needed, that part of the report of the commission was submitted confidentially to the president as a separate document. He didn't have time to do much with it, with all that was going on campus, and for a few years, I wondered what he ever did with it. Later my friend Paul Gray told me. He said, "Oh, that's simple; he gave it to me when I became the chancellor." He said, "I put it on my nightstand. I went down that list, and I checked those things off one by one as I did them."

As you implied, the Commission was an enormous thing for an unknown faculty member to take on. Quite apart from the reception of the report, how did the experience affect you personally?

It was in many ways the greatest thing that ever happened to me. My God, what I learned. I learned enormous amounts about MIT and what makes it tick, through talking to people and meetings with people and *reading*—oh, my God, did I read.

When this commission started, one of the members was Louis Smullin, who was then head of electrical engineering and computer science and an old-time MIT person. He would talk to me about the institution and what makes it tick, and I listened to him for hours. For example, one of the first things he told me that you've got to know about MIT: "It's a sweaty place."

I said, "Excuse me?"

"It's a sweaty place. It believes in the work ethic."

He said, "If you don't think that's true, here's how you can test it. You go take a walk around through the corridors of MIT and see how often you pass a faculty member's office, and the door is open, and you see the faculty member in there reading a book. I can tell you in advance you will never see this.

"Now, what are you supposed to conclude? That the faculty members don't read books? Or that they don't read books in their offices? That can't be true. It's just that when they do, they get up and close the door."

He said, "You know, for an assistant professor at MIT, the work week is minimally eighty hours. It's the minimum. You don't have time to do anything else, if you take all the teaching seriously and the other things that you have to do and you're working hard on research. That's just the way it is."

He said to me, "Caltech is a graduate school of science. MIT is an undergraduate school of engineering. It always has been; it always will be. We have the most powerful science departments you could imagine, but in its *soul*—it's still a school of engineering."

Then I started looking at the math department and noticing that there are a lot of faculty members in the MIT math department who could just as easily be in some other institution. They could be at Harvard, for example—some of the really strong ones. What makes them choose MIT? It's because they like being part of an institution that's connected to the real world. It's not that they go over and talk to the engineers or anything. But they just feel more comfortable being in a place that's less ivory-tower. And they're very proud of the fact that MIT gets into practical world affairs. So there's a self-selection of that kind that goes on in the faculty.

Things like that, you see, I would learn from Smullin about what is going on beneath the surface. And then, of course, I read histories of the place. I don't remember a great deal of it now, of course. But it was a very valuable learning.

Valuable in what sense? How did these things help you?

They helped mainly so that you keep a proper perspective. You have a proper respect for the institution and why it does certain things the way it does. That was, I think, mainly the thing I got out of it. So I can't identify specific things that I did more wisely because I knew that; I just knew that I had gotten a valuable education. Jerry Wiesner, who was provost then, told me I got the most expensive education MIT ever gave anyone. Which is his way of saying, we spent all this money on this Commission, and about all we got out of it was we educated you! [Laughs.] He didn't say it that way; but I knew what he meant.

This must have put quite a dent in your own work as a mathematician.

Oh, the minute I said "yes" to chairing that Commission, the dent was big right there. I had to take my PhD students and give them to Singer, because I knew that this was a twenty-four-hour-a-day job. And it was.

Were you ever able to get back to work as a mathematician?

No, no. I got back to writing books and so on; but I never got back to mathematical research.

Do you regret that?

Oh, yes and no. I don't know. I'm different than many people. I don't look back. When things are done, they're done: I don't sit around and moan, you know. You say, don't you miss teaching? Yes, I suppose. On the other hand, I'm doing other things. It doesn't break me up.

Department Head

In 1971, shortly after finishing your work with the Commission, you became the first department head in the post–Martin/Levinson era. What was the biggest change you implemented when you came in?

The one big innovation is, I created the undergraduate mathematics committee and a role for its chairman, Arthur Mattuck, who handled all the affairs relative to the teaching and administering of undergraduate math programs. There always had been, I think, some kind of graduate chairman. But we created the undergraduate chairman out of whole cloth.

Why did you create that position?

The sense that we had to give greater focus to it; that it shouldn't just be the department heads paying attention. You needed somebody doing it who was closer to it and to the students, and who could handle all those affairs.

And the first undergraduate chairman was Arthur Mattuck.

Oh, yes—wouldn't have created it otherwise. Because he's an unusual man, with a rare gift for sensitivity to what was going on in the undergraduate level and a very deep interest in all sorts of things. Very skilled in all those affairs. Arthur is very astute, so I put him in charge of all that.

He said, "What does it mean that I'm in charge?"

I said, "It means the following. You do what you think is necessary. If I see something you're doing that I don't like, we'll discuss it. And after we're through discussing it, if we still disagree about it, then I will support your position."

"Now, of course," I said, "you understand: I am the head of the department. And so I *could,* under some extreme circumstances, violate that," I said. "But if I ever do, you should resign immediately, citing breach of contract."

And Arthur said, "That's clear enough."

In an article for The Founders of Index Theory, a festschrift for Singer, you wrote, "I sought … to make [the department] as strong as any department in the country." [29]

---

[29] Hoffman, K. (2003). Fifty years of friendship and collaboration with Singer. In S. T. Yau (Ed.), *The founders of index theory* (p. 232). Somerville, MA: International Press.

And we did. That of course is in many ways the main job of the department head. It's an enormous task, making sure that you use very high standards for the promotion and tenure and recruiting and hiring. And there, of course, you have to listen to your advisors. I do not have the kind of insight into future talent that somebody like a Levinson or a Singer has.

In the same piece you quoted Dan Quillen's saying that Singer's talent identification process is "an astonishing thing to watch." What is it like to watch? What does he do?

He just knows! There's nothing you can say about it. He spotted Quillen himself, as an undergraduate, and he said to me, "We're going to hire that guy." See what I mean? He just knows who's going to be the really whizzy person. We're talking about a very high level—we're not talking about who is a promising student. But it's an essentially *infallible* instinct. I don't have that. So when you're making choices and decisions, X versus Y, you have to have people like that to listen to.

Presumably hiring and keeping good people is not just a matter of giving them more money.

No. It's the care and feeding. As a wise person once said, "When a faculty member leaves, rarely is money the first reason. But it's always the second reason." It is a factor, but it's not the only factor. They feel more comfortable at another place; they have some family circumstances which make them more content to be in another part of the country. Just an admixture of all kinds of things. It's hard to say. I can fill a book with just what it took to get John Milnor to MIT. He was one hundred percent Princeton, through and through. And to get him to move here—my goodness, what an undertaking. And you have wives or husbands involved in these things—oy! It takes great care and thoughtfulness and attention to build and maintain the quality of a faculty. It's a full-time job.

What's an example of the kinds of things that a department head would have to do in order to attract a person like that?

Well, of course, you have to convince them that they would play an important role in your department or the development of a certain part of mathematics. Another thing you have to take into account, frequently, is what the spouse is going to do, because there are a lot of couples where there are two professional people involved, and the spouse works in industry or in academe. It's very complicated, first to understand what the needs are, and then to understand how to meet them. People are people, you know? They have personal wants, needs, desires; and you have to attend to those. You do what you have to do.

For example, we stood MIT on its ear to hire Dennis Sullivan when he was young. He was a *super*-phenomenon, but he didn't like ordinary teaching. So we had to create a special position for him, which didn't exist at MIT. We just went to Jerry Wiesner and said, "This is what we're up against." So he invented a position, right there on the spot. I think we called it the Sloan Professorship and used Alfred P. Sloan money.

It would have been great had you had the opportunity to talk to Jerry Wiesner. This man was so smart, he was frightening. When he was provost, he made all the decisions about academic programs. You'd call him; and he'd call back—because he was never there when you called him—and the first thing he would say to you was always the same. He'd say [briskly], "What's up?" Then you'd explain what you were calling about, and you'd get halfway through the second sentence, and he'd take over and finish the sentence. Then he'd explain to you the rest of what you were worried about.

When we went to hire John Milnor, I called Jerry from Is Singer's office, and we told him that we probably had an opportunity to hire Jack Milnor.

"Really?!"

First of all, our provost knew who that was immediately. "That's fantastic," he said. He asked a little bit about the circumstances, and we explained.

He said "Okay! Make up a salary." I forgot what salary we quoted to him—some outlandish figure. He thought a moment, then said, "Okay. Go! I'll talk to the president later." That's it. There's no paperwork. "Go!"

That's a wonderful person to work with. He was not only a president and a provost of MIT but one of the most *brilliant* people I ever knew in my life, just an utterly amazing man. He was a communications engineer, like Ed David: they were a little bit more on the mathematical side, and he understood the great importance of mathematics to MIT and would have talked to you about it, I think, rather eloquently.

You wrote in your piece on Singer that you sought to give the department a greater role in the affairs of MIT. How so?

You have to be present in the councils of war! The research and education role of the department in MIT is governed by just a few people from the department who are involved in institutional affairs. If you're involved in these institutional affairs and in constant running dialogue with the higher-level decision makers, then nobody can ever forget that there is a math department. And if they like what you contribute, then this reflects well on the math department doing its share, helping run institutional affairs. Levinson was always consulted by the upper administration people at MIT. So we got more people out and around the institution: Hartley Rogers going off to the provost's office. Ted Martin actually became chairman of the faculty. All of these things give the math department some presence in the affairs and the institution. That's all. I don't think it gets you any more money.

You (and other heads, from Martin on) must have done something right in building up the department, as the MIT math department was recently ranked number one in the country, for what it's worth.

Those rankings are really opinion polls of knowledgeable people around the profession. It's based upon the general sense of quality of the faculty and the PhD students that come out. Very imperfect, and very approximate; but you get pretty good convergence.

Sammy Eilenberg, the great mathematician at Columbia, once said the following: "You should not try to identify the ten strongest math departments in the country. You have to discover, for each field of science, the magic number N such that it's possible to identify the best N out of $N + 1$ departments. These are the N departments that will appear on *everyone*'s list of the top $N + 1$." You have to leave room for the person to put their favorite department on the list, you see. That was Sammy's wisdom.

Washington and the David Committee

You later went to Washington, D.C., where you served as executive director for the Ad Hoc Committee on Resources for the Mathematical Sciences of the National Research Council, popularly known as the David Committee. I was interested to learn that the committee's chair, Edward E. David, Jr., had done his PhD under Jerry Wiesner.

Yes, in communications engineering. Ed David was the chairman of the Visiting Committee of the MIT math department, one of the great inventions of James R. Killian, another outstanding president of MIT. Killian introduced this system of visiting committees for the departments, such that each was chaired by a member of the MIT Corporation, the trustee body. The Visiting Committee also has alumni on it and professionals from the discipline. It visits the department every other year—and visits not at budget time, when contentious issues may be afoot. The agenda for its meeting with the faculty or the department is made up by the department head and the chairman of the Visiting Committee, to ensure that what is important to the department is what is discussed.

When they're through, the Visiting Committee reports back to the MIT Corporation, so that there is information about the department flowing back to the trustees that doesn't come through the upper administration. Though the upper administration people will attend the meetings of these visiting committees—essentially all of them. They learn a lot this way.

One great feature of this idea is, it assures that every member of the MIT Corporation knows one or two departments in the institution intimately. This promotes incredible loyalty of the Corporation to the institution. The minute there is any problem at MIT—outside political forces try to impinge on the institution or something—that corporation will run their wagons around in a circle and stand by the place, because they feel that *they* are part of the institution. Now, that's a very clever administrative idea by Jim Killian.

The David Committee played a central role in increasing government funding of mathematics. Why and how was it set up?

We had problems about research funding in the country for mathematics—federal support of mathematical research. It wasn't just financial support; it was other support as well. At the Academy you couldn't say you were setting up a committee to look at the *funding* for this field. That is not allowed. You had to claim that you were looking at a whole spectrum of things—"resources." I don't even remember what we described. It was just a smoke screen. Everybody knew we were really looking at the funding [laughs].

Bill Browder and Is Singer and a couple of others were influential down at the National Academy of Sciences in getting the National Academy—which calls itself the National Research Council when it advises the public—to set up a committee to look into the support of mathematical research. And they wanted to set up this committee. I had just come off being the department head at MIT, so Bill Browder asked me if I would consider becoming the executive director of that committee. I said I would, providing we could get Ed David to chair it. Ed David was chairman of our Visiting Committee for years, so he knew lots of things about the math department. I knew him very well. Respected him greatly.

Browder said, “Oh, that isn’t going to work; I already tried that twice; and he said no.”

So I said, “Well, let me talk to him.”

And I did. I said, “Ed, there is the question of the chairmanship of this committee. And I know that Browder has talked to you and you said no. I assume that the reason you said no is because you’re just too busy to do it?”

“That’s right.”

I said, “Well, let me ask you a question. Don’t you think that anybody that we would want to chair the committee is too busy to do it?”

“Yup.”

I said, “Then that argument is irrelevant.”

This brought about a stunned silence on the other end of the phone, whereupon he started talking about “We” this and “we” that. I got him! He agreed to chair that committee, and it ran for three years.

But it was not a committee that just made a lot of recommendations. He made two conditions on agreeing to chair it. One condition was that I would do all the work. That I expected. Second condition was that every basic recommendation that we make will have been set in motion before the report is issued. He said, “I’m tired of having committees recommend into the wind and write reports that sit on shelves and gather dust.” So we had an action committee. We started implementing from day one what we were going to eventually recommend. We worked very closely with the presidential administration team, and it turned out to be a rather effective committee.

This was the Reagan administration at this time, right?

Yes. It was a wonderful administration to work with, because it was staffed with clear-thinking and decisive people, and you could get yes/no answers bing-bang-boom, like that. They understood why the federal government had to support pure mathematical research: because it was not plausible that any other sector of society would do it, and yet, it was absolutely essential to the whole engine of science and technology. There was no more to the argument. That’s it! End of the discussion. Reagan’s science advisor, Jay Keyworth, had that clarity. He was the one, absolutely.

What was the committee's effect on the funding for mathematics?

It got the funding for mathematical research doubled—but then the mathematicians let it erode away again. Because they didn't keep up the effort, you see. The math community has to *keep* a strong presence in Washington that watches what goes on and keeps working with people like the White House Science Advisor to make sure it doesn't slip backward again.

What we discovered was that there was a point in history going back to the mid-sixties where mathematical research was funded very well by the federal government, but the mathematicians just let all the money go away. We lost *half* of the federal research funding for mathematics in a span of four or five years. This was connected with the Mansfield Amendment, which was only in effect for one year; but it was an amendment that put strict strictures on the Department of Defense about what it could fund on the research end. It could only funds things directly connected to the immediate mission of the Defense Department.[30] As a result of that, the funding of all these things like the Air Force Office of Scientific Research that I talked about and research grants to numerous mathematicians—that funding was all moved from DOD to NSF.

The only trouble is that it never got to the mathematicians, because the mathematicians were asleep at the switch, see? So it all went into physics and chemistry, because physicists and chemists ran everything at NSF! That's how when you're asleep, you pay a big price. I said to Ed David once that the mathematicians believe in the theory that God will provide. He said, "That's correct. And God does provide—but for the physicists."

The setting up of this committee, by the way, was strongly supported by the physicists and chemists in the Academy. They talked to Bill Browder and Is Singer; and they said, "My God, that's a terrible situation mathematics is in. Somebody has to fix that."

Yet every million that goes to mathematics takes it away from physics and chemistry.

Yes, but the math amount is *trivial,* you see? At that time, the whole thing—federal support for mathematics—was less than a hundred million dollars

[30] Protests on college campuses against the Vietnam War and university involvement in military research led Senate Majority Leader Mike Mansfield, a liberal Democrat, to propose an amendment to the Defense Procurement Authorization Act of 1970, limiting Department of Defense support to basic research "with a direct and apparent relationship to a specific military function or operation."

a year. To those people that's just chicken feed. That's the other thing that made the physicists and chemists mad. How could this be allowed to go on when it costs so little to support the math work? Early on, when we set up the David Committee, I was in the office of the Undersecretary of Defense for Research and Engineering. That's the big-wheel scientific guy in the Department of Defense. He was a wonderful man named Dick DeLauer, who had been executive vice-president of TRW. I was in his office the day after he was confirmed by the Senate, and I talked to him about this funding problem. Finally he asked me, "How much money are we talking about here?"

I said it's a bit less than a hundred million a year.

He looked at me, utterly baffled. "We don't *deal* in sums of money that small!"

"I'll tell you what," he said. "I'll give you a B-1 bomber and you can invest it and live off the interest. Or no, better yet," he said, "go to the Exxon Corporation. If you look at the published annual budget of the Exxon Corporation, the one—the unit—in the budget is a hundred million dollars a year. So you would be below the discernible level! Less than one!"

So was he telling you, "Get out of my office, don't bother me with these details"?

No, no. Because he also said, "Look, if the DOD needs to do something, write the memos for me, and I will sign them." Dick DeLauer was a wonderful man. Smart as a whip, and very supportive.

But you had to *be* there. You have to be part of what's happening, you see. So we would tell the story to people and say, this is what happened and we've got to try to fix it. Well, we did do some fixing. But the mechanisms weren't put there to try to keep up the good work. What people don't understand is that your community has to be present in Washington, and you have to know—very well—people at very high levels. And they have to *trust* you. They have to trust *you*. So I did that job for years.

So what actually happens in Washington? What do you actually do?

What you do is be in the right place at the right time when key decisions are made. And so a great deal of the time you don't very much at all! But if, for instance, some new program starts up, it's going to require new research funding. You've got to make sure that your field gets its proper share of it. Now, they will all *tell* you that it's going to get its proper share, and they'll argue and talk this publicly and so on. But at the upper level, when the final decisions are really made about how to hand out the money—if you're not there, you'll get pushed out.

In a couple of cases I know, I wasn't in the room; but I went to a very high-level person in the government and told him what happened. I said, "Fix this!"

He said, "Okay!" and they just go fix it.

But mathematicians don't like this kind of work, you see? It's too political. It's the kind of work in which certain people in the government have to know not only that they can trust you; they have to know they can *count* on you, see? *Count* on you. And that means you have to do things to *help* them with one thing or another.

Count on you for what? They're not going to come to you with equations to solve.

Well, for example, if you have a lot to do with the Department of Defense because your field is quite heavily dependent on it, which mathematics was, then you can tell the White House what's going on in the Department of Defense. Because they don't know! Things like that.

So it really is Washington politics, leaks and relationships and all that stuff.

Yes, exactly. Just as your department has to have people who are part of MIT politics, your field—your discipline—has to be part of Washington politics.

Last question. You've worked on math or for math at almost every conceivable level: student, teacher, academic, administrator, Washington lobbyist. You have served at state levels and national levels. What do you see as the primary engine of mathematical innovation in this country? Is it graduate students going for PhDs? The junior faculty gunning for tenure? The senior faculty working free of the pressures in the marketplace? The department heads with their eyes on the deanship? The members of government committees?

The mathematicians themselves, and the fact that their subject is inherently important. When I was in Washington we got the Congress to have Mathematics Awareness Week, which tickled the funny bone of a lot of people in Washington. The proclamation was written up by the White House. Now, what are they going to say about mathematics? I read only the opening clause. It said "Since its beginnings in ancient Mesopotamia seven thousand years ago..."

I said, "We're in!" [Laughs.]

Because the White House knew that mathematics is extremely important and always has been! It's been around a long time, and it's going to be. I just think what drives it is the curiosity of the people who pursue it. In the end, you see, it's going to turn out that mathematics *is* reality. You watch the way particle physics goes, or theoretical physics. What's reality, huh? Steven Weinberg, the great particle physicist, once said it may turn out that in the end the abstract patterns of mathematics are the only reality there is. Now, I don't want to make a big fuss about that, because I don't know enough. But I do have the feeling that mathematicians *know* that. They know a contradiction in the logical structure of mathematics could be found tomorrow; nevertheless, they believe that it's the best truth we have going, the truest that any discipline can be. Understanding mathematics is the driving force in understanding everything about science, technology, and so on. It's fundamental.

~ March 21 and 26 and June 25, 2007

Editor's note: Kenneth Hoffman died suddenly at the age of 77 as this book was going to press. Hoffman contributed many hours of conversation and correspondence to this project. He was an excellent storyteller, immensely proud of MIT and its history, and devoted to mathematics at every level. The mathematical community as a whole, and this project in particular, have lost a generous and enthusiastic friend and advocate.

Alar Toomre

From Estonia to Long Island

J.S.: *Before World War II, most of the world's prominent mathematicians were European born and trained. Your own generation was the first in which that center of gravity shifted to the United States. You and Michael Artin are in fact among the few MIT mathematicians of your generation who were born in Europe, Artin in Hamburg and you in Rakvere, Estonia, though both of you moved here as children. Do you have memories from Rakvere?*

A.T.: Not of that little town, because I think one doesn't remember two- or three-year-old things. But I certainly remember our apartment in Tallinn, the Estonian capital, which we left in a hurry in September of '44, actually for Hitler's Germany. Stalin's army was approaching, so we had a choice between the devil and the deep blue sea. My father, who was quite enterprising, picked the deep blue sea. He had figured out that going to a small village about 50 miles west of Leipzig in Germany was pretty safe in the wartime sense. Going to a city would have been ridiculous.

My father had been a lawyer before the war. From that point on, he was just scrapping, one job after another. Maybe he worked for the buses and trams administration in Tallinn during the wartime, something like that. My uncle, my father's brother, younger by two years, actually had gotten drafted into the Red Army, when the Soviets were in there in '40–'41. Because remember, when Hitler and Stalin split up Poland between them, the Russians moved into the Baltics. And then, of course, they had a prompt election: it was 99.6 percent voting to join the Soviet Union [laughs]. Well, very soon the Russians deported some ten percent or more of the educated types. My father was very lucky not to be deported. My uncle, I think, was a little Socialist in his leaning, so he probably saw which side the thing was painted on and said, oh, joining the Red Army is probably safer than joining Siberia. It wasn't as hideous as the German depredations, because there weren't too many killings, but they were obviously going into very rough circumstances, the gulags and so on. I think maybe 70 or 80 percent of the people eventually trickled back, ten years later, after the war. But, you know: not a nice way to greet your new satellite republic [laughs].

My American wife and I are going to Tallinn this August for my third time back, her second time ever. I've been reading guide books again, and I'm just reminded: there was one severe bombing by the Russians in March '44. It scared the hell out of this little kid of course, especially since several places nearby were burning. But we were survivors. We survived that, and we also survived an attack on our ship afterwards.

You were sailing from where to where?

From Tallinn to Danzig—Gdansk—on some crummy German freighter loaded with refugees, maybe a five thousand–ton freighter. And it was attacked; maybe it wasn't torpedoes but bombs, anyway, which fortunately missed, but a couple of other ships got sunk, and there were two thousand deaths each time. Not a healthy time. But obviously we were not as badly off as the Jews, because the Germans had the silly idea that the further north they went the more Aryan it became, so therefore we were reasonably safe in that sense.

How did the war end for you?

As I said, we ended up in a little village near Leipzig, which was going to be East Germany. Of course, we didn't know it then. The American troops finally came pouring in, tanks and candy and whatever else, in mid-April of '45, roughly I would say within a week of when FDR died. The war was almost over. I saw my first black man ever, an American soldier. The Americans were quite nice to us. As a matter of policy the Baltic countries were singled out as having been unfairly absorbed into the "Great Union." Thus the Americans soon moved us some distance south to towns like Erlangen and Würzburg in Germany, which were to be in the American zone.

I went to school with German kids. My parents were sensible enough to send me there rather than go to camp school where I'd learn more about the glories of Estonia. We lived five years in Germany, but in 1949, finally, the Immigration Act in this country opened up the doors enough to let displaced persons like us come in. There's a funny story with that: We're nominally Lutheran, although thoroughly nonpracticing in my case. But we had the help of the Catholic Relief Organization to find a sponsor for my father, somebody who'd guarantee him a job for a year. And so these Lutherans, helped by the Catholics, found a Jehovah's Witness in Somerset, Ohio, having a small farm, who kindly employed my father for the year. It's only in America, this kind of mish-mash of religions [laughs].

Eight months later, we got the hell out of there. It was a pleasant enough place in the country, but no prospects or anything. So we ended up, as so many immigrants do, where there are others of your own kind. Turned out there were plenty of Estonians in the New York area, where the going was reasonably good, so we moved to Long Island, about ten miles east from Queens. I started high school there and within a year met my future wife—I was more visible than she was, in the sense that I was an exotic foreigner.

How was your English when you came here?

That's a funny story. My spoken English was pretty crummy, so I got placed in a rather low grade in Ohio, because they were worried that my English wasn't too good. Then soon came a standardized test—they called it the "Every Pupil Test." I was the top kid in math; that was not surprising. But I was also the top kid in English! Because I'd already started two years of grammar in German school, and the Germans had at least gotten around to teaching us nouns and verbs and so forth. The Ohio farm kids were only getting around to it.

You were hurriedly placed in a higher grade, I presume.

Yes, yes. And it was a rapid progression from then on. Less than four years later, at sixteen and a half, something like that, I was a freshman at MIT.

Undergrad at MIT

You mentioned an early aptitude for math.

Right, but I was afraid to go into mathematics itself. As an undergraduate here at MIT, I was in aeronautical engineering and also earned a second degree in physics. Maybe five or ten percent of the students got double degrees even then. Nowadays, I think it's become quite popular: maybe fifteen or twenty percent of our students get double degrees at graduation. You see: I was already hedging my bets. I was not quite believing that aeronautical engineering was going to be enough for me, so maybe physics. I still like aviation to this day: I'm basically an aerodynamicist, even though that morphed into stellar dynamics and so on. But I guess it's the old story with immigrants: you look for jobs which are likely remunerative or at least safe. You're obviously not going to go into humanities or politics or anything else very chancy. And math struck me at the time as something that was for the really brainy ones, a few geniuses. Everyone else was to avoid it, because it seemed too theoretical and probably too much like opera: you got paid well only if you were terrific.

What did aeronautical engineering mean to you at that time? Airplanes?

Airplanes, yes. The department, soon after Sputnik, got renamed into "Aero/Astro"—Aeronautics/Astronautics—but in my time it was simply called Aeronautical Engineering. That department was quite a distinguished one. I think it was the oldest aero degree-granting department in the country. For instance, the famous General Doolittle of Tokyo-bombing fame in World War II, and who later headed the Eighth Air Force in England, had gotten a PhD in aeronautical engineering from MIT back in 1925.

What made you decide to come to MIT in particular?

That's a good question. I'm often asked why didn't I think about Princeton or Harvard or something like that, and the answer probably is the engineering side. I must have had some counselors in high school who said MIT is very famous. They also urged me to apply to RPI, which I did. I applied to Johns Hopkins; the University of Rochester was chasing me, because they chased various bright students in New York high schools. I guess I also applied to Brooklyn Poly, just as my safe school.

But, curiously enough, nobody urged me to apply to Harvard or Yale or Princeton: these were just too much like rich man's schools. As an immigrant kid, you just didn't apply to them. And nobody also had told me about Caltech, which would have been many miles away.

So four years after arriving in the United States, here you are at MIT. What was undergraduate life like for you?

I was living in East Campus—again, this is a cautious immigrant type: picks the dormitory closest to the class! And again, as a safe thing, I decided to room with a distant friend from high school, Kent Mercer. I've never seen him again, more or less. But it turned out that he was a chemistry nut, and he loved to play tricks on other people. So there I was one evening, maybe two months into the term, alone studying in our room, and some brown liquid came under the door, and it stank like a skunk! It was clearly a retaliatory raid by some of his victims.

East Campus didn't have much of a social life, but as it happened, I worked in the neighboring Walker Memorial, which was the Student Union, as student staff, clearing tables and serving in the evenings. I eventually became Head Captain there. I also was a local student politician in that East Campus dormitory. In fact I became House Chairman of the whole blooming lot: for one year, I had five hundred constituents, so to speak [laughs]. I guess I must have felt that it looked good on my resume. I wasn't particularly cut out to be a politician, but it was nice to know, "Oh, yeah, I could do it." So I was a busy, busy lad on East Campus, chasing two degrees and so on.

Your English must have improved considerably by this time.
Well, it's never gotten good because it's always got this crazy accent—accents—laid under it, some mish-mash of Estonian, German, and later even British, I guess. But I was quite facile by then.

Marshall Scholar

In 1957, in your senior year, you were chosen for a Marshall Scholarship. What were these scholarships?

These were essentially Rhodes Scholarships, but given for a much nobler purpose. Rhodes was a rich man but a real stinker: segregationist and so forth, had Rhodesia named after him.[1] Marshall Scholarships were instead given by the British government—they still are—as a "thank you" for the Marshall aid. And of course that Marshall was a much more commendable person, who was our chief general, and a very thoughtful general, in World War II. Later he was also our Secretary of State.[2] It must have been com-

[1]Cecil Rhodes (1853–1902) established De Beers, which dominated the world diamond trade, and colonized the settler-state of Rhodesia (today Zambia and Zimbabwe). Segregation and the conviction of African inferiority, while real in his time, became more rigid and formalized after his death. Rhodes's will, which declared his desire for "the extension of British rule throughout the world," established the Rhodes scholarships so students currently or formerly under British rule could study at Oxford.

[2]George C. Marshall, U.S. Army Chief of Staff (1939–1945) and Secretary of State (1947–1949).

mencement in 1947 at Harvard where he announced what became the Marshall Plan. We essentially dumped something like ten or fifteen billion dollars, which was real money then, into the European economies. And the British, by 1953, were so thankful for the fact that we didn't let them collapse that they set up this scholarship, which selected a dozen new scholars each year. Eventually there were, I think, three or four dozen annually. At the coronation of Queen Elizabeth in 1953, it is said that when Marshall walked into Westminster Abbey, the rest of the congregation stood up. Now, that was a mark of real respect, because it was not for Marshall alone but of course for the U.S., which was highly regarded then.

Marshalls, even now, fifty years later, still don't have the track record of Rhodes, because the Rhodes have a hundred years of being in the business. But Rhodes scholarships are only to Oxford, and in the sciences this is not a particular blessing: Cambridge is surely superior, broadly speaking, to Oxford in the sciences. Likewise you might prefer to go to Imperial College, London, or Manchester or Edinburgh: it's not always clear that Oxford is the number one place.

How did you hear of these scholarships? What led you to apply for them?

Leon Trilling was a wonderful advisor for me as a senior in Aero. He was a Jewish American who had gotten his PhD from Caltech maybe ten years earlier, so he was not too much older than I. He lives still in Boston, somewhere on Beacon Street. I just wrote to him, after walking back across the bridge on a sunny Monday a week ago, that my own fiftieth MIT anniversary was coming up, and I wanted to say thank you again. He replied to me nicely this morning.

Holt Ashley was the other faculty member. Holt was a big hulking guy, six foot eight: I was always impressed with anybody much taller than me, at six foot three, and Holt Ashley certainly was. He left MIT after a while and went to Stanford. He died last year.

Between those two, they gave me very good advice. They said, of course go to Caltech if you want to—which I almost did—but even more impressive would be to go to England, because they knew about Professor Lighthill.[3]

[3]Sir James Michael Lighthill (1924–1998) was a distinguished British fluid dynamicist. He was professor at Manchester and later at Cambridge, director of the Royal Aircraft Establishment, and fellow of the Royal Society, among other positions. Lighthill's papers defined two new fields in fluid mechanics, aero-acoustics and nonlinear acoustics.

So I did. Obviously, at that level, you should probably go for the person more than the place. It's a mixed blessing, of course, because the person may move, as it happened in my case.

The Tech wrote at the time that you didn't want to go to Caltech because you were anxious to start research right away.

Yes, I guess I had a reaction to the fact that I'd been working myself to death with all the courses I had to take here at MIT. I didn't want to take any more courses then. And the appealing thing about the English scheme was that you became a research student, as they called it, right away, so you didn't have to take any more courses. I was a cocky son of a gun, but I had some of this kicked out of me pretty damn soon, because soon after arriving in Manchester, I decided, "Well, I ought to go listen to some theoretical physics courses also"—particle physics and so on. Oh, no! After two weeks or a month of that I gave up in disgust, because I just couldn't handle it. I didn't even know what they were talking about!

What was James Lighthill like?

He was probably the single most brilliant guy I've been close to, in the sense of someone I've seen or talked with several times a month. James Lighthill and Freeman Dyson were classmates at Winchester College and went on to Cambridge afterwards. The war intervened, and so their careers were mixed up with working for the Royal Air Force and whatever else. But he and Freeman Dyson remained peas in the pod in a sense. Freeman Dyson of course went on to a great career as a theoretical physicist; he's been a professor at the Institute of Advanced Study in Princeton. A very bright guy, many talents, maybe a tad less mathematical than Lighthill but perhaps even more imaginative. Lighthill was surely the more flamboyant: he had a loud voice and was eccentric in various ways. But he was not just plain show. This was his personality. He had a theatrical way of talking, and my wife accused me for years afterwards of imitating him as a teacher, unconsciously or otherwise. I would do some of his tricks: loud voice, or colored chalk, or trying to startle the students with something ridiculous you say; things like that.

Lighthill died about a decade ago at the end of a swim around a small island, Sark, in Great Britain. He loved swimming, and he apparently made a point of swimming around this island every five years or so, to show that he could still do it. And probably he could in this case, but then a heart attack or something overwhelmed him near the end of this long swim. He had scared people before. He disappeared, I remember, on one astronomical trip to the French Riviera, where he went swimming and he didn't return for three hours. We wondered if the poor man had expired. No, he hadn't—he just loved to go to this next place and swam back gradually. Fortunately, there were no sharks in the Mediterranean [laughs].

So you worked closely with Lighthill, but he didn't end up being your thesis advisor.

Right, well, I went to Manchester chasing Lighthill; I didn't actually *catch* him, in the sense of him becoming my official advisor, because halfway through he got offered the directorship of the Royal Aircraft Establishment in Farnborough. So off he went, and Manchester suddenly became a lot duller, for my taste. The special kick left when Lighthill left. On the other hand, we remained lifelong friends, which was wonderful. Of course, there were people whom he had attracted there as faculty members and who were still there: it wasn't all lost. But I could have kicked myself. "Why the hell didn't I go to Cambridge instead, where there would have been a larger number of fine people?"

Who was your advisor?

Paul Owen, a decent fluid mechanics person, who did a lot of committee service with the British Aeronautical Research Council and eventually became a Fellow of the Royal Society himself. I did partly an experimental thesis, wind tunnels and so on,[4] in a little department of fluid mechanics, not strictly in math but very much under the wing of the applied mathematicians. For this, you didn't have to take graduate courses; you could get your PhD essentially for just a research thesis—which is okay, but I don't look back on it with any great pride. At twenty-three, I had my "el cheapo PhD," as I call it, but I was dumb as hell [laughs].

Manchester had a real applied math and fluid dynamics history. I read that during applied math chairman Sydney Goldstein's tenure there, the influence of Osborne Reynolds and Horace Lamb, both fluid dynamicists who were born in the 1840s, was still palpable.

Oh, you could drip with history there if you wanted to. It's true. I guess also Rutherford, in physics. Sydney Goldstein had been professor at Manchester.

[4] Toomre's thesis was titled "Some Studies of Flows with Vorticity."

Just as Barrow resigned his position to make room for Newton,[5] there's a joke that Goldstein resigned early from Manchester because it was known that Lighthill was on the way.[6] Goldstein then went to the Technion in Israel and later to Harvard.

Manchester used to be, and maybe still is, a way station for subsequently famous people on their way to Cambridge. It is sometimes said to be the most reputable of the red-brick universities. Red brick meant they were built in the 1880s, Victorian things with brick on the outside. They were essentially in the same tradition as MIT or RPI: you founded polytechnic or technical institutes, in this country and many, many other places, because these were believed to be practical and useful. The University of Manchester was almost like having a Harvard and MIT next to each other, because they had UMIST, as they called it, the University of Manchester Institute of Science and Technology, right next door. Now they've essentially merged.

Why is it that fluid dynamics was so central to applied mathematics that early? Was it because the equations were tractable, was it driven by practical applications, or was it a combination of both?

Well, certainly hyperbolic equations for high-speed flight were tricky as all hell, especially in the days when you didn't have computers and you had to do it more with paper and pencil. Transonic flight was even worse, because it was partly hyperbolic and partly elliptical. Equally clear was that, especially with the pressure of the post-war competition with the Russians, you wanted to have fast airplanes; and the military obviously was after these things. Jet propulsion and rocket propulsion were also coming. These were all tough, tough fluid mechanics cases, where mathematics of the applied kind was really essential. So you could see why a Goldstein or a Lighthill or a George Batchelor were in great demand. Or their subject, broadly speaking; and of course also theoretical meteorology and oceanography. It's not that applied mathematicians did very much of the latter, but some of it they did. Not exactly weather forecasting but certainly trying to understand the weather motions—in the days, again, before big-time computation.

---

[5] Isaac Barrow resigned the Lucasian chair at Cambridge in favor of his student, Isaac Newton, in 1669.

[6] James Lighthill (at age 26) succeeded Sydney Goldstein as the Beyer Professor of Applied Mathematics at Manchester University in 1950.

I find it a little hard to explain it even to myself, looking back. But if you think of Courant and Hilbert, what the heck were they doing? They were doing hyperbolic equations. That kind of stuff had a long usefulness. It was definitely not an artificial sort of construct. Pure mathematics may occasionally get into that trap, here or there, but it's clear that fluid mechanics had much relevance, especially during and after World War II. These days I would have to admit that the great glory days of fluid mechanics are probably over, but it sure did not seem so then.

Applied Mathematics Instructor at MIT

In 1960, at age 23, you came back to MIT. Who brought you back here?

Probably Lighthill's best student ever was Gerry Whitham, and Whitham was a professor here at MIT.[7] It must have been the January or so of my third year that I'd heard from some Manchester types that MIT was hiring applied mathematics. I remember writing very timidly to Whitham, "Is there any possibility that I might get an instructorship?" and by essentially return air mail he said, "Oh, yes, a very good chance." I suspect now, looking back, that some expensive transatlantic phone call was probably made to Lighthill, and also that my local reputation here at MIT may have been pretty good.

I came here to this department for what looked like it was going to be just a two-year transitional job, and that was going to be it. So I laughed: "What the hell am I doing in the math department?" But you see, my British experience made being formally in applied math at least plausible.

Whitham wasn't here for all that long: he went to lead the new applied mathematics program at Caltech.

Yes, Gerry was gone by '61, unfortunately. He was a wonderful guy, a lively guy. We still treasure his excellent 1974 book on wave mechanics,[8] discussing Lagrangian and Hamiltonian dynamics of interacting waves and group velocity in umpteen ways. Also, he and Lighthill had a famous paper in the mid-fifties on traffic flow, using characteristics and all that sort of stuff.[9]

---

[7] Gerald B. Whitham was a member of the MIT faculty from 1959–1962.

[8] Whitham, G. B. (1974). *Linear and nonlinear waves.* New York: Wiley.

[9] Lighthill, M. J. & Whitham, G. B. (1955). On kinematic waves. I. Flood movement in long rivers. *Proceedings of the Royal Society of London, A229*(1178), 281–316.

Lighthill, M. J. & Whitham, G. B. (1955). On kinematic waves. II. A theory of traffic flow on long crowded roads. *Proceedings of the Royal Society of London, A229* (1178), 317–345.

Toomre leaning over his desk, shaking a hanging chain whose behavior is explained by the mathematics on the blackboard at rear.

It looks trivial in retrospect, and of course by their standards was all sort of easy, but his professional life was built largely around hyperbolic equations and characteristics.

The other person whom I actually was hoping to meet here but never did—because, dammit, he also left!—was George Backus, the geophysicist. He was already famous then for his efforts with dynamo theory in conducting media. Just before I arrived, he went to La Jolla, the University of California at San Diego. Him I actually knew more about than Gerry Whitham. Gerry was highly regarded, but I thought I was going to do some geophysics at the time. I realized that I didn't want to just do aerodynamics, because that was getting to be rather well covered, it seemed. I wanted to hedge my bets, I think, looking back. I was interested in astronomical things; I was interested in geophysical things. I also realized that theoretical oceanography was not a silly thing to think about, especially with the Woods Hole [Oceanographic] Institution nearby. Even back in Manchester when I was involved with my wind tunnel experiments, I remember reading with great enthusiasm about the dynamo work of Edward Bullard and Walter M. Elsasser. Bullard went on to be a major figure in geophysics in Cambridge, in England. Elsasser was a German Jew who immigrated to America and made his reputation here in dynamo theory. And then George Backus was the young whippersnapper coming up and cleaning up the field. So I was hoping to actually work with George Backus.

Harvey Greenspan, in his interview, attributes the MIT department's loss of both George Backus and Gerry Whitham within a few years to the lack of support for applied mathematics within the department back then. You had come from a very well established applied math situation, whereas at MIT the applied math people were still very much under the thumb of the pure mathematicians.

Yes. I'm thinking of this right now, as I'm applied math chairman again, and so I'm professionally concerned with the health of things. So let me just philosophize about this problem.

American math departments, in the middle of the twentieth century, let's say from 1920 until 1940 or '50, got into this terrible trap of becoming very, very pure. American math departments generally, in the best places—not even counting MIT among the best places then—had forgotten their origins to some extent. The Brits, to their credit, had not fallen into that trap nearly as much, and so the idea that under the same roof you'd have pure mathematicians dealing with topology and algebra and so on, and also applied mathematicians dealing with fluid mechanics, maybe theoretical physics, was quite common, quite sensible. Not every place did it as well as Cambridge or Manchester, but enough places wouldn't have laughed at it.

MIT, to its great credit, said no, you must not act like that. I think our administration and some senior faculty members like Levinson and Wiener must have realized: better for the health of the department and health of the field if you make sure that something like one-quarter or one-third of it is genuinely applied. It's just *silly* to do otherwise, because for undergraduates it's wonderful to see that mathematics can be used in various ways, and then they don't think, "It's only mathematics." It opens so many doors, to see that mathematicians and users of mathematics can coexist in the same place. Even the teaching of, say, differential equations or numerical analysis or complex variables is all enriched by the fact that these are people who actually *use* stuff rather than just have some pure person do it under protest because there's this mob of students clamoring to take courses.

I think it was under this general impetus that people like Backus and Whitham and Lou Howard and Harvey Greenspan were hired. Lou was wonderful, a geophysical fluid-dynamics kind of guy. He is probably the most mathematical of us all—or was: he left us after a while. He's got a proper pure math degree: he was a Swarthmore undergraduate and went on to get a Princeton PhD with Donald Spencer.[10] I think he was a mathematician who fell more and more in love with fluids puzzles and the instabilities of parallel shear flows, that sort of stuff. So Lou was one of the rare guys who was essentially, I think, a pure mathematician at heart, because he knew a hell of a lot of it, like a Mike Artin might. Yet Lou also had more than just a protective coloration in applied: he actually knew his science as well. His best collaborations were with geophysics types and oceanographers at Woods Hole.

We also had Francis Hildebrand. He and Prescott Crout were both in the numerical analysis end of things.[11] But I think the reason that the brash newcomers like Greenspan or Whitham never paid too much attention to them was that they were essentially developers of methods, rather than scientists in their own rights.

C. C. Lin

You didn't end up working with George Backus, your initial interest, or with Gerry Whitham. Instead, the senior MIT mathematician who turned out to have the most influence on your subsequent career was C. C. Lin. He was also one of the principal architects of the applied group here. What were your first impressions of Lin?

Put it this way: he was the first Chinese scientist I knew whose English was terrific. I remember thinking that when I came here as an instructor: the one thing that stood out immediately was "My *goodness*, what a cultured and articulate guy." In those days, we did not have many Chinese or Chinese-American colleagues in academia, although at MIT there were already several good ones in electrical engineering and physics. But even compared with those guys, I claim that C. C.'s English was superior.

[10] Louis N. Howard earned his PhD from Princeton in 1953. The title of his thesis was "Constant Speed Flows." He served on the MIT math faculty for 26 years, from 1955 until he left for Florida State University in 1981.

[11] Francis Hildebrand received his PhD at MIT in 1940. Prescott Crout was his advisor.

C. C. Lin.

He got his undergraduate degree in China, at Tsinghua University.

Yes. It must have been something like '38 or '39, just before World War Two. He had the good sense to go to Caltech, where he was von Kármán's student.[12] It must have been still wartime: maybe he was not having to do war work because he was an alien, so he was safe from the draft, and he could study. He and George Carrier, who later moved to Harvard, were both faculty members of Brown University before MIT had decided to strengthen its mathematics by hiring applied types.

How did your own connection to Lin begin?

I didn't even know C. C. very well the first half year or so. He was a major figure, but to me somewhat intimidating with his expertise, so I didn't at once talk with him at any length. C. C.'s wonderful reputation was based in part on a very nice monograph, summing up his own work and that of

[12] Lin received his PhD at Caltech in 1944, the year that his advisor, Theodore von Kármán, became director of Caltech's Jet Propulsion Laboratory. Von Kármán had already been director of the Aeronautical Laboratory and founder of the U.S. Institute of Aeronautical Sciences. He was awarded the first National Medal of Science by President John F. Kennedy in 1962.

others on the hydrodynamic instabilities of parallel shear flows.[13] It showed his mathematical aptitude very well. So he had an excellent reputation as a theorist who knew about instabilities of parallel flows and boundary layers, which are very important for airplanes and so on. On the other hand, he *must* have been more broadly interested than just that. All of us have to make our push in something so we can really establish credence and say we've accomplished something, but I think he must have been eager to do something else.

By the time I showed up, he had a phase where he was saying, very much contrary to Feynman and several other noted physicists, that quantized vortices in liquid helium were probably baloney and that the motion of this helium was really explained by continuum reasoning. In retrospect, C. C. turned out to be completely wrong there [laughs]. But I remember thinking that "My gosh, here's a man who's already in his mid-forties"—an ancient, ancient man—"who's broad enough and courageous enough to take on the physicists on these quantized vortices." I also felt rather good that C. C. was paying attention to the George Backus types. George was gone, seduced away to the West Coast, but C. C. was reasonably knowledgeable about geophysics, too, I thought—for a mathematician, anyway.

So Lin had established a reputation in fluid dynamics, but he also had a side interest in astrophysics. He had taken a course with Fritz Zwicky, the famous astronomer, when he was still at Caltech, and he'd read Subrahmanyan Chandrasekhar and Martin Schwarzschild. But then apparently in 1961, shortly after you arrived, he went to a conference in Princeton, where his interest really took off.[14]

Yes. Lin was, as it turned out, something of an amateur astronomer from way back in his career. Zwicky was a wonderful nut of a physics/astronomy person back at Caltech. So C. C. probably did take a course with Zwicky. And it's absolutely true that he had gotten invited to a conference in Princeton, held at the Institute for Advanced Study in March or April of 1961. I showed up as an instructor in the fall of 1960, so six or eight months after I arrived, C. C. went off to this conference on galaxies and learned how they were full

[13] Lin, C. C. (1955). *The theory of hydrodynamic stability*. Cambridge, UK: Cambridge University Press.

[14] The proceedings for the conference, which took place at the Institute for Advanced Study, April 10–20, 1961, are reported in Woltjer, L. (Ed). (1962). *The distribution and motion of interstellar matter in galaxies*. W. A. Benjamin.

of these instability problems, ones that nobody had really tackled in a big way. That was the theme.

The good theoretical astronomers knew that this was an area worth exploring, both for the spiral structure and for various distortions that galaxies exhibit. They are wonderful objects. The famous Richard Feynman himself had some quote in one of his freshman books that if you're looking for a good problem, look at spiral structures. It was clear that they cried out for an explanation. It was like the modern photographs of hurricane cloud bands from above: they look clearly spiral, but what causes it? Nobody really worked on that very much. "It's just cloud bands!" But spirals are of course self-gravitating, and there must be some great reasons for why they look so spiral most of the time.

With plasma physics going strong and fusion research also under way, it reeked to high heaven that the kinds of gravitational plasmas which represented the stellar systems were also worthy of much further study than they had gotten so far. The poor old observer types and even Zwickys could only ruminate, but it was clear that a time was coming when theorists should have a real go with these collective instabilities. So C. C. had gotten a dose of this medicine from good people like Lodewijk Woltjer and Bengt Strömgren, and maybe Martin Schwarzschild or Lyman Spitzer, at this conference in Princeton. He came back all enthusing about this, and he promptly invited two of these big shots, Woltjer and Lüst, to come to visit MIT during the summer of '61. Lo Woltjer was a great catch. He went on to a wonderful career, including becoming the director of the European Southern Observatory, headquartered in Munich. And Reimar Lüst, in due course, went on to become the president of Max Planck Gesellschaft in Germany. So C. C. showed excellent taste by chasing down people like that so promptly and early.

I think it's interesting how success in one area opens doors into something else. C. C. was nothing in astronomy, but he was a major figure in hydrodynamics, so therefore the door was already open a crack: maybe a man of this sort might actually have something to contribute. And so he felt very much appreciated.

What other steps did Lin take to get this new enthusiasm off the ground in the department?

Partly to learn things better himself, C. C. started a reading group, or sort of internal seminar, so we could teach ourselves why the stars are called O stars, B stars—elementary stuff, but we were just doing catch-up. And certainly we read chapters of Schwarzschild's book on stellar structure. I'd actually attended some lectures in Manchester on it, so I had a bit of a head start. It was lovely. I guess it must have been in the fall of '61, the beginning of my second year here. I remember Chris Hunter was there. And in the second year our department hired Norm Lebovitz, who had just gotten his degree from Chandrasekhar in Chicago. Norm actually ended up as my office mate for my second year. We had office 2-278, right next to Norbert Wiener's. It was awesome to be so close to him. Wiener more or less looked through us, when he noticed us at all [laughs]. He was a barely grown-up child prodigy, very self-centered. But we were dealing with genius, so you can put up with that sort of curious behavior from a genius.

I wasn't in C. C.'s group in a literal sense; I was just an instructor there. But C. C., to his great credit, was very generous towards me and Chris Hunter and some of the other youngsters around at that time, encouraging instructors to learn other things and do other things than just their theses, and through his enthusiasm for this stuff made me realize, "Oh, it's okay to enthuse about galaxies!" I was somewhat disgusted with my thesis: I didn't want to do any more wind tunnel experiments and shear flows, so I was delighted to have this welcoming atmosphere. And this is exactly what I try to do even now with our instructors. We hire them for two or three years—now it's more like three years—in the hopes that they'll catch on to *something*. And of course, most stories don't turn out as well as my story with Lin: the senior professor doesn't come back enthusiastic from a conference and bring in Lüst and Woltjer to lecture to you or have a seminar series the next year about the things, just because he's himself trying to learn and therefore lets the others learn also. In that respect C. C. was just wonderful, as a senior figure who stirred the pot by going someplace and then sharing his enthusiasms. Mind you, they were only enthusiasms; they weren't any more than that, at that point.

How did the first problem that you yourself chose to work on crystallize from this initial enthusiasm?

Lo Woltjer came, it must have been June of '61, to give maybe two or three weeks of courses, and I remember listening with big ears to him telling us about the gas in the galaxy, about how thin a pancake it is: it's something like ten thousand parsecs across to the outside—three light years to the parsec, more or less—and it's flat within a hundred parsecs. Just one part in a hundred! And yet this disk is distinctly bent or warped near its outer edge. That made a big, big impression on me, and the fact that the Dutch astronomers, especially, had inferred from the Doppler velocities what the distribution of gas was, and that they already had spiral pictures. This was all old stuff by then, I guess by five years—it was done in '57–'58, something like that—but it was news to me!

After I'd heard Woltjer and Lüst, I started thinking about the instabilities of a disk. I knew about Jeans instabilities, the gravitational instabilities of a homogeneous three-dimensional medium. But I was thinking, "Well, what about a thin disk? Supposing it rotates in equilibrium, under its own gravity; what does it take to keep it stable against condensation?" C. C. hadn't told me that I should study this; it just seemed like a very reasonable thing to do. And then to my great astonishment, I discovered that this simple question had not been asked or answered very well before, by anybody. So my earliest contribution to astronomy stemmed essentially from my second year here at MIT in the mathematics department as a result of having listened to Woltjer and C. C., especially Woltjer, talk about this thin gas layer amidst our Milky Way disk of stars. I started asking, "Is it possible for a thin disk to remain smooth and stable, when everything in it is simply orbiting in concentric circles?" The answer was no: it doesn't seem possible, not if all that stuff is merely circling about a common center, as in a so-called "cold disk." To avoid serious gravitational instability or clumping, we also need a good deal of noncircular or random motion within a massive galaxy disk like ours.

And that, of course, was very relevant to what C. C. himself was trying to do, to find that a smooth distribution doesn't like to remain so for very long; it must have random velocities in it to keep it mixed up. You can't just rotate in concentric circles: that's a no-no. Unless you have very little mass, like in Saturn's rings, but our galaxy has a lot of mass in its disk.

Princeton

After your second year here as an instructor, you went to the Institute for Advanced Study, a very prestigious placement. How did that come about?

I remember going in, early in my second year, to Dick Schafer, who was the associate chairman of the department—a rather cautious soul compared with the very warm and optimistic personality of our department head, Ted Martin. "By the way, is there any chance of staying on longer at MIT?"

Schafer's answer was "Oh, we never do that." But of course the answer was that they can do it by having you go away.

I'd already done this nice stuff on the instabilities—the realization that, hey, these cold disks aren't stable and that you need random velocities and so on. It didn't explain spirals at all, but at least it said, "Wait a minute, this is very promising; gravity plays a large role." Anyway, that work had been known only locally. C. C. was the one who, when my instructorship was coming to an end, stirred the pot for me with Strömgren at the Institute for Advanced Study. So essentially all I had to do was write to Strömgren to ask, could I possibly get a position there for a year? And that's where I went to for eight months, in '62–'63.

Was there some sort of understanding that this was essentially to get around the math department rule against hiring instructors?

Well, I was unaware then of what the understandings might be. I went off to Princeton thinking that halfway through I'd really have to start looking for a regular job. But the lovely thing is that I'd hardly been in Princeton for a month or two, when I think probably C. C. or Harvey called me up and said, "Would you like to come back as an assistant professor?" It was pretty damn quick.

Part of the vision for the IAS was as a seeding ground, a place where the brightest young researchers, early in their careers, could come and work in the company of world-class scientists, then take that experience with them back out into the world. What stood out for you at Princeton?

I remember that Strömgren himself had the office which was previously occupied by Einstein: I was very impressed by that. I was also very impressed by the fact that the famous Oppenheimer was director. Oppenheimer came to one talk of mine, asked a *pro forma* question at the end, just to show that he'd been awake. I had hardly any contact with him otherwise, and he died about four years later, from throat cancer. Down the road from our apartment was Paul Dirac, the famous theoretical physicist, also visiting. So how could one not be impressed by that place? It was lovely, a green oasis. Oppenheimer was around, Dirac was around. The astronomers had the custom of having weekly lunches, Tuesdays. Bengt Strömgren was very shy and retiring, so he let Lyman Spitzer from the nearby university run them.

I remember in February '63, when I was there, Spitzer came in, very serious, and said, "There's important news from California." A "star" had been discovered with a red shift of fifteen percent.[15] This was the announcement of 3C273, which was the first quasar discovered. It looked like a star—quasi-stellar in that sense—but its red shift was terrific!

Anyway, I got to know Spitzer, and Martin Schwarzschild himself, the one whose book we had been studying barely a year before. Also Otto Struve from Yerkes Observatory was there, but he died soon afterwards, because he was an old man. It was a wonderful, formative time for me, because here was this young whippersnapper who was just working on instabilities of galaxies, and thought he contributed something, and was already mixing it up with these big-time folks. I was delighted that I could hold my own—not terribly well, but at least I wasn't laughable during my year there at Princeton.

MIT *and the Spiral Galaxy Controversy*

You came back to MIT in 1963, where you and Lin began to part ways, professionally, as you took up different positions on how spiral galaxies can be explained. With applied mathematics, unlike pure, the test of a theory is not the rigorous proof but a match with physical reality: this creates the possibility of serious disagreement as to how well a given model explains the phenomenon it attempts to describe. In your case the disagreement was within the department, so it seems worth discussing. How did it start?

I had just come back from Princeton in '63, beginning of my assistant professorship, when C. C. suggested that maybe he and I should write up this instability thing together. I thought this a little pushy on his part and also unwise for me. If I'm going to be giving my best work away under the general umbrella of Lin, then what will this do for *my* reputation? It will be thought of as Lin, Lin, Lin.

And yet, in the broader sense, it *was* Lin. Because Lin with his great enthusiasm, back from that Princeton conference, had clearly turned me on in that direction. I was enough of an amateur astronomer that I might have done it anyway, but I didn't. It was Lin who brought Woltjer, who brought in Lüst, and so on. Honestly, in any great scheme of things, we were *leaning* on each other. As the younger person, I was leaning on him wonderfully

[15] That is, an object that was not relatively close and faint, but that was speeding away from Earth at fifteen percent the speed of light and was therefore much farther (about 1.5 billion light years, more distant than anything yet seen) and brighter (thus "quasi-stellar") than was then believed possible. The discovery of quasars allowed great advances in understanding the origins and expansion of the universe.

much, because when a senior person says something is doable, he's probably right. Although there's also this famous saying that if a senior person says something is *im*possible, he's probably wrong! C. C. was very influential in encouraging me to think about these things, sending me to Princeton, hiring me back, and so forth. Conversely, my realization about the instability and the need for random motions was, I'm sure, a very helpful thing for C. C., because at least it gave hope that, yes, here might be the mechanism to get these wonderful waves, these patterns, started. Certainly his early work leaned rather heavily on my stability criterion, which stated that such-and-such a minimum random or noncircular speed was needed among disk stars of a given total mass and rate of rotation to avoid at least the axisymmetric sorts of gravitational instability. He was, I think, very sincere in recommending me to Strömgren.

Around that time, C. C. may well have felt that instabilities of some sort are needed to begin with, but then they may saturate: they may stabilize at some final amplitude. Again, like hurricane cloud bands: clearly, the atmosphere is unstable, and yet when you look at it from above, it looks like a whirlpool. Well, C. C. went after the galactic whirlpools. He was hoping—quite plausibly, if somewhat optimistically—that the end result of his or his group's efforts would be to show that the spirals are everlasting, or long-lasting, at any rate. They're not just instabilities, he must have hoped, but instabilities which somehow develop to a finite amplitude and then stay that way and rotate rigidly more or less ever after. This was the basis of his so-called quasi-steady spiral theory. C. C. got sucked into this hope rather early in the game, but I didn't believe it was particularly likely that most spiral shapes were long-lasting—rather than ever-changing, like a kaleidoscope image. So I slightly stiff-armed him, as politely as I could. I said, "No, let's keep the stories apart"—my story, with the need to avoid gross instability, versus his story, with the long-lived structures based possibly on much subtler spiral instabilities. I think that was probably a parting point for us, because it was clear that he wanted to do his spiral waves. So I guess the friction was simply, I think, a disappointment on his part that I wasn't pulled into his enterprise more or wasn't more immediately helpful to him in terms of fulfilling his hopes. And the answer is that I might have been, if those hopes had been more substantial.

I found a good deal of information on how all this unfolded in I. I. Pasha's account, "Density-Wave Spiral Theories in the 1960s," published on a Russian site called Astronet.[16]

[16] I. I. Pasha. Density-wave spiral theories in the 1960s, arXiv astro-ph/0406142 and 0406143 or http://www.astronet.ru/db/msg/1183369/eindex.html.

Yes, that's a very good historical review by Pasha, quite accurate. He's Russian, in his late forties. He was Alex Fridman's PhD student in the early eighties in Moscow and Irkutsk, but in the employment situation in Russia, which is now miserable, he sort of washed out of astronomy and plasma physics. He doesn't have a regular faculty position, so he maintains himself I think mainly by tutoring students in Moscow for university entry, that sort of stuff. On the other hand, he has a great interest in the history of astronomy, and he also remains a *keen* photographer. He went to the Caucasus for the recent solar eclipse. He has some beautiful pictures from Mount Elbrus, the tallest mountain in the Caucasus, which had that eclipse come right over it.

I originally corresponded with Fridman and Pasha back in the eighties briefly and then forgot it. Pasha lives in Moscow, and he contacted me back in 2000 or so, saying he was trying to write the history of the density-wave theories in the 1960s. He interviewed everyone in sight he could get hold of, and two quite nice and very detailed papers came out of it. I turned out to be probably his most prolific correspondent, in the sense that I opened up my old files and dug up old papers and old letters and so forth. He's actually coming to my house here in just three days. In the fall of 2000, I think, he also came to visit. It was his first time in America. I introduced him, by prearrangement, to Lin, and he had a good hour or two alone with Lin to talk.

In a nutshell, then, how would you describe your respective positions on the spiral structures of galaxies? Where do you disagree with Lin?

C. C. fell too much in love with the idea that one could demonstrate long-lived perfect structures other than the so-called bars near the centers of galaxies—which he has not; and nobody else has either. My eventual emphasis was rather that, no, it's much easier to demonstrate that an orbiting lump would cause waves: also, a galaxy going past will cause big waves. Even in retrospect, there's a lot of truth to those claims. For instance, I was one of the first people to show, with Julian back in '66, that if you do have a lump, you can have tremendous responsiveness.[17] The galaxy as a whole—the disk—responds very strongly to having such a little "ship" traveling within it. It develops a *two-sided* wake, because this "ship" is almost at rest, although in orbit, but the "water" flows past one way on the inside and the other way on the outside. So it's a shear flow. These gravitating shear flows

[17] Julian, W. H. & Toomre, A. (1966). Non-axisymmetric responses of differentially rotating disks of stars. *Journal of Astrophysics, 146*, 810–830.

are very responsive in a spiral fashion: in that sense, we do have a unifying principle that massive shearing disks are very responsive and tend to produce spiral structure, at least for a while.

Is there a general consensus now about spiral arms? Where does it stand?

Alas, it's a *long* story. There's a lot to spiral structure that is difficult, even to this day. There's no question that much of the spiral stuff in fact involves density waves. That is to say, they're vibrations, like waves on the water. The big issue still is: are some of them really long-lived single patterns? Or do they all come and go? There, I'm afraid, the jury is still very much out. The joke is that the most beautiful ones, the ones which C. C. was hoping and dreaming about—reasonably enough—were almost certainly made from the outside, relatively recently. The Whirlpool Galaxy, M51, as it's called, is so striking. I would think it 95 percent certain that its neighbor did it: there's indeed a neighbor galaxy behind it that has incriminating tidal streamers coming from it. The famous Zwicky had already pictures of those streamers. My brother and I[18] in '72 simulated that interplay; there was no question that an encounter was involved. It seemed likely—although the proof is *still* not quite there, even to my mind—that one can reconstruct the whole tidal experience. But it looks very probable, twenty to one odds, anyway, that the density wave there is a *transient* density wave. It's one hell of a good density wave, but it's not the long-lived one that C. C. was hoping for.

Thus my only real quarrel with C. C. is that he should have warned the astronomers early on, saying, "I'm not really *sure* if they're long lived." They probably are waves; there's no question about that. On that we're all pretty much agreed. But are any of them really *long*-lived? This question has still not been settled. As I said, his best example, M51, one that he and Frank Shu were claiming that they had explained, turns out not to have been explained by them at all. Very likely, as I said, that damn thing was stirred up by the passage of the neighbor galaxy only a hundred million years ago or so. That means it's one percent of time ago since the Big Bang. You want to call this long-lived? No. Because wait twice as long—we can't, of course, but if you *could* wait twice as long, or three times as long—it would essentially be a pretty dull galaxy again, or at least so I continue to suspect.

You see lots of fragmentary arcs; most spirals are rather ragged, like hurricanes are mostly ragged. You hardly ever see clear, two-armed hurri-

---

[18] Alar and Juri Toomre, in their joint paper, discussed further below.

canes, right? You think you do, because the human mind tends to simplify matters, but look at it: their spirals are three-and-a-half arms or something like that. Messy spirals are considerably more common in the sky than picture-book perfect ones. The jury's still out on *exactly* how you make them. Is it from bombardment by clumps of unseen dark matter? This idea is not as ridiculous as it sounds. Is it the instabilities of my old kind in the gas disks, which create lumps, and they in turn create wakes? Probably some truth to that, too. Or is it some new and trickier disk instability? It's just one of these messy situations that is not clear to this day. But it is almost certainly not a superposition of the hoped-for permanent waves that Lin had in mind.

So here you are, with these very different viewpoints, roaming the same halls every day. Did your disagreement become personal in any way? Did it affect your relationship at all?

Not especially. I winced when C. C. said some things, and C. C. must have winced when I said other things. But no, no, we were politeness itself, as things go. C. C. and I became coprincipal investigators on NSF grants—this remained true for about fifteen years until he more or less retired in the mid-eighties.

C. C. and his wife are very graceful: just the right kind of behavior that you'd expect from a decent sub-department head, group head. I feel very much indebted to C. C. in many ways. As far as the outside world was concerned, although all the experts knew that there was a lot of scientific disagreement between Lin and Toomre, nonetheless it was a fight within the family. We both believe that gravity does it. And we both continue to believe that each of us had a lot to do with explaining it [laughs]. So it's a friendly friction, let's put it this way.

Galactic Bridges and Tails

In 1972 you and your brother, Juri Toomre, published "Galactic Bridges and Tails,"[19] *which a later article in* Natural History *magazine called a "monumental paper" that "provided a comprehensive new way of looking at the data on galaxies that had accumulated over the course of the century." Scientific collaborations between brothers must surely be relatively rare in modern times. But Juri had really followed in your footsteps to an amazing degree.*

[19] Toomre, A. & Toomre, J. Galactic tails and bridges. (1972). *Journal of Astrophysics, 178*, 623–666.

Yes, in a very flattering way. My brother is younger—three and a half years literally younger and maybe five years academically. I went to MIT; he came to MIT. I went to aero; he went to aero. I lived in East Campus; he lived in East Campus. I worked for the student staff and became the head guy; he became the same thing. I was student politician in East Campus; he was student politician! And then, lastly, I got myself a Marshall scholarship to England; he got himself a Marshall scholarship. But there he did me one better. He realized there was no Lighthill to chase in academia, so he just went to Cambridge. And of course he was happy ever after. He came back to the country in 1967 or so, to New York, NYU and NASA's Goddard Institute, and he was saying, "As brothers we should do something together!" I remember at that time I was thinking, maybe we should do something about the motion of the earth's core. We tried a little bit, but it wasn't going anywhere.

You ended up writing about the mergers of spiral galaxies in a process now known as the Toomre sequence. How did this come about?

I'd then been at MIT '63 to '69, and in the seventh year, you get a sabbatical. So I decided it was time to go to Caltech, which astronomically

Alar Toomre with his children in a publicity photo for a children's lecture, "Twenty Jumps from Here to the Galaxies," 1973.

was a very big Mecca, even more so than Princeton, in this country. I got myself a Guggenheim—partly, again, with C. C.'s encouragement: even if we had friction, I'm sure he wrote for me. So MIT paid for half of it, the Guggenheim paid for the other half.

I had an excellent experience at Caltech. It was essentially during that time that I had my eyes opened—not by Zwicky, who was still there, though I didn't see very much of him; but by Chip Arp of the peculiar galaxies book. Arp, from 1966, had this whole atlas of pictures.[20] He may have been a little silly in asserting that these weird-looking galaxies implied unusual forces or explosions, but his plate collection was wonderful. Just like C. C. had learned from his Princeton visit that spiral structures were well worth investigating, I realized from Caltech that the peculiar galaxies are quite numerous, the funny-looking galaxies. It was this in 1969–'70 that opened my eyes to the realization: We have a growth industry here. Let's do some more with colliding galaxies.

I had already realized, at MIT before I even left, that you could make some *amazing* shapes by just flying by. I had then found a nice, long-armed spiral from a fly-by, more or less what appears in the 1972 paper of ours. So I was ripe in that sense when I met Arp at Caltech. In addition, Juri was again asking long-distance, "Well, should we do something?" He had good computer connections there at NASA's [Goddard] Institute for Space Studies near Columbia. So I said, "Yes, let's do something. Let's work on these colliding galaxies." And that's what we did.

Is working with your brother different than working with another academic colleague?

It's different, yes. I guess we're both perfectionists, but brothers can get into horrible fights and yet make up for it repeatedly. The actual paper-writing and drawing duties were three-to-one my burden: I wanted to do it, because I was prouder of this and because galaxies are more my bag than his. But we

[20] Arp, H. (1966). Atlas of peculiar galaxies. *Astrophysical Journal Supplement,* 14, 1–20. In *The Arp atlas of peculiar galaxies* (2006, Richmond, VA: Willman-Bell), published to commemorate the original's 40th anniversary, the authors' note by Jeff Kanipe and Dennis Webb explains the importance of Arp's work. "Few of the galaxies in Arp's *Atlas* corresponded to the symmetrical forms specified in the Hubble sequence of spirals and ellipticals, the most prevalent galaxy classification scheme in use at the time.... [P]rofessional astronomers...were accustomed to ignoring the occasional pathological galaxy. The sight of so many variations in one place, however, was startling. Obviously, quite a few galaxies could not be facilely placed into the elliptical- or spiral-shaped bins that astronomers had fashioned for most of the galaxies then known."

got into these horrible fights. "You lazy bum, you haven't written anything on this paragraph, this chapter or section or whatever, for a whole fortnight now." And yet we pulled together, because we had decided this was kind of a nice thing to do. It was a lovely collaboration. And very influential, fortunately.

How do you explain that?

We did a number of things right, let's put it this way. First thing, we were very careful not to over-claim priority. Just two weeks ago, some German science writer, writing about interacting galaxies, wrote to me: "Weren't you and your brother the first to do that?" And I promptly replied to him, "Oh, no. We were not the first at all. In fact, since you're writing in Germany, you might as well mention that there was this rather nice German work: even in our own paper, we pointed out that Pfleiderer and Siedentopf had done something very relevant already in 1961 and '63."

One section of your paper cites "the ingenious and still astonishingly relevant optical-numerical N-body computations of Holmberg" from 1941. This was before computers were in use, of course, and I read that he did a computer simulation with light bulbs.

Oh, it's a charming story. Erik Holmberg was a Swedish astronomer, I guess maybe thirty years older than myself, but he's been dead now for the last ten or fifteen years anyway. He was an observational astronomer from Uppsala in Sweden—not a really famous observer, because Swedish telescopes were nothing like California telescopes, but very sensible and thoughtful. Back in 1941, he had the brilliant, nutty idea to actually play a game on a table, so to speak, with the best standard light bulbs he could get from a Swedish light bulb company. He went on to use photoelectric cells to add up the forces that a bunch of equal masses exert on each other, imitating the inverse-square gravity of stars in one or two galaxies by the inverse-square dropoff of brightness of light with distance! He commanded his army from the readings made by photometers. Photometer looked this way, photometer looked that way, so he got the two vector components of the force at each point. He would read out these things probably to one percent accuracy, which is kind of crappy; and then, from these readings, he would move his seventy-five or so particles and also keep on changing their velocities. When you look at it from our era of fast computers, it was a mildly ridiculous analog computation, about as ludicrous a way as you could do it. Because even in those days, you probably would have been better off if you just used an adding machine to add up the forces. But he had this charming idea to use light bulbs to add up these forces in analog fashion.

If Holmberg's work had been at all accurate, he would have discovered the fierce instabilities of disks. If you were doing it on a PC these days, you would find this disk of fewer than a hundred particles would be unstable in just one revolution, essentially due to the instabilities which I was predicting theoretically in 1961. But doing it by hand, and probably dismissing any deviating tendencies as mere numerical errors, these disks of particles would not have seemed horrendously unstable. So Holmberg never saw them go bananas, as I said, because he was just moving them smoothly with his computational hand. Nonetheless, it was done very early, and it had good enough accuracy that he got some two-armed shapes out of it: from passing another galaxy by, he found that, yes, he could get this vaguely two armed-reaction. If you irritate what is originally a circular ring by a passing body, then at intermediate times afterwards, the ring is more and more distorted into this shape.

We have some elementary examples like this in our 1972 paper. But Holmberg had some as well, correctly so. So that was encouraging, and it is a charming story indeed that he used those light bulbs [laughs].[21] Anyway, he published this paper in 1941, and my brother and I are proud of the fact that we cited him. We were careful to cite all our predecessors. As calculations go, there were plenty of calculations before us. But as insight goes, our explanations of these wonderful tails were distinctly the best. By the time we got done, I think we were much more convincing about those tails than anyone else had been before us.

Judging by reactions I've read, the paper seems unusually strong, not only in what you say but in how you say it, as an example of scientific writing. Can you describe some of the considerations that went into writing it?

---

[21]For a detailed explanation of Holmberg's contributions, see Rood, H. J. (1987). The remarkable extragalactic research of Erik Holmberg. *Astronomical Society of the Pacific, Publications*, 99, 921–951. The author, then at the School of Natural Sciences at the Institute for Advanced Study, wrote, "As my own research capabilities continually developed, I have come to realize ever more fully that the caliber of my own work could have been significantly enhanced if in graduate school I had only been aware of and studied Holmberg's works."

The paper is damn well illustrated. Just as any good movie should leave a lot of footage on the cutting room floor, we had *tons* of trial diagrams, pictures and simulations and "try this, try that," which we essentially junked. And we deliberately didn't put any equations in there because there was no particular reason to show off in that trivial way. We had a good story, and we told it.

But what really made that paper was that we had some interesting conjectures at the end, what we called "Broader Issues." First we said, look, it seems to be that galaxies can induce trouble, not because they're orbiting in circles, but because they've essentially been in long orbits and fallen back. That's absolutely true.

Secondly, we said that because several of the interacting galaxies seem to have two tails sticking out of what appears to be a single pile, it looks to us like two galaxies are actually merging as a result of this process. Could it be that elliptical galaxies are made this way? Absolutely true.

Thirdly, we said that often these interacting galaxies seem to have very peculiar star formations near their middles. Could it be that the gas got dumped into the middles and formed lots of new stars there? Absolutely true.

And lastly, we asked, could it be true that some of the most exceptional spirals, like M51 or NGC 7753, and even M81, are tidally made? Probably true.

So we had a good piece of work there, which leaned on other people and was careful about praising them, did its own drawings and explored lots of situations and viewing angles, and then had conjectures of that caliber, all of which were only speculations at the time, but which turned out to be outrageously successful.

When you say these conjectures have turned out to be true: isn't it hard to authenticate them? The time scale is so vast. How are these things established?

Ah, that's always a very good question, because how do you establish something when you can only take a snapshot? The answer is, there's probably safety in numbers. You catch different things in different phases: these days, the Hubble Space Telescope—among others—often manages to photograph galaxies at distances so vast that they were only one-half or one-third as old, when that light left them, as the age of our own galaxy now. So you can look back in time, and you find that these interactions are more numerous when the universe is half as old. This has to be all part of the bigger picture.

Little by little, the evidence has come pouring in. For instance: we didn't know, and we couldn't dare to claim at the time, why galaxies seem to decelerate markedly when they fly past or through each other, or how a galaxy could then almost come to a screeching halt. We didn't know, back in '72, that galaxies are embedded in dark halos. When our paper was written, dark halos were barely dreamt about, and we didn't even dare to mention them. And that was beside the point for us. Our model galaxies were artificial little constructs: just two disks of test particles, each surrounding a point-like central mass. So our model galaxies themselves could not by *definition* slow down and merge. And yet it looked as if they were doing so, somehow.

But the main thing we left out, which turned out to be wonderfully true for other reasons, are the dark halos.[22] Because they're big things, and also very massive. You see, we were trying to make two galaxies—disk here, disk there—come grazing past each other to make the wonderful shapes that they do. But if each disk is near the middle of an unseen, massive puffball about ten times larger in radius, then the puffballs themselves go whizzing through each other. The galaxy, the visible part of the things, represents only a small fraction of the total mass. So it's primarily the puffballs, these big spherical things, that do the bulk of the braking and merging. Thus it has become dead clear now, from many observations and much experimental evidence, that mergers occur all over the place.

Did you suspect the existence of these dark matter halos at that time?

No. We didn't. But on the other hand, we were looking at the evidence, so that the galaxies *looked* like they're merging. We dared to say orbital decay, but that was just a cautious way to stress that good brakes were needed. We also did our numbers: we said that at the moment we see this many galaxies in the most violent phases of their tail making. How long will it last? Not too long, maybe five percent of time. Therefore, if things were going on at the present rate, there must be twenty times more relics than the ones that are currently obvious, thanks to the tails. What could the relics possibly be? They could be the ellipticals. The numbers there were about right. We said this all in the old paper. As a simple insight, that was damn good for the time. We certainly hit a home run there.

[22] These dark matter halos refer to the need—if current gravitational theories are correct—for a great deal of nonvisible mass to account for spiral galaxies' unexpectedly high rotational velocity far from its visible center and also for the large-scale arrangement of the galaxies themselves into various clusters and voids in our expanding universe.

The other thing—one more brag—was this third little section, which we called "Stoking the Furnace." You know, if you have a small fire in a furnace, you can stoke the coals, stir them up and make the fire a lot bigger, especially if you toss in some fresh fuel at the same time. Juri and I thought this was a wonderful title for the little section, where we simply pointed out that we had not forgotten that here we were dealing not only with existing stars, but also with the interstellar gas. Interstellar gas would likely end up in the middles and cause prolific star formation, or so we conjectured. Well, as luck had it, just ten, twelve years later came along an infrared satellite: IRAS, infrared astronomical satellite, a Dutch-American thing; NASA put it up. And they discovered there were some galaxies glowing in infrared radiation a hundred times more strongly than they were in light. That is to say, ninety-nine percent of the energy of those galaxies was coming out in heat.

It soon turned out that these IRAS galaxies had to involve lots of bright new stars deep inside a pile of gas and dust so dense that no ordinary light could get out very easily, only the heat from that intense pile of stars warming up the place. Then, when people got around to imaging them, just a year or two later, practically all of these things, which had been picked up by their infrared glow, turned out to be interacting piles, interacting galaxies, each with incriminating tails of one sort or another. Merger relics. Ooh—now *that* made it much more interesting! Just as we had no knowledge that dark halos would turn out to be so abundant and useful, we also had no foreknowledge of this very pleasant development about the IRAS galaxies, or ULIRGs, ultra-luminous infrared galaxies, as people soon called them. Horrible acronyms. You asked me, what's the evidence? Well, those are two items. They were things we really did not foresee, but which turned out to provide wonderful support.

I want to touch on the background to these advances in theoretical astrophysics: the changes in observational science and computational technology that made the theoretical advances possible. My understanding is that after Chandrasekhar's Principles of Stellar Dynamics,[23] *which gave the mathematical underpinnings to establish stellar dynamics as a branch of classical dynamics, astrophysics was relatively quiet until plasma physics and radio astronomy came along. Number-crunching, while in its infancy by today's standards, was starting to make analysis of all the data that resulted more feasible. So the time was ripe for a blossoming of the more theoretical side, the explanations of what astronomers were seeing.*

---

23 Chandrasekhar, S. (1942, 2005). *Principles of stellar dynamics.* New York: Dover.

Yes. Every one of us is a product of our age; you can only do what others have made possible, or what's becoming possible. As I said, a good physicist like Feynman could smell from miles away that one should send bright young people to work on the structure of galaxies. It's just fascinating. What the hell causes them? Feynman was not too far from Zwicky, and Zwicky had a pile of examples an arm long, already by the fifties, of weird, weird things out there. They turned out to be gravitationally easy, but they sure looked weird at first. And it's clear that this cried out for explanation, or study—for data mining.

Likewise, it's absolutely true that, in the late fifties, success in the plasma physics was just coming and was going to be very beneficial, since collisionless plasmas are rather similar to stellar systems. Computational stuff was coming, and also people with time on their hands were coming. World War II was over, and you weren't sending out soldiers or other war workers any more. Europe was recovering, and people in the States were building big departments. Clearly the time was ripe for a lot of this stuff. Computation and imaging also overtook oceanography, and atmospheric sciences benefited a hell of a lot. Planetary exploration benefited hugely from rocketry. So lots of things became possible.

You said, "computational stuff was coming." Your work in the 1960s started roughly the same time as the math department was putting in a computer lab for the first time. How did your own use of computers progress?

We had a room across the hall from my later office with eight to ten desk calculators, even before I arrived there, thanks to Hildebrand and Crout and people like that. Marchants and Monroes, electrically driven things. Some of them even took square roots, would you believe. But you couldn't do any sines and cosines; for those, one had to go looking in the tables.

My first use of an electronic computer was actually '62, right here. IBM had given MIT a computer: I remember submitting decks of cards. A little bit of that showed up in my 1964 paper on the axisymmetric instabilities of a disk.[24] By the late 1960s, the terminals were coming into existence: in the same room which had all these calculators, there were probably two terminals connected to our main frame, the Compatible Time-Sharing System, CTSS. MIT, after all, benefited from the fact that we had top-notch com-

[24] Toomre, A. On the gravitational stability of a disk of stars. (1964). *Journal of Astrophysics, 139*, 1217.

puter scientists already at that time, and they were absolutely clamoring to do something with time-sharing.

Nineteen sixty-six—from Woods Hole, not yet MIT itself—was my first use of the remote access rather than cards, using an electric teletypewriter, which connected to a distant computer. Back at MIT, during the very same week when the Russians were invading Czechoslovakia in '68, I recall that I was making plots from the data—not computer plotting, but from data coming in over similar wires—into diagrams for my group velocity paper, which was published in 1969.[25]

My finding in that '69 paper that one can obtain a fine reaction from the close passage of the LMC, Large Magellanic Cloud, was likewise done here at MIT, essentially by calculating with a hundred bodies and plotting manually where the hell they had gotten to. It was not computer graphics exactly, but if you were patient and you were eager, and you wanted to see where a hundred bodies went with a computation in 1969, you sure could do it.

By the time I was back in MIT from my Caltech experience, all throughout the seventies, I was having better and better connections to MIT's main frames through teletypewriters, remotely connected terminals—later on, even with XY plotters with colored pens—in Harvey Greenspan's lab. Harvey was very nice and let me have a corner in there.

Irving Segal and the Red-Shift Controversy

Astrophysics seems like a controversial subject, as evidenced by a later controversy involving another department member, Irving Segal.

Yes. For me, it wasn't much of a controversy, honestly. He was controversial, but I personally haven't had much controversy with him. Segal was a very bright man. He was somebody *clearly* deserving of an MIT position. In Chicago, when Irving Segal was one of the graduate students, the famous von Neumann is supposed to have said, "There's the brightest young man that I know."

Wow.

Wow. Yes, indeed. So Irving Segal was fine in his own business—I think analysis and group theory was his main business. On the other hand, he was

---

[25] Toomre, A. Group velocity of spiral waves in galactic disks. (1969). *Journal of Astrophysics, 155*, 747.

rather a headstrong guy, in many ways. He apparently told his pure colleagues over and over exactly what they should do and whom they should hire and so forth, but certainly Irving was excellent in his own business. Then he, too, like C. C., did the commendable thing of trying to apply his knowledge or his enthusiasm to do something else. He tried to apply it to cosmology, and he had this very unusual theory that predicted that there should not be a linear-law red shift—there should be a quadratic-law red shift instead. I forget what he called it. It was some fancy topological stuff, which was probably quite impressive in its own right. But it, unfortunately, ran smack against what is probably the number one cosmological observation of the twentieth century, that there seems to be a red shift that increases linearly with distance. In the last twenty years of his life he was constantly pushing his cosmology.

Being a very bright man, he learned the weaknesses of the cosmologists' own thinking, and he was having one fight after another with noted observers like Allan Sandage and Gérard de Vaucouleurs who were claiming to measure the Hubble constant[26] to remarkable accuracy, although their separate estimates kept on differing by almost a factor of two. But they were just measuring the linearity. Segal would say no, no, they're faking the data, they're fooling themselves, it's really quadratic. He got to be an embarrassment after a while. But he was a member of the National Academy of Sciences, very highly reputed as a mathematician, so you could not disregard him completely. Segal was an interesting character, but nearly every department has one.

Some controversy erupted within the department about a course that he wanted to give.

Yes. It must have been roughly in 1990 when he suddenly proposed to give a summer course on his kind of cosmology. And it did raise some eyebrows and hackles of various people, including those of our department head, David Benney: why is this being peddled now to the outside world? If you want to be silly about yourself to our own students—go ahead, was the attitude. But this was being advertised as a summer course to the wider public, and that drew some big *harrumphs* from people. I remember getting a really annoyed letter from Ed Salpeter, a big-time astrophysicist in Cornell: "What is MIT doing teaching this as a summer course and advertising it broadly?" I

[26] The Hubble constant gives the proportionality between the speed at which a distant galaxy is receding (which is calculated linearly from the red shift) and the distance of that galaxy.

felt it was a dubious course to give, but that MIT would survive it. Benney told Segal "no" at first, but in the end, I think either our dean or provost intervened to overrule him, after hearing a loud outcry from colleagues all over MIT that one must not trample on Segal's academic freedom. So his course remained in the summer catalog, but almost nobody signed up for it. Either way, he did not make many converts.

Yet some people still seem to think that Segal's theory hasn't been disproved and that he didn't get a fair hearing. An article in the Notices called "Einstein's Static Universe: An Idea Whose Time Has Come Back?" asks, "Can Segal be right against almost the entire astronomical community?" [27]

Look, journalistically and underdog-wise it's wonderful to ask whether Segal might be right, and it makes for good articles, but scientifically speaking, it's largely junk. Occasionally even Segal seemed to admit that his theories "violate causality," crashing into the big motherhood issue that you just cannot have later events influencing earlier life. That's what causality means.

I'm not one to judge whether it was a blunder or not: all I can judge is that Segal *butchered* his treatment of astronomical evidence. Maybe he fooled himself into thinking that these statistics, which astronomers were using to establish the red shifts, were very weak: he butchered them by mixing up essentially these kinds of galaxies and those kinds of galaxies, and these kinds of clusters and yet others. After he got through stirring those five or six different things into the pot, I, too, would have to agree that the statistics looked very dubious, based on that. But if you limit yourself to just the great clusters of galaxies, which [Edwin] Hubble and [Milton] Humason had been doing from the 1930s onward—the great cluster in Coma, the cluster in Leo, and several other great clusters of galaxies—you would need to examine only a few of them to see with your own eyes: do they look smaller, in inverse proportion to their red shifts, or do they look bigger? And Segal was just completely disregarding that. Some of the statisticians have merely revealed their own weaknesses when claiming, "Segal had very good statistics." This is the reason why statisticians usually ought to learn more about the subject to which they are applying their stuff. Because here they simply got taken by Segal. Segal's knowledge of statistics may have seemed impressive, but his reading of the data was not.

[27] Daigneault, A. & Sangalli., A. (2001). Einstein's static universe: An idea whose time has come back? *Notices of the AMS*, *48*(1), 9–16.

Research Directions

You're known as a person who chooses good problems to work on. What do you think young applied mathematicians should be looking into now, and how has that changed since you were in their position?

You want to go look for gold where there are a few yellow specks; you don't go looking for gold in just any old desert. These days I would think that a bright young person would be a fool not to look into biology. Or at least some corners of biology: even with biology, you want to develop a taste. You can't do every moth and every butterfly. There must be some combinatorial or structural or geometric problems which are more obviously "mathematicable," or more likely to yield to your limited talents than others. If I were fifty years younger, I'd be looking into biology like mad! The human genome things: How is the control actually exercised? How is it passed on? I would probably look into the mathematics of evolution. Evolution is blathered about so much, but is it *really* plausible that selection will do so much, or does there have to be some more active feedback from your experiences back into your genes? These are perhaps somewhat silly questions, but as a bright young person, I would imagine people would be looking at biology.

Maybe even financial math, damn it. With that, I hold my nose most of the time, but there must be corners of econometrics, corners of statistics, where it looks like what you've learned already might spill over, and just by the luck of the draw, if you're clever enough, you might actually stumble onto a connection there which hadn't occurred to others that were too much "nose to the grindstone."

But turn the clock back forty, fifty years: a bright young person with physical sciences interests would then be looking at the geo- and astrophysical things, all sorts. Supersonic flows were still very interesting; certainly the rocketry or [other] things that were possible. I remember from my high school's Westinghouse Science Talent Search contest—which I never won, but when I was at least proposing—they asked me, what are you hoping to do? And I said, I was hoping that I would maybe take part in the moon program. There wasn't any moon program. Even the man-made satellites didn't exist yet. Sputnik went up in '57, when I had just gotten to England. It was quite a culture shock to have left this country and have found suddenly, "My gosh, these backward Russians have launched something." We were *going* to launch something, but we hadn't done it yet. Yet even before Sputnik, it was clearer and clearer to bright kids, even high

Toomre working in the mid-1960s; the sketches on the wall eventually became some diagrams in a 1966 paper.

school students, from all the gushing about V-2 [rockets], and Wernher von Braun, and the coming of space flight, and so on, that something like trips to the moon were not just a dream: they were going to happen. Same with supersonic flight: it was perfectly obvious, certainly by the time I was an MIT undergraduate, that we were going to be flying supersonically.

And so I could foresee that my career might in some sense tie in with going to the moon. Not me going to the moon, but making it possible for other people. It happened sooner than I would have dared to predict. I predicted in twenty or thirty years: from '53, I would have predicted it by '83, but it happened by '69.

A minute ago you said that what you've learned already might spill over into another field, giving you a different perspective than those who have their noses closer to the grindstone. Harvey Greenspan said something similar, that one advantage applied mathematicians may have over more specialized colleagues is the ability to take a general mathematical concept—he gave the example of wave phenomena—and apply it in different situations.

Yes. Last week I was writing that thank-you note to Leon Trilling, my advisor from undergraduate days, as I mentioned to you. I was thanking him partly because he had introduced me to what's called free molecular flows, flows way up in the atmosphere where molecules hardly bump into each other. A high-speed object comes flying through them, and the gas then doesn't act like gas anymore, because it's just simply particles. It's so rarified that those molecules mostly miss each other. Well, it's sort of like what I got into in my stellar dynamics, the stars and like that. They fly past, you know. And I told you I did these rather dull wind tunnel experiments, with shear flows and so on—dull in retrospect, certainly—yet in a typical disk galaxy, where the orbital periods vary markedly with radius, there is also a lot of shear flow! It didn't carry over very closely, since a gravitating fluid is a rather different thing. But nonetheless, the idea of shearing is not alien to you. You know what kind of math might be needed or that you can do. My math has never been very subtle, but at least I know what can be done, numerically, and what can't be done. So intellectually, in some sense, it all was useful: don't throw away the old stuff! Much of it does connect, somehow.

You've received some of the scientific community's highest accolades for your work, including membership in the National Academy of Sciences and a MacArthur Fellowship from 1984 to '89.

Yes. I had those two nice pats on the back. The first one didn't give me any money, but it gave me stature. In '83 I got elected to the National Academy of Sciences, which was kind of nice. And then, just a year and a half later, a MacArthur. My joke about this is that one was applause for the work I'd already done. The other one was saying, you have hope yet!

~ May 29, 2007

Steven L. Kleiman

History of an Office

S.L.K.: This room[1] has special memories because it was the chairman's secretary's office when I was a student, and there are some leftovers from that period. There's a floor button over there, you see, and it still says, "W. T. Martin." That went under the secretary's desk. She stepped on it, and then a buzzer rang in his office next door. Somehow it stayed here for all those years. There was a doorway right behind this bookcase: you can see the hinges. There's a door frame there, and the door, but it's been covered up on both sides. I think [Richard D.] Schafer was here for one year when he took over from Martin. Levinson, I think, already moved the department headquarters, but I'm not a hundred percent sure. For many years it was Sy Friedman's office, and before that it was Ken Hoffman's. Ken really had the office refurbished. He knew he would move in there after he was chairman, so he had the place spiffed up.

Undergrad at MIT

J.S.: *You arrived here in what year?*

As a student? In '58. I turned sixteen at the end of March of that year.

When I entered MIT as a freshman that fall, the linear algebra course was numbered M-62—M for mathematics. Nowadays all the departments have numbers, but at that time there were two tiers of courses. The main ones had numbers, and then the service departments had letters. So mathematics was M. That was already different my second year. It went over from being M to 18.

[1] Kleiman's office, 2-278.

What brought you to MIT specifically?

Oh, it was always in the family. My dad went to MIT. My uncle, his brother, went to MIT. My dad studied electrical engineering. I'm not sure about my uncle. It was understood: my mother made it very clear from early on that that's where I would be going. My mother sort of guided my early education. She insisted on having me registered as an applicant when I was thirteen. I couldn't apply for a few years, but that was something that was important to her, so we did it somehow. This was the only college that I applied to, so I would have been in a lot of trouble if I hadn't been accepted. I was lucky.

I went to a private school. I wasn't doing very well in the local junior high. I would have gone on to high school in Lynn [Massachusetts], but my parents realized that that course was probably not best for me. Most of the public school students fooled around: there wasn't much interest in learning per se. At the time, there was a private high school in Copley Square in Boston called Chauncey Hall School. It's not the same school today that it was back then: they had trouble attracting students, so it merged and moved. Back then, it was an excellent preparatory school. Many of the students were there on the GI Bill, back from the Korean War. They wanted to refresh their knowledge of technical subjects and mathematics and to prepare to go to college. That made the place really special for me, because those students were mature. They were older than most high school students, and they were serious about their studies—they were there to learn, and they knew what learning meant.

The school had a lot of foreign students. It advertised that it was preparing people for schools like MIT, technical schools. So families sent their children to study in the U.S. and begin their technical education. I had a friend from Bombay, India, and another one from the West Indies. Somebody else from Greece. So it was cosmopolitan as well. It was really terrific for me, a perfect match. There were a couple of others like myself. One of them went to Yale afterwards, not MIT, but then he showed up at Harvard when I did. We were first-year graduate students together: Martin Schultz. He went into applied math. He's on the [computer science] faculty at Yale now.

When did you first become interested in studying math?

Well, I've always had a special interest and a special ability in it. That was pretty obvious. My early interest in mathematics came first of all from my father. He had some of his books around. One was a book on differential equations written by Phillips[2] here at the math department at MIT that my father had used when he was a student here. He had it around the house, and I looked at it. He helped me work my way through it.

But I entered in electrical engineering, because my dad did. I switched from electrical engineering to mathematics halfway through my stay here. One problem I had was with the laboratories: the Institute has a laboratory requirement, and the electrical engineering department has laboratories that were important, and they gave me trouble. So I began to think twice about going on in electrical engineering. That was one thing.

The other consideration was that I was going to get two degrees, but if I dropped one of the two, I could graduate a year earlier. That idea was attractive, in part, because my mother was paying for my education—I mean, my parents, but it came from my mother's salary. She was a bookkeeper in my father's business. He had a small machine shop. So it seemed nice to be able to save them that money, because in those days no graduate student paid for his education. I guess that's still true. But in those days there were lots of fellowships. Everyone had one. So I knew that if I went on to graduate school a year earlier I would save my family a year's tuition.

So it was a transition. I started out in electrical engineering and had my advisor there. Then I became a double major and was assigned an advisor in the math department, and that was Arthur Mattuck. That might have been my sophomore year—I was only here for three years. And then, probably at the end of that year, I became a full math major and dropped the electrical engineering. I remember talking to Arthur about what would that mean [laughs].

I had a problem, actually, that Arthur Mattuck solved for me. One of the Institute requirements is to have a couple of years of calculus: it was four terms when I was a student. But I hadn't taken that much calculus at MIT. In the fall of my senior year in high school, I took Chauncey's only calculus course. But I wanted to learn more. So that spring, I took the second term of freshman calculus as a special student, commuting to MIT twice a week from Copley Square and back. That summer I continued by taking the fourth term of calculus, differential equations, as a special student. I simply skipped the third term. So I didn't have formal credit for two terms: the matter had just fallen through the cracks. When it was time for me to

---

2 Phillips, H. B. (1922) *Differential equations*. New York: John Wiley & Sons. Henry Bayard Phillips served as head of the mathematics department from 1934 to 1947.

graduate, there was a problem, because I hadn't satisfied this requirement. So Arthur found a way out and arranged for me to get some sort of transfer credit for it.

Do you remember any particular courses that you took here as an undergraduate?

My first term I took linear algebra and a freshman elective in elementary number theory from Kenneth Hoffman and enjoyed his teaching immensely. He was just a Moore Instructor my first year, and then he got promoted and eventually became chairman of the department. At the time, he and Ray Kunze, who was another Moore Instructor, were writing a linear algebra book, and they kind of test-drove it in our class. [Hoffman] was just a terrific teacher. He had his own personality, and it came through. His notes and discussions were clear, and he somehow captured the attention of the audience in ways others didn't. I found it remarkable how he could so smoothly make the transition from the advanced course to the freshman course in the pace of his lecturing, and the kinds of details he presented—it was just so natural for him, in both places. You'd never guess that this was the same instructor, except that his personality was the same. He was always my favorite instructor or lecturer. I chatted with him about points of mathematics after that on a couple of occasions, and I have fond memories of those discussions.

I enjoyed his courses so much that in my senior year I made a special point of driving in from the North Shore—my family had moved to Marblehead—by myself, just for his lectures. Most of the time I commuted with a number of others from the area, one of whom drove. The course was on advanced complex analysis: H^p spaces in the half-plane. Hoffman was writing another book, in his own field this time. I don't think I took it for credit, although I may have, but I enjoyed it so much that I drove in. I still have the notes I took in some of those courses. Here they are.

They're amazingly neat!

I rewrote them; that's the way I learned mathematics. Most of my courses were somewhat different in those days, in that the lecturers wrote out *everything* on the blackboard. It was like a book: no abbreviations. Students used to say you could just send your secretary to class [laughs]. There weren't really many textbooks then, not like today. The professors all had their own lecture notes, which they wrote out on the board in detail. So we'd spend all our time copying them down. There really wasn't much time to think about it: it was hard enough to keep up with the copying. I'd think them through as I copied them again at home. I found that much more effective for myself than just reading them. It took a lot of time: probably about three hours for each hour of class, depending on the subject. The more advanced subjects took more, the elementary subjects took less.

I liked algebra, so I took Arthur Mattuck's course on modern algebra, as they called it in those days, in the second term of my freshman year, although it was a graduate course. That was then: we now have honors undergraduate algebra courses. Arthur had a special reputation. He wrote entertaining calculus exams. He had a character that he created, like a cartoon character, and he would have little stories along with his exams. So, there was little Egbert who walked along the x-axis and you had to solve a problem related to what he was doing. I never took one of those exams, but I heard about them from my fellow students. My friends told me that they used to spend the first five minutes of an exam laughing. He stopped writing those kinds of exams by the time I came back here, for sure. But he was always a special lecturer. And of course he won the Big Screw several times for his lecturing in calculus.[3] He had a big reputation.

I remember [Kenkichi] Iwasawa's graduate course on algebra. That was a serious first-year graduate algebra course that spanned the whole year of my senior year. I think I still have the notes from it as well. His command of English was rather poor, and so he just wrote on the blackboard. Fast, but complete. We just came in and copied. He didn't talk much, but it was a terrific course, with terrific notes. Iwasawa had a reputation for being completely on top of the subject, and it was clear from his lectures that he was. I remember one time when someone spoke up and asked a question about whether a statement was true. Iwasawa thought for a moment and then wrote out a counterexample. It was an impressive performance.

I took Whitehead's course, too, in homotopy theory, my senior year. That was a famous course. He turned his notes into a book, compiled by his PhD student Robert Aumann, who won the Nobel Prize in Economics last year. The book was so successful that Whitehead published a second edition, twice the size of the first.[4] And then I remember Dan Kan, another long-

---

3 See interview with Arthur Mattuck, page 69.

4 The first edition, compiled by Robert J. Aumann, was Whitehead, G. W. (1966), *Homotopy theory*. Cambridge, Mass.: MIT. The second edition is Whitehead, G. W. (1978). *Elements of homotopy theory*. Graduate Texts in Mathematics, 61. New York: Springer.

time member of the department. He was a topologist, too. I was in his topology course in a crucial year, because—well, he lectured, and things seemed to go well enough. I was an undergraduate, but it was a graduate course, and he didn't give much homework, if any, and no exams except for one at the end. And people did *very* poorly on this exam, to his surprise. It was because there was a disconnect between the theory that he lectured on, and that we learned, and the *use* of that theory in examples. Well, he was so upset that he switched format the next term and had *us* lecture on the material. And that developed into a seminar format for graduate topology that has continued to this day.

I used to come and chat with Kan in the mornings. He was an early person. The first two years, I commuted with several others. One of them was the driver; he had his own car, and the rest of us rode with him. My senior year was different because the driver had graduated the year before, so I drove, and I probably insisted that we arrive early enough so that I could talk with Kan. It was partly to give him the chance to get to know me so he could write a reference for Harvard, but we chatted about different things. He's an emeritus professor: he hasn't shown up at the department for many years.

I heard that Whitehead hired so many topologists that people used to call this the Massachusetts Institute of Topology.

Well, that's a half-truth, I guess. I don't know if he was responsible for hiring. There were certainly a lot of topologists. And when I came back in 1969, one of the secretaries had mistyped the name of the school on somebody's paper, and that name stuck. That's where the name came from. But it fit, because the topology group was the most active at the time, and it was already active when I was a student. Certainly Kan and Whitehead were the two most known. And [Franklin] Peterson. They formed the nucleus. I didn't have much contact with them then. John Milnor[5] from Princeton spent a year here, along with some young mathematicians.

Did you have any contact with Norbert Wiener?

He was around when I was a student. I went to one of his lectures at the math club, which was something for me in those days, because it meant I

---

5 By this time John Milnor had received a Fields Medal for his work in opening up the new field of differential topology, as well as the National Medal of Science.

had to give up my ride home and take the train. But I wanted to hear him because he had a reputation: there were lots of stories floating around about him, and we'd see him in the halls. He didn't prepare at all. He just came in, and he said he thought it would be nice for us in the math club to hear from an active mathematician about what he was doing. And he just went into a discussion of his work. I don't think, though, that he tried to take his audience into account and explain things in a more elementary fashion or bring us into the subject at all. Just, "This is what I'm thinking about nowadays." He lectured to us the way he would to a seminar of specialists.

Some people actually seem to get more out of a lecture that has some loose ends in it.

It's funny. My own style has evolved. I started out doing what I think all my teachers here did, which was to write out everything. I still think I prepare as carefully, except that I don't write out everything. I use more abbreviations. I talk more to the audience. So the style is different. On the other hand, I remember when I was an undergraduate here, I had one teacher who was sloppy about his presentation. And I really was angry about it, because I felt he was wasting my time. I felt I learned by thinking through the arguments that were presented: I had to fill in the gaps and straighten it out for myself. But I didn't want to be the first to think through the standard theory. I mean, I didn't want to do this fellow's preparation: that was his job! The result was that I had to spend *so* much more time on his subject, and I didn't feel I learned anything more from it than I did from my other courses. So I was just angry about that.

Were there other undergraduates you studied with who later went on to become mathematicians in their field?

I remember one other—Steve Orszag. He joined the MIT faculty in applied math, but because of the social atmosphere at the time, there was much more mixing. He was my only classmate on the faculty in the math department for many years, until he went to Princeton. He worked on weather forecasting, developed a computer model of how the atmosphere behaved, and used it to forecast the weather. Fluid dynamics of some sort. Of course, Alar Toomre and Gil Strang were also MIT undergraduates, although some years before me. Alar did not major in math; I'm not sure about Gil.

Were there other people who stand out from your undergraduate days? What was it like being a student here back then? How was it different than now?

Well, I was a commuter, which makes student life very different than when you live on campus. There were some 600 commuters at the time, in all four years. And we were given a locker room, which happened to be in the basement down here. They're now graduate student offices. But I have memories of eating my lunch, which I brought there from home. And one of the other people there, who was not a mathematician but he's also on the faculty now, was Fred Salvucci, the father of the Big Dig project. He was an undergraduate at the same time as I was here, in civil engineering. He used to eat lunch at the same time as I would, and he told stories of all kinds of civil engineering projects that he knew about. I really enjoyed that.

At Harvard with Oscar Zariski

And then you went from here to Harvard and studied with Zariski as your supervisor. How did you connect with Zariski?

I went with the idea of going into topology, actually. But I happened to be standing around in the common room the first day of classes chatting with some other first-year students from Columbia, which had three or four students at Harvard that year. And then all of a sudden they said, "Aren't you coming along to Zariski's class?" I hadn't even heard of him before, but I didn't have anything else on my schedule. So I tagged along, and one thing led to another. It was the force of his personality that changed me to study algebraic geometry, first of all, and secondly, with him. In fact, apparently I got permission from him to photocopy his notes from that course. Here they are, in his handwriting, with places where he crossed out the text. That's a special treasure for me.

They're beautifully done.

Yes. But I wasn't ready to appreciate the mathematics fully. It was an advanced graduate course on algebraic surfaces, and I hadn't taken the introductory graduate courses in algebraic geometry yet. So I and some other friends from Columbia came back to MIT that fall for Arthur Mattuck's course in algebraic geometry. So at the same time I was taking Zariski's course in algebraic geometry, I used to bike back to MIT to take Arthur Mattuck's course as well. It was easy—still is easy — to cross-register.

And then I came back in the fall of 1962 for a second introductory course in algebraic geometry from Bert Kostant. Kostant was more of a representation theorist, an algebraic group person. He lectured the first term, but in the

Oscar and Yole Zariski (seated) with some of his PhD students, 1981; from left to right (standing): Heisuke Hironaka, David Mumford, Steven Kleiman, Michael Artin.

spring he ran the course as a seminar, so we students did the lecturing. He had some notes from a course in Paris by [Claude] Chevalley, so we took turns, essentially, translating the notes from French to English at the black board.

Algebraic geometry is a big subject, so it's helpful to take a lot of introductory courses from different people. They have their own points of view. There was a small community of algebraic geometers in the Cambridge area, so there were seminars. Arthur used to come to the seminars at Harvard and so on.

What do you feel you learned from Zariski?

Well, mostly philosophy. Relatively little technical mathematics, although I made it a point to try to read his papers, which I enjoyed. But he was old, about to retire. Algebraic geometry had moved on to a new phase, new techniques, not the ones he grew up with. He, to his credit, helped promote these new ideas—he brought the principal architects of that theory to Harvard. And he lectured on some of these techniques himself at a Summer Institute, wrote it up for the *Bulletin,* and lectured on it in his algebraic functions course, which I took in the fall of '62.

Zariski's course was actually more modern than Arthur's course, which was more old-fashioned. But it was funny: in a private discussion, when a certain mathematical argument would require this technique of cohomology of coherent sheaves that he taught in his course, its use was not natural for him. He would fall back on the earlier version that his teachers had taught him in Italy: formulas. Whereas we would write down exact sequences and derive from them certain relations among dimensions of cohomology groups, he just knew the dimensions as certain variants of the situation, and he remembered relations among them. That was just what was natural for him in these situations.

I was always struck by that. It was surprising because, first of all, nobody else did it that way. I mean, it wasn't natural for us, because we had the general theory, coherent sheaf theory, that would apply in this case and in others. And because he lectured on the general theory in the course. That's where I learned about it! Not only in the course but also in his report in the *Bulletin*. When the theory was first developed, they had this conference in Boulder, Colorado, sponsored by the American Math Society, and one of the main topics was this new development by [Jean-Pierre] Serre in France. They had gotten hold of the manuscript before it was completed and published, a thin outline, and they worked through it. There was a list of results with a few hints, and they had to find all their own proofs, which oftentimes are different from what appeared in print later. And then he wrote up a lovely report on the results of the investigation.[6] It was fun to read, I remember, as a student. He lectured on the theory. And yet when you came to his office, he would do it differently.

What struck me most about Zariski's final exam was one of the problems: the question was to write an essay on the sheaf-theoretic proof of the Riemann-Roch theorem. Just a whole new point of view! Zariski grew up mathematically in Italy, where this tradition of exposition of ideas was promoted as much as technique and detail. And also that's just sort of the Harvard approach, somehow. In my earlier work at MIT there were always problems: "Prove it." Or: use the Riemann-Roch theorem to do this, that, or the other thing. Make a computation. But here we were supposed to write an essay. I was completely free to describe whatever I wanted.

---

[6] Zariski, O. (1956). Scientific report on the Second Summer Institute: Several complex variables [1954, Boulder (Col.)] Part III: Algebraic sheaf theory. *Bulletin of the American Mathematical Society, 62,* 117–141.

What was Zariski like to work with?

I remember one thing that struck me about him was that he was always enthusiastic and helpful if your work was going well. But if it wasn't, he was just indifferent [laughs]. At that time, the faculty at Harvard had a reputation of being distant—just letting the students sink or swim. Although I personally never had that problem: I always found it easy enough to talk to anybody I wanted to—just go up and talk to them! The idea of a distant faculty was foreign to me, because at MIT that idea just didn't exist. The faculty was always there and accessible. There was no feeling of distance. Secondly, Zariski, again, came from Italy, and you'll read in his biography that Enriques, his teacher, used to have him at home for discussions.[7] And Zariski to a certain extent maintained that tradition, in that he often invited people to *his* home. It was a short walk from the math department. So I think this tradition of welcoming [students] to his home and doing mathematics at home, rather than a separation between work and home, probably comes from his time in Italy.

I think this story is in the biography, but I remember hearing it from Zariski. Zariski went to Rome and didn't know anybody. He went to [Guido] Castelnuovo's course on algebraic functions and then went up to him afterwards to ask if he could work with him. So Castelnuovo invited him to walk and just chatted with him, and Zariski realized when it was over that Castelnuovo had been asking key questions to find out how much Zariski knew and how well. By the end of the walk Castelnuovo was satisfied, and told him to come to his class, and how to register, and so on.

Zariski used to do something similar. He'd ask me to walk with him on errands, when he wanted to pick up his car at the garage where it was being serviced, for example. And we chatted mathematics. He certainly wasn't testing me, but that was a way that he could do two things at once, so to speak: run his errands and meet with his students. But I think he really enjoyed the companionship and opportunity to chat while he was walking on his errands. He had a lot to do, so he had somehow to combine his jobs.

On a couple of occasions, he invited me over to his house just to chat. And others, too. I remember one time in particular, one of his other students, [Shreeram] Abhyankar, was visiting for the year—a former student, about ten

---

7 "Unlike Castelnuovo, who in six years of close association saw Zariski only by appointment and never anywhere but his study, Enriques often invited Zariski for drinks or Sunday dinner with his wife and daughter." Parikh, C. (1991). *The unreal life of Oscar Zariski* (p. 207). New York: Academic Press.

years earlier, before even David Mumford and Mike Artin.[8] I had an appointment to meet with Zariski at Harvard to talk about my progress, and I showed up at the office, and I was told by the secretary to go to his house because Zariski was sick. So I went. I found him in bed, talking math with Abhyankar.

Any other Zariski stories?

At the memorial service for Zariski, [Heisuke] Hironaka said that when he was Zariski's student, he couldn't afford a typewriter to type his thesis on, having come from Japan with no money, nothing. Then he met Zariski, and Zariski had just gotten his salary in cash. So he pulled out the money and gave Hironaka the money for a typewriter. And his wife, Yole, helped Hironaka with Harvard's foreign language requirement. Hironaka figured English was enough of a foreign language, but she taught him French.

Another student in Zariski's introductory course was Joe Lipman. We've maintained contact ever since. Joe was from Toronto, and his senior year he went back for the holidays and was given a job at Purdue. So there was some transition going on with his visa. Well, when he got to the border to come back to the U.S. for the spring term, they wouldn't let him in! But he was due to graduate in June with me and others, and one requirement at Harvard is that you have to be on campus to defend your thesis. So there was a problem.

Zariski went to work on both fronts: he tried to get the rules waived at Harvard, and at the same time he tried to get the immigration rules waived. It so happened that the U.S. Secretary of State, McGeorge Bundy, had been dean of the Faculty of Arts and Sciences when Zariski was chairman of the math department a few years before, and Zariski got to know him, so he wrote to him personally to have an exception made. Harvard voted to allow Lipman to graduate without the defense here, if there was no other way. And the State Department agreed to let him in just for the defense, provided there was no other way he could graduate! So in the end he had his choice and came back, but he had to leave again until his new visa was approved.

Another Harvard graduate student who was a contemporary of mine, Dan Quillen, came to MIT from Harvard after graduating. He left in the early eighties and went to Oxford. He was a student of [Raoul] Bott's at

---

8 Shreeram Abhyankar earned his PhD under Zariski in 1956, Michael Artin in 1960, and David Mumford in 1961.

Harvard, a topologist. He was very good.[9] I learned a lot from him, which reminds me of another point: it was very definitely the feeling among students at the time, certainly at Harvard, probably at MIT, that we learned as much from our fellow students as we did from the faculty, partly because of the student seminars and partly, we just went to each other with questions.

So the atmosphere was more helpful than competitive?

It was both. It was a very productive kind of competitive atmosphere, which brought out the best in us and pushed us to do even better.

Besides doing your PhD with Zariski, you met some other seminal figures in algebraic geometry at Harvard.

Yes. I sat in on Serre's course on Lie algebras and Lie groups in the fall of 1964. Actually, I first met him in December 1963 at a conference at Purdue University, which a number of us from Harvard, students and faculty, went to together. And I talked with him again at the [Woods Hole] summer institute in algebraic geometry, which Zariski organized for the AMS.

And Grothendieck was at Harvard my first year: he had come even before I went to Harvard, so he was there for two terms, a year apart. His course was too advanced for me my first year. He lectured on Part IV, which was actually four thick volumes in itself, of his *Elements of Algebraic Geometry*.

Grothendieck had tremendous insight. Really remarkable insight. He could see similarities in results in what would appear to be *very* different fields, but he somehow sensed a common element and could outline a program of research. He would write a couple of paragraphs and have students develop them into theories of hundreds of pages. He saw the essence in a way that nobody else did. It was remarkable, because he saw it in so many different lines of theory within algebraic geometry.

I mean, ideas always develop, and there's never really an earth-shaking break. Serre was a member of Henri Cartan's school in Paris. They developed, notably, some ideas that had been conceived by [Jean] Leray in a prisoner of war camp during War World II; they abstracted the ideas and turned them into powerful tools. Serre wrote his thesis in homotopy theory, using one of these tools, spectral sequences, in a remarkable new way. It was great

---

9 Quillen received a Fields medal in 1978 as the prime architect of the higher algebraic *K*-theory.

work: it won the Fields Medal. But another tool that came out of Leray's work was the cohomology of sheaves. That tool was developed for use in several complex variables in the Cartan seminar, but Serre saw that it could be algebraized and used in abstract algebraic geometry.

We had a summer school in Trieste a few years ago on Grothendieck's *Fundamentals of Algebraic Geometry,* a collection of some of his Bourbaki talks.[10] His original talks were around 1960, so forty-five years later we found them still of enough interest that we had a summer school based on them. We expanded on them, but they were the basis. As part of my lecture, I gave a history of the subject, up until Grothendieck developed it. Part of the point was to show how the ideas developed and just what Grothendieck's contributions were to the field, but in the historical context of what had come before. So I included a history of algebraic geometry, because this was the central topic in algebraic geometry.

I sent this history to Serre for his comments, and he commented on a number of things. In particular, I had said that Serre's development of the cohomology of sheaves was revolutionary for algebraic geometry, and he objected to my wording. He felt there was no revolution, because he hadn't destroyed what had come before and replaced it with something new and very different, but was simply providing another tool. Nevertheless, this tool has taken over, because it is so much more powerful than the older tools; but it is still a tool, another way to understand the geometry of the solutions of polynomial equations.

Serre always saw a limit. He would abstract to a certain point, beyond what was already in use, but he didn't want it to go over the edge. Whereas Grothendieck felt differently. Grothendieck had deeper insight. Serre would work over a base field—it didn't have to be the complex numbers anymore, but it was still an algebraically closed field, and it had arbitrary characteristic, because he knew that case was important. But Grothendieck saw the importance of working over an arbitrary *ring.* There were already signs that you should work, say, over the integers, and then reduce modulo a prime, and others had started to develop theories of algebraic geometry over rings like the integers, but it was still somewhat restrained. Grothendieck saw the right general context. It was remarkable, because there's a fine line there between generalization

10 The 2003 Advanced School in Basic Algebraic Geometry (Trieste, Italy).

for its own sake and fruitful generalization that's really driven by a deep understanding of what's needed. And time after time, Grothendieck had the latter, though to a casual observer, it might appear that he was making an idle generalization for its own sake: kind of formal, rather than useful.

Grothendieck's ideas in his Bourbaki talks culminated in the theory of the Picard scheme, which I wrote about, but a lot of supporting ideas came before. The AMS published our write-up as a book, *Fundamental Algebraic Geometry: Grothendieck's FGA Explained* (2005). My piece about the theory of the Picard scheme is at the end.

Doctoral Work

Carol Parikh's biography of Zariski says that in the spring of 1965 you "proved some important results" for your thesis, "including a conjecture of Chevalley."

It's true that I proved a conjecture of Chevalley's in my thesis, which I always felt was the main result. Zariski mentioned Chevalley's conjecture in his course my second year, but he never gave any sense about how difficult it was. He was always proud of that fact later.

There's always some discussion of this matter in pedagogy, whether an instructor should give students an idea of the difficulty of problems before they devote too much time to them. My own feeling is that people don't make enough of a distinction between challenging problems and "exercises for one's mathematical health," as Grothendieck put it to me. When you're learning a subject, you need this kind of exercises to check on your knowledge, so they shouldn't be difficult and involved. If students find difficulty with them, that is a sign that they haven't understood something. But students who understand the principles and techniques that have been explained in class should have a fairly easy time with them.

That's one kind of problem. But there are some textbook authors and course lecturers who mix in, with that kind of problem, problems that are difficult or unsolved, without distinguishing very well between the two kinds. I think that's a mistake. It's not necessary to say "this is a very difficult problem," but you should make clear that this is not an exercise for your health, where you check your understanding; this is an interesting, important problem, which is difficult or unsolved.

Anyway, so Zariski mentioned this conjecture of Chevalley's, and it sounded interesting to me, so I worked on it. There were some theorems in

David Mumford's course that I saw how to develop and apply. It was a slow process, but it was one focus of my research while I was a graduate student, so that was good.

I tried to solve it in various cases. In the first case it occurs, for curves, it's fairly easy; that's sort of an exercise for checking your understanding of the tools at hand. Then for surfaces, that case was a little more difficult: with more advanced methods, you could do it. But after that, it was unknown, and it was always sort of in the back of my mind, and finally I had some success.

But to actually prove the conjecture, I had to develop some other tools, notably a criterion for when an abstract variety could be embedded into projective space. That criterion came first, and that part of my thesis has been the most significant in the sense that it's been used by other geometers over the years. It's now forty-one years old, but that's a part of the thesis that's attributed to me and used. Whereas Chevalley's theorem, it seems to me, is far deeper. There's an equally difficult step to go from that first criterion to establish the conjecture. To me it always seemed not just difficult, but it required some—I called it a trick; I'd say some intelligent work. But it didn't lead anywhere. Nobody really cared about Chevalley's conjecture, per se. That's often the case in mathematics, that these conjectures are of interest because they focus the development of mathematics, and tools that are developed to solve this particular problem are then used to solve other problems.

It's not the end point that's useful, but the way of getting there.

Right. The same thing is true about Fermat's Last Theorem. Gauss was supposed to have said that he never worked on it; he didn't see that it was any more interesting than a whole lot of other, similar problems. It was only later that that particular Diophantine equation led a lot of serious mathematicians to develop a lot of theory, and it's the theory that's of importance, not this particular conjecture.

There was a second topic in my thesis. I developed an important set of tools that David Mumford had introduced. It's now known as Castelnuovo-Mumford regularity, and it's really become a subject unto itself. I found it helpful in connection with my study of the Chevalley conjecture; although it was not helpful directly with the conjecture, it was helpful with related questions that I studied in my thesis. That work was never really picked up by anyone else, but Grothendieck asked me about questions on the Picard scheme when I was a post-doc with him in Paris, and I saw that these techniques would help. I pushed them a little bit further to prove some finiteness results for the Picard scheme, and Grothendieck liked that approach and asked me to write it up. Grothendieck had listed a number of such theorems in his FGA, and he had his proofs, but he never wrote them up. So in a way I wrote them up, except that I provided my own proofs; I didn't develop his ideas for proofs.

I learned much later that actually David Mumford had originally developed this theory of Castelnuovo-Mumford regularity for that problem in a seminar that they had at Harvard with Grothendieck. He developed it to some extent in the seminar, and then he lectured on that in his course, but it was in a special case. I carried the ideas a little further, and that's what I wrote up in the late sixties for Grothendieck. It appeared in one of the Grothendieck seminars, SGA 6.

At Columbia with Heisuke Hironaka

You mentioned Columbia, which was your next move from Harvard.

Yes. Hironaka went to Columbia. After he graduated Harvard, he went to Brandeis for a few years. And actually at that point, the whole algebraic geometry group went to Brandeis from time to time. Hironaka was special because he was Zariski's student. Zariski made at least one trip to Japan, and he brought Hironaka back. Japan had developed a school of algebraic geometry: I guess they had an older school in algebra and number theory. They read Weil's work very carefully; they didn't at that time pick up the French approach. A number of Japanese mathematicians left. [Jun-ichi] Igusa went to Johns Hopkins.[11] [Teruhisa] Matsusaka left Japan shortly afterwards. In 1954, Weil invited him to the University of Chicago. He spent three years there, three years at Northwestern, then a year at the Institute for Advanced Studies in Princeton. In 1961, he went to Brandeis and spent the rest of his

11 Zariski was responsible for bringing Igusa to the U.S. as well, according to his biographer. "In September 1952, Zariski received a note expressing interest in his work from a young mathematician in Kyoto who had already published several papers, Jun-ichi Igusa.... In November, Igusa wrote, 'I believe one can solve any problem when one decides firmly to solve and work persistently. I hope that I could get a conviction for the method in future developments of algebraic geometry under your direct leading and suggestions.'

"Zariski replied immediately with an invitation to Harvard, and Igusa arrived in Cambridge in the fall of 1953. Two years later he accepted a position at Johns Hopkins.... An eclectic mathematician, Igusa has made substantial contributions not only to algebraic geometry, but also to number theory and several complex variables" (Parikh, 121).

career there. He developed algebraic geometry at Brandeis, but he never really embraced Grothendieck's approach. To some extent he did embrace Serre's techniques. And he's had some good students, notably Marc Levine and János Kollár, who are probably around fifty now. Marc is at Northeastern. János is now at Princeton.

So there was a long tradition in algebra at Brandeis, but Hironaka was the first modern algebraic geometer. David Eisenbud was the second, although he developed while he was at Brandeis from a noncommutative algebraist to a commutative algebraist and then to an algebraic geometer.

So what took you to Columbia?

Oh, Hironaka went from Brandeis to Columbia. He attracted a dozen students there, and he needed some help teaching them algebraic geometry. I was in my fourth year that year, and we knew each other. So he called me up in late December—in my dorm, actually; we had a phone in the hall, I remember—and asked me if I would be interested in going to Columbia. I think Zariski must have arranged it somehow. At that time, I wasn't even sure I was going to have a thesis and graduate.

How long were you at Columbia?

Four years, except that I spent one year in Paris on a NATO post-doctoral fellowship. The situation at Columbia was very exciting because of Hironaka and some of the others. It was a smaller department, and I really had the sense that the school was trying to develop. There were some older members that were going to retire soon. When I was first there, especially with Hironaka's personality and stature, it seemed that we would build up a really strong group in algebraic geometry. So I wanted to stay there. The faculty was a small group. We went to lunch together at the Faculty Club often, and we had colloquium dinners once a week after the colloquium. I ran the colloquium one of the years, and that meant having people back in my apartment afterwards for more chat. There wasn't a large group of people, a dozen at most, usually fewer. But it was a very nice feeling on the faculty there, of working together to develop the department. MIT, in particular, but other schools, as well, tried to attract me my whole stay there, every year. But it was just too hard for me to leave, because the sense of building something of value at Columbia was very strong, whereas other schools were already established. And larger.

Steven Kleiman at his desk.

But then Hironaka left, so the situation changed. He came to Harvard, making Columbia less interesting and MIT more attractive. The next year I was made undergraduate chairman; the department chairman was an older fellow, and he had different views of things, so there were some minor disagreements, as I see it now [laughs]. I wouldn't say those disagreements were the primary reason for my leaving Columbia or my coming to MIT, but they were a factor.

Student Unrest: Columbia and MIT

Your wife, Beverly, told me that this was also a difficult time at Columbia. They had perhaps the first major student sit-ins and protests after Berkeley. The divisions extended not only between students and the administration, but also between older and younger faculty members—not a good academic environment. You also mentioned an incident that got your picture in the paper.

That incident occurred in the spring of '68. The students were protesting expansion plans of the University into Morningside Park, in particular, but more generally the Vietnam War. Morningside was a public park on the edge of the campus that was used by the local population, which was largely black and poor. The students didn't feel that it was appropriate for the campus to expand in that way and gave this reason for the protest. The expansion was a serious concern, but larger social issues were involved. The protest was just another reflection of the division between the people in power and idealistic students.

So the students took over a number of buildings and occupied them for a number of days. And some of the faculty took portable blackboards out and held classes on benches outside, to show that we still believed in education. I did so a couple of times, and one of the times I was out with an honors calculus class, a photographer happened by and took my picture. My memory is that the picture appeared in *Time* magazine: I think we still have a copy in the attic. The photo was actually picked up and published in one of the Paris newspapers, as well. The secretary at the Institute in Paris where I had been a postdoc saw my picture and recognized it, and sent a copy to us.

Anyway, the buildings were going to be cleaned out the first night by the tactical police, who had assembled in the tunnels underneath the campus. The idea that something was going to happen spread among the faculty, and we cordoned off the library where the administration was. The president was in there as well. We surrounded the library building arm to arm; some of the senior faculty got the president's ear, and the police action was called off at the last minute. However, not before the first edition of *The New York Times* went to press: the 11:00 p.m. edition described the police action as having taken place.

You returned here in the fall of 1969. There was a certain amount of unrest at MIT that year as well, right? The November events.

Oh, yes. In fact Beverly and I went to the medical department to get a blood test before we got married, which happened to be across the corridor from the president's office, and we had trouble getting in to have our blood test, because of the demonstration. The medical department has since gotten larger and moved a couple of times, but that's where it was then. So yes, there was some protest, though never as extensive as at Columbia. One main reason, beyond the war, was that MIT was going to help the Shah of Iran develop a nuclear program: the nuclear engineering department was going to accept a large number of Iranian Master's students, and the undergraduates didn't like the idea. It's sort of funny, because now it's come around again.[12]

---

12 When this interview took place, President Mahmoud Ahmadinejad's determination to pursue Iran's nuclear program was very much in the news.

MIT

Was there someone at MIT who was the principal person in getting you to come here?

Yes, Norman Levinson was the principal person. He lived a few doors away from Zariski, and they were very good friends. I remember Levinson told me that Zariski used to say to him, "At MIT you have good undergraduates!" although I think I was the only one to have come from MIT to Harvard in those days [laughs].

Levinson had a strong personality. He was in a different field, but he was a statesman, and he appreciated mathematics in a way that a lot of the younger people didn't. They wanted to advance their field, but they had a very narrow kind of an appreciation, cultural appreciation, for the development of ideas and where their work fit in with other branches of mathematics, and also for how their own field developed and where would it go.

Did you have contact with Levinson as an undergraduate here?

I don't think so. He wasn't chairman when I was an undergraduate, but he became chairman. Actually, my first contact was with Schafer, who was acting chairman at the time. Maybe it was my first year at Columbia, and he looked me up. There was some meeting of the American Math Society in New York that he came to, and that I went to, and he sought me out and asked me how I would feel about coming back. But it was really Levinson, with Artin and Mattuck, too, who was responsible my returning.

Mike Artin came to MIT right from Harvard: he spent three years at Harvard as an instructor and then came to MIT. He had been the assistant for the course I took with Zariski my second year, so that's one way I got to know him. Mike had a large group of students in algebraic geometry when I first came to MIT. It was a little bit like coming to Columbia, where I helped out Hironaka with his large group of students, but I came from Columbia with six students of my own. So the algebraic geometry group was a very large group—and active, seminars, and so forth. The Artins were very social in those days, too. They had parties, something like the Zariskis' parties, although they were more limited in their scope. It was largely MIT people. Maybe the party would be in honor of some visitor. And there were a lot more students.

When Beverly and I came to MIT in '69, we went down to the Cambridge City Hall one day, just the two of us, and got married. That evening, we invited the graduate students that came with me from Columbia to celebrate at Anthony's Pier 4 restaurant. We hadn't yet gotten to know the MIT students and faculty very well. But the marriages at the city hall were registered in the *Cambridge Chronicle,* and less than three days later we found a wedding present on the doorstep from Fagi Levinson.

The Levinsons used to have dinner parties, and after our daughter was born a year later, they told us to bring her along, assuming we wouldn't want to leave her with a babysitter. So she just slept on the bed.

The growth of the department must have made it harder to maintain such social interactions.

Well, the department head used to invite faculty members to his home for dinners, usually in smaller groups—six or ten people, a couple of tables. Wives, and children sometimes, particularly small babies. But then, while Hoffman was department head, his marriage broke up, and then of course he stopped doing it. Danny Kleitman became chairman after Hoffman; he had a few parties in his home, I remember. In addition, at least the pure math chairman, notably Frank Peterson—he was pure math chairman for many years—used to have a party or two every year; maybe one a term. Those parties were big. Everybody in the pure math group was invited, plus spouses. There was a buffet supper, and it was very nice. That practice continued for many years. But the subsequent heads didn't maintain the tradition, which is kind of a shame. There's a party every year held at Endicott House. It's nice, and it's worth continuing, but it's not the same. There's a certain warmth to being in a home.

You've mentioned also that there were teas?

Yes. Those were wonderful, too. There still are teas, but they've become less and less popular over the years. In the beginning they were more formal. The department had two large samovars: one had concentrated tea and the other hot water. And the department had fine china, and cakes, and a woman who worked for the department and served. So we would line up, and pick up our cup and saucer, and go over and get the tea, and then some cakes. It was a regular event, every day at 3:30: I think it's always been down the hall in the common room. Everyone who was around always went to these teas, and then there was an opportunity to chat informally. Not only the faculty among itself, but the graduate students, as well. There were some chairs, but most people stood around, and there's always been a blackboard or two. Recently, Mike Sipser renovated the room, so now there's a lot more blackboard space.

So those teas were really nice. Then, when the woman retired, the department tried to maintain the tradition. They got different people to serve the tea. I think Arthur Mattuck asked graduate students to take turns, but even that practice has pretty much disappeared. Still, somebody comes in and makes the tea ahead of time. There's just a large aluminum teapot there, and you can fill up a Styrofoam cup and take some cookies if you want.

Algebraic Geometry

Algebraic geometry is a vast field today; it was obviously much smaller back then.

Oh, yes. In fact, there were only two places in the world where you could learn the subject from Grothendieck's point of view: Harvard and Paris. Notably, the founder, Serre, was there, and of course, Grothendieck too, also Chevalley,[13] Samuel, and others. There were other places, in the U.S. and elsewhere, where there were distinguished algebraic geometers, but really only two places where Grothendieck's ideas were embraced.

Zariski regularly had parties, kind of math parties, where he invited all the active algebraic geometers in the Boston area, which meant mostly Harvard, MIT, and Brandeis. But there was a good group there. There was a lot of math chat: it was the point of the party. And he included the students, myself and a couple of the other graduate students.

So here was the Boston algebraic geometry world in one living room. How many people were there?

Gee, probably twenty to thirty—including wives, and children sometimes. It was somewhat more than just algebraic geometry; it was probably the larger algebra world. So, other people like Richard Brauer, John Tate, and others.

I think you were the one who started the algebraic geometry seminars at MIT. Can you talk about the growth of the Boston-area algebraic geometry scene?

In my student days, I think all the seminars were at Harvard; I don't remember coming to any seminars at MIT. We had student seminars where we used to go through papers, the literature, and Zariski ran a topic seminar where people would speak about their research. So that practice was one thing I

---

13 Claude Chevalley, one of the youngest of the Bourbaki mathematicians, moved to Princeton in 1939 because of the war, then to Columbia in 1949, but was able to return to Paris in 1957.

learned there and maintained: when I went to Columbia, we started student seminars, and when I came back here, we started both student and research seminars in algebraic geometry, which I ran for 25 years. In Cambridge, we used to have two research seminars in algebraic geometry, one at Harvard and one at MIT, but they've merged now. We take turns. One week it's at MIT, and the next week it's at Harvard.

For a number of years, there were seminars at other schools. Five years ago or so, they had a regular seminar at Brandeis, one at BU, and one at Northeastern, and I would go to all these schools—not regularly, but often. It happens less now. Some key faculty have left Brandeis. David Buchsbaum has retired. He wasn't an algebraic geometer but a commutative algebraist. David Eisenbud was also first of all an algebraist but then developed an interest in algebraic geometry. He had an active seminar at Brandeis, which I went to a lot. I used to go often to the geometry seminar—algebra and geometry—at Northeastern. That's still going on, but it's not quite as attractive to me as it used to be. And BU has lost its algebraic geometry strength. Dan Abramovich left a few years ago and went to Brown, so he's still in the neighborhood and comes up from time to time for the seminars at Harvard or MIT, but less so every year. Even Tufts has a few algebraic geometers, who come to the seminars, but there's not very much activity per se—talks, seminars—there.

How do you see the MIT group's role in the history of algebraic geometry?

Mike Artin played a major role in developing many of the tools. As I mentioned before, Grothendieck came to Harvard, and Mike Artin joined that school and helped develop some of Grothendieck's ideas, particularly with respect to étale cohomology, which was later developed and used by [Pierre] Deligne and by others as well. Mike did much of that work when he was at Harvard and then as a postdoc in Paris. When he came here, he developed some of his own ideas, but they grew out of Grothendieck's theory; they concern algebraic spaces, which involve another way of constructing abstract algebro-geometric loci by patching together pieces of varieties. He had a graduate student who worked on that subject.

But then the development of the subject kind of quieted down. Six students came with me from Columbia, as I mentioned, and Mike Artin had an equal number, but the group of algebraic geometry students has never been up at that size again. We all did our own research in the field, but it wasn't the same intense period of development of tools and then applications of the tools to study some of the classical problems in the subject. So I don't think MIT ever played the same role that, say, Harvard did in the development of algebraic geometry. The subject was in a different stage then.

Were you disappointed in some sense, looking back on how things developed?

No, I wouldn't say I have ever been disappointed. On a personal level, I've always felt behind in my research work in the sense that I've had more work to do than I ever had time for. I'm just so behind in my writing: I have to fight off the temptation to get into a new project before the old one is finished.

Enumerative Geometry

In these papers that analyze the networks of mathematicians in a given field, you are the central person in enumerative geometry. How did you get into the field?

One thing led to another. First, at Columbia, Hironaka asked me about a problem on algebraic cycles, and I saw that the Schubert varieties would help with that problem. Hermann Cäsar Hannibal Schubert was a German mathematician, who worked mostly in Hamburg. He's best known for just part of what he actually did. His work was done for the purposes of enumerative geometry, but then a part of it was picked up in the first half of the twentieth century and has been generalized. And I had learned about his work on the Grassmannian when I was a graduate student. I worked on this problem that Hironaka had mentioned, and I used the geometry of Schubert varieties in a paper related to this problem that appeared in the Zariski volume, so my name was connected with Schubert's a little bit.

Then the *American Math Monthly* wanted to have an article on Schubert calculus. I think they asked Phil Griffiths first, and he wasn't able to write it for one reason or another. So they asked me to write this article. I didn't know much about enumerative geometry at the time. When I was in grad school, I had heard the name. And I'd seen it mentioned, notably in André Weil's *Foundations,* but Weil didn't know what enumerative geometry really was either, as I discovered later. But sure, I said, I'd be happy to write something. I got my graduate student, Dan Laksov, to help me, and we wrote an article, which is still of interest. It's an introduction to the subject, as I knew it then, which was intersection theory on the Grassmannian, and the applications to some enumerative problems. But I didn't know the full scope of Schubert's work then, and I didn't have much sense for the history of mathematics at that point.

The *Monthly* article led me to do some research into an issue concerning transversality that came up and that hadn't really been addressed, so I wrote a little paper about that issue, and my work has turned out to be of use to people since then. Robin Hartshorne of Berkeley—another student at Harvard when I was a graduate student there; Zariski is one of his teachers—wrote a very popular introductory graduate text in algebraic geometry. So when he wanted to talk about this transversality issue in there, he included my treatment of it.

A couple of years later, the AMS held a symposium on the seventy-fifth anniversary of the Hilbert problems. It turns out—I had no idea—that Hilbert's fifteenth problem called for the rigorous development of Schubert calculus. So the symposium organizers invited me to speak about the Fifteenth, because I had already written about Schubert calculus. At that point, I started to do some more serious historical research on it. I actually *looked* at Schubert's book [laughs] and other enumerative geometers' works from the late nineteenth century—and it turned out to be a much richer subject than I had realized. When I started, I thought my report would be a simple further development of my *Monthly* article, but it turned out there was just so much more to the subject, much of which had been completely forgotten.

Enumerative geometry was taken very seriously by well-known mathematicians in the first part of the twentieth century. Their names were certainly known to everybody, but this particular aspect of their work was not. Severi, a famous Italian geometer and one of Zariski's teachers, devoted a lot of work toward developing the foundations of intersection theory for applications to enumerative geometry and wrote monographs about the applications to some problems. And van der Waerden. And then finally Weil, although Weil had no real interest in enumerative geometry and had *no* idea whatsoever, I'd say on the basis of what he wrote, of what enumerative geometry was really about. He made this boast in his *Foundations* that a reader could find in his book everything that was needed to solve Hilbert's fifteenth problem, which is far from the truth.

So it had attracted some serious interest. Part of it, the intersection theory, had continued to be developed, but the original problems in enumerative geometry, the geometric problems, like finding the number of lines or the number of conics tangent to five given conics—that aspect of it had been forgotten. And it was beautiful mathematics that needed modern development. I used the subject as a source of thesis problems for a number of students, and I did some work on various aspects of it myself.

With the emergence of string theory in physics around 1990, enumerative geometry has become more popular, and my transversality result is often cited in this connection. By mathematicians—the physicists don't care [laughs].

How do enumerative geometry and theoretical physics intersect?

These strings are one-dimensional real objects. Real as opposed to one complex dimension, which is two-dimensional real. But the string develops over time; it evolves and sometimes two strings come together and so forth. And if you look in space-time, then you'll see a compact Riemann surface—in other words, a complex projective algebraic curve. So it was of interest in the physics community to understand what kinds of algebraic curves could be in space-time. They had a more advanced view of space-time: it didn't have simply the usual four unbounded dimensions, but also some dimensions that kind of wrapped around themselves, so that altogether there were, I think, ten dimensions, the four unbounded dimensions and another six.

And what's the structure of the six? It had to have certain properties that were reflective of the world, and these properties are those of an abstract geometric structure called a Calabi-Yau manifold, which had been studied already by two differential geometers, Calabi and Yau. There are various kinds of these manifolds. The simplest is defined by a polynomial of degree five in four-space. It's then a complex three-dimensional object, a real six-dimensional object. This was one candidate for a space with the extra dimensions. So there was some interest in knowing about the kinds of images that the Riemann sphere has inside of this space. And that was a problem that had been studied to some extent in the nineteenth century. In 1885, Schubert—the same Schubert—had found the number of straight lines on this space.

Graduate Chairman

How did you come to serve as graduate chairman, and what did you try to accomplish?

Mike Artin was graduate chairman for many years, and then he suggested that I take over, because I had a lot of students myself.

I tried really hard to recruit especially qualified students. Because we're in competition with the other major departments, we need to give them a little more incentive to come to MIT. So one thing I did, the first year I was graduate chairman, was to get a graduate fellowship for the department. Money was tight in those days, and the federal government had cut back on support. In my day, everybody had an NSF doctoral fellowship. That practice continued until the early seventies, when the government cut back, so by the time I was graduate chairman, around 1980, there wasn't a lot of money. The size of the student body was shrinking, and there was a question about whether it would really get too small to maintain an active graduate program. So we needed money to support graduate students.

A visiting committee meeting was to be held in the spring, and I had to prepare a report on the graduate program. Kenneth Hoffman was in the office next door to me, and I went over and chatted with him about what to say, and he gave me some good advice. He said that the Visiting Committee has a direct line. They speak directly to the higher levels of the administration, who have special funds and could take money for a fellowship directly out of these funds without going through the usual channels. So he suggested that I write to certain key members of the visiting committee who understood the problem and ask for their support. They took up the cause and spoke to the provost, notably, who took money from his special funds to give us a graduate fellowship, which I then named after Levinson. The department still awards the fellowship every year. Now it has money to attract a *number* of talented students.

I also tried to recruit women students. I had one of my own woman graduate students make phone calls and maybe arrange for entertaining these students when they came to visit. Those practices have been kept up and expanded by later graduate chairmen. I believe the graduate chairman assigns a mentor to take the visiting student around, and the department now asks various faculty members to meet for a short time with the students in the areas of special interest to the students.

At the end of my second year as graduate chairman, I had a sabbatical year, and somebody else was appointed. Before I left, there was a little friction with Danny Kleitman, though. There had been from time to time some friction between the pure group and the applied group. I don't think there's anything similar nowadays, but I got caught up in it in one respect: the graduate program. In principle, as far as the Institute was concerned, there is one graduate office. But in practice, the applied group did their own admissions and had their own fellowships, and so forth. I was never really told what the conventions were, except to some extent by Ken Hoffman. In some of the advice he gave me, I think his own background kind of came in, so I had a rather one-sided view of things. I probably heard something from other people, like Frank Peterson, but not a lot. That second spring of my tenure as graduate chairman, Kleitman was chairman, and he wanted some combinatorics students admitted as pure students and supported by pure math funds, even though he and the other combinatorialists in the department are in applied. It wasn't clear who had the responsibility: he was head of the department, but I think that formally, as graduate chairman, the decision as to whether and how to admit them was mine. So there was some disagreement about how to handle that matter, and there was some tension between us. It didn't develop into anything serious. But it was hard for me that spring.

There was also some issue going back to this matter of funding of graduate students. The question was how many students we could admit. There were many people in the department who felt that you couldn't oversell the airplane. And that wasn't good, because quite clearly, we were in competition with Harvard and Princeton, Columbia, and other places, and the top graduate students have to go somewhere. They're not all going to go to MIT! But to what extent can you make more offers than you can fund? The difficulty is that if you admit only your first choices, and some of those people decline, then you'd like to go on to your second choices. But meanwhile, those students have decided to go somewhere else, and so you're down to your third choices. Whereas, if you could make offers both to your first choices and your second choices, then there's a much better chance that you're going to get those second choices.

So there was some disagreement on strategy. I felt we could make more offers than had been made traditionally. I did so, and there was no problem. But the fellow who took over the job after me, Nesmith Ankeny, did so too, and the department got into trouble, because too many students accepted.

Mathematics Writing

How did you get into teaching writing for mathematicians?

Arthur Mattuck asked me. The Institute started a program, a writing requirement, in the early eighties. The Institute has all these different kinds of requirements. There is a swimming requirement. If you know how to swim four laps, then that's fine. But if you don't, then you have to take a course on how to swim. Well, there's sort of an intermediate ground: maybe you need a little coaching so you can make that fourth lap. So this writing requirement slowly evolved from a requirement into a form of teaching. The question was what to do to encourage students to work on their papers. There's always more work involved than they realize. One idea that evolved, which originated outside our department, was to require the paper to be due before the last term; if the students hadn't written an acceptable paper in time, then they would have to take a writing course. There were general science-writing courses offered by the Institute's writing professionals.

So after a while I was helping the students edit their papers, and every year it got to be more and more work. There's always a fine line separating acceptable papers from unacceptable ones, and our standards went up.

Actually, Ken Hoffman was one of the original architects of the Institute's writing requirement. And the Institute had some structure for it. The first couple of years it was run largely by the writing faculty. But then John Deutch—*the* John Deutch; he was dean of science at that time, and later director of the CIA—he decided that the requirement would have two phases. The first would be in general expository writing; the second would be specialized in the writing style of the individual departments, under the control of the departments. It was supposed to be a sample of professional writing in the field. If you're a physics major, then you were supposed to write about physics. If you were a mathematics major, then you'd write a math paper.

That policy was somewhat contrary to the original spirit, and some of the original people left the program. I was there from the beginning, even before it changed, but I took on more responsibility when it became the department's responsibility. Each department needed somebody to oversee the program in that department. Arthur Mattuck was chairman at the time, and he asked me.

I had run an undergraduate seminar on several occasions before then. The students did the lecturing, but in addition I asked them to write a term paper, an essay, on some topic that wasn't covered in the class. In those days I didn't like calling them essays, but that's what they were. A term paper seemed like a good requirement to me at the time. That requirement might have been one reason why Arthur thought of asking me.

Twice, early on, I also gave a course on mathematical writing: I ran this course in the department for our majors. So I had already studied writing; I read the literature on organization, style, and formatting, everything from the fine points of typography—when to display lines, how much space to put between symbols, and so forth—to the importance of conceptual proofs and how to develop them.

So at that point, the job became more serious, and I felt more responsibility for making sure that the papers looked like a sample of good mathemati-

Steven Kleiman lecturing.

cal writing. The program went on for about fifteen years, and every year it got a little more serious, until eight years ago we started an undergraduate journal.[14] That idea was promoted by the people who were in charge of the program for the Institute.

Publishing the journal was a good move because the students liked the idea. Before then, they would put in a lot of work on their papers, and when the papers were done, the process ended. Now, others see the results, published in a professional-looking journal. Publishing gave the students more incentive to work hard on their papers. It also gave *me* more incentive to make sure that the papers were as good as possible.

Anyway, the writing requirement ended two years ago and has been replaced by a communications requirement, which is supposed to be broader. My own Phase Two writing program has evolved into a course in mathemat-

---

[14] Volume 1 of the *MIT Undergraduate Journal of Mathematics* was published in June, 1999.

ical presentation. The students come in with a paper that they've written for another course or special program, and then we edit it. It goes through some six rounds of serious editing, revision of the content, the organization, the style, and the format. About half of the students have to work on the mathematics; I find technical problems in what they say. Completeness and correctness are big problems for beginners. They're writing on material they just learned about. And they make mistakes: they don't know the whole context, for one thing; and for another, they don't always have the details correct. So I do spend a lot of time helping them with the mathematical content of the papers, not simply in organizing it logically and presenting it clearly. Then I help them with the various mechanics—formatting and punctuating properly, using simple and concise sentences. There are a lot of problems [laughs].

I added an oral component, in keeping with the goal of the new communications requirement, and more writing exercises as well: the students have to lecture and to write a short review of each others' lectures. I give some lectures at the beginning, on mathematical writing and the use of LaTeX to produce high-quality papers, but otherwise they do most of the lecturing. At the same time, a lot of the other math courses, seminars and so forth, have taken on a larger component of writing and presentation. So the new communications requirement isn't as focused on writing as the old writing requirement, but that change might be appropriate anyway. There's now a lot of opportunity for our majors to take these various courses, to spend most of their time learning new mathematics and just a little time writing. When our majors take these courses, they don't get the same kind of intensive training in professional editing as they would have under the old requirement or as they would in my new course, although two courses, 18.091 and 18.097, do offer somewhat more instruction in writing mathematics than the others.

What mathematicians have you encountered whose writing you enjoyed?

Serre, first and foremost. But Zariski too. His command of English was incredible, and English was definitely not his first language. Nor his second language. In fact, most their contemporaries wrote well. It's a matter of generation, too, I think. Their generation took writing more seriously; later generations, far less so.

Your "Writing a Math Phase Two Paper" must have been one of the first things ever written on the subject.

Steven Kleiman talking with Dirk Struik and Mike Artin on the occasion of Struik's 100th birthday celebration, 1994. In the background are MIT faculty member Daniel Freedman and Domina Spencer, the third woman to receive a PhD in mathematics from MIT.

That essay was written before the journal. The students wanted to have some idea of what was expected of them. So I wrote that paper to show them the format and style, first of all, and at the same time, to give them some tips on writing. It's not widely distributed, but colleagues at other schools have asked me for copies to use in their own teaching. I put a copy of it in Volume 1 of *The Undergraduate Journal.*[15]

We'll see if my course—18.096, "Principles of Mathematical Exposition"—and the *Journal* continue. They might, but they might not, because a lot of the math majors might prefer to learn more mathematics and spend less time learning writing. Undergraduates don't have the background—the experience reading other people's writing, both good and bad—to know firsthand how important writing is.

---

15 Kleiman, S. L. and Tessler, G. P. (1999). Writing a math Phase Two paper. *MIT Undergraduate Journal of Mathematics, 1,* 195–206.

The opening Donald Knuth epigraph really captures the motivation. "Word-smithing is a much greater percentage of what I am supposed to be doing in life than I ever would have thought."

He's a genius.

~ December 15, 2005, and May 15, 2006

Harvey P. Greenspan

The Growth of Applied Mathematics

J.S.: *Let's start with the growth of applied mathematics as an American discipline. In your 1961 article "Applied Mathematics as a Science" for the American Mathematical Monthly you wrote, "The great majority of the senior faculty in applied mathematics are products of a European education"—mostly Germany and England—"men like Friedrichs, Courant, Goldstein, and Lin."*[1]

H.P.G.: That goes back to the late forties. From Germany, NYU had Richard Courant and Kurt Friedrichs, and Brown University had William Prager. We had Eric Reissner, who was a pre-war émigré but had done his doctorate at MIT. And Caltech had Theodore von Kármán and a first-rate group in aerodynamics.

In England, applied mathematics had a rich history from Newton, Kelvin, Rayleigh, and Taylor, and right up to the present time Cambridge University remains a powerhouse. Sydney Goldstein created Manchester's applied math group, which included James Lighthill, a real luminary, as well as Gerry Whitham, who later settled at Cal Tech. Goldstein went to the Technion [in Haifa, Israel] and then left to come to Harvard, where he presented a comprehensive plan for the development of applied math but in such specificity that it was hacked to death. "Week two we'll do this. Week four...." But he did influence Harvard. And, of course, Harvard had the Aiken computing center early on.

Applied mathematics in England was never subordinate to pure mathematics, as it was in the United States, where a tradition of pure math was

---

1 Greenspan, H. (1961). Applied mathematics as a science. *American Mathematical Monthly, 68*(9), 872–880, from an address presented to Mathematics Association of America by Greenspan, January 27, 1961.

strongly influenced by Harvard and George David Birkhoff. Every university came to have a pure math faculty, and the mathematics establishment produced a lot of people without balance or constraint. Applied math at Harvard developed in the Division of Engineering and Applied Physics with Howard Emmons,[2] George Carrier,[3] Bernard Budiansky from Brown University.[4] That's the second generation already. NYU had Joseph Keller, Peter Lax, and many others who were first rate, mainly students from New York. Cal Tech had Julian Cole; George Batchelor took the reins at Cambridge.

So that's, broadly, the stream. It was developing in different places in different ways in response to many new and urgent problems in fluid and aerodynamics and the need for theoretical analysis in these and other fields.

What role did Norbert Wiener play?

He really didn't play a role at that late time; I did not have any contact with him. Wiener was instrumental in getting Levinson here; they both had difficulties being appointed because of anti-Semitism at MIT. That's all before my time.

So when places like Caltech, MIT, NYU, Brown, were starting to have applied groups, what was the conception regarding how to go about this?

There was no common conception. Absolutely no conception. But applied math originated from theoreticians in continuum mechanics, and the Cambridge tradition, personified by G. I. Taylor, must have had a strong influence. Taylor was, I think, the leading applied mathematician in continuum mechanics in the world, deserving of a Nobel Prize and genuinely so, in the best of the English tradition, like Faraday and Rayleigh.

The domination by pure mathematics meant that applied math usually had to develop separately. Harvard operated as a separate group, in the Division of Applied Science; at Brown and RPI they were mostly separate, and with weak math departments. I don't know of another applied group

[2] Emmons was professor of mechanical engineering and engineering at Harvard from 1940 to 1983.

[3] Carrier was at Brown from 1946 to 1952. He then moved to Harvard, where he was appointed Professor of Mechanical Engineering, then Professor of Applied Mathematics in 1972.

[4] Budiansky enrolled in Brown's newly established graduate program in applied mathematics in 1947, and completed his doctorate in 1950. His advisor was William Prager.

within a strong math department. We were the only one. I don't think it would have been possible, and I think the evidence bears that out. I had a conversation with the dean at UCLA in the late 1960s, who wanted to start applied math at UCLA, within the math department. I detailed for him all the unsolvable problems that would occur this way to make the effort fail [laughs]. Because it really depended on having a spearhead to break through resistance. Otherwise you couldn't do it. Two from our group, Victor Barcilon and Alan Newell, one of Benney's students, who did attempt this task, failed. Victor migrated over to geophysics where Julian Cole was then a professor and subsequently moved to the University of Chicago; Newell left for Arizona, where his efforts were rewarded. Later I heard that the dean, visiting Stanford University, again stated that he would like to start applied math at UCLA—"but not the kind Greenspan wants." He wanted a different applied math, something that would *fit in,* you see?

The Applied Mathematics Committee

How did you get into applied mathematics?

Well, I went to Harvard as a graduate student to study math and physics, and the only way I could do that—given that there were no real applied-math programs—was to join the division of engineering and applied physics. I did my degree with George Carrier, who was among the very best applied mathematicians. After I graduated they made me an instructor and a research associate, and then I became an assistant professor. But I was told they were really top-heavy and not to expect too much, since they had recently appointed Goldstein and a number of other more senior people. I said I didn't mind that, but not to expect me to stay around longer than my opportunities would permit me.

How did you get to MIT, and who was here in applied when you got here?

I left Harvard in 1960. I had met Lin at Avco-Everett: I went there for the summer as a consultant, and Lin was also a consultant. I worked on a stability problem, and he came in weekly and we talked about that. I gather Carrier had told him about me, and he was interested in my coming to MIT. Lin had succeeded in bringing several people to MIT in the late fifties–early sixties. The department really did need a representation in applied math because such a curriculum was very important to MIT. So he had George Backus, who's now at the University of California San Diego,[5] and Lou Howard appointed. And he also convinced Gerry Whitham to come. I knew and liked Gerry Whitham; he had offered me a job at NYU, but when he went to MIT, I just followed. Eric Reissner, in solid mechanics, was here, and he preceded Lin. Levinson himself had been trained in electrical engineering.[6]

So these people came; they were all first-rate. It was a most successful start in the formation of the group, a credit to Lin's efforts, but it lasted only two years after I arrived because of the pressures within the math department. Gerry Whitham decided to go to Caltech,[7] which shows you how smart he was: Gerry had correctly assessed the difficulties of doing anything in the math department at MIT and left immediately. And George Backus left to go to San Diego. There's a normal level of disagreement in a department, which is natural. And then there is an abnormal level of disagreement, in which people leave. So the group was collapsing.

Seems like one of those junctures that's either a crisis or an opportunity. What happened then?

Well, something had to be done. It wasn't clear *what* could be done. So Lin and I talked the matter over, and we established a theoretical, philosophical basis for the field, which said enough about what the structure of applied math *should* be—and no more than that, because you could be picked to death. It was, essentially, a study of commonality among different disciplines. We argued for a restructuring of the department and a committee to be organized that would enable us to appoint more applied mathematicians and to circumvent some of the difficulties that we had. The difficulties were mainly that the pure mathematicians recognized the need for a *service* group in applied math, not for a research group. They listed applied math as one of perhaps nine areas in pure math, and they had rules which said that every applied mathematician had to have a mathematics degree in his background.

---

[5] George Backus is Research Professor of Geophysics, Emeritus, at Scripps Institution of Oceanography at UC San Diego.

[6] Norman Levinson was awarded a Bachelor's and Master's degree in electrical engineering from MIT in 1934. MIT also determined that during this time he had done sufficient work for a PhD. in mathematics, since he had already taken almost all the graduate courses the mathematics department had to offer.

[7] Caltech established an applied mathematics program under Whitham's leadership in 1965.

C. C. Lin.

Which is not what we meant. We treated applied math as a science, grounded mainly in the elucidation of scientific principles that were common to different fields—and not as basically a mathematical enterprise; although it was that, in the creation and study of the mathematics that had been produced by physical problems for the *solution* of physical problems. It had to encompass basic ideas common to many different fields. It had to include statistics. It had to include probability and computation as well as continuum mechanics, stability, wave theory.

And so I wrote up a document. Lin was a very important voice in setting the philosophical basis, but I always did the writing. Lin and Levinson presented this to the dean. It was looked upon favorably, and in 1964, a committee was established on applied mathematics, which was supradepartmental. It was really an Institute committee.

What do you mean by "Institute committee"?

The dean, Jerry Wiesner, was the head of it. Under Alberty it was the same way. And we had very distinguished faculty from other departments on the committee: Jule Charney from meteorology, Marvin Minsky and Claude Shannon from electrical engineering, Bob Solow from economics. So we had a first-rate committee to start. We really wanted to get people recommended and validated by these experts, and it worked well.

The success rate for getting a senior person was only about one in ten at best. We decided we would try to build an excellent group; that excellence was the only way we could succeed, not just adding people. We could have easily developed a first-rate fluid mechanics group, but we aimed at something broader and to serve MIT's interests by serving our own—in other words to do it naturally. And we began to restructure the department.

Well, the pure mathematicians thought this was divisive—and it was, obviously. After all, we wanted a separate group to decide on appointment qualifications. We wanted to be able to get a person who was a physicist (or even an experimenter) into applied math, yet it was very hard to find a physicist who also had a mathematics degree. We immediately set out to appoint new people and to construct a curriculum that mirrored these basic precepts. In my eye, we really set out to create a separate department, or something that had the structure of a separate department, so that new people coming in wouldn't feel like aliens, and those students applying for admission would get a very clear idea as to what we were offering—because it didn't yet exist here.

"Wouldn't feel as aliens" among the pure mathematicians?

Yes. The pure mathematicians were highly rated, but one could chop off the group and put it on the other side of the Charles River and not many people at MIT would recognize they were gone. They weren't intrinsically involved with the technical community. They are now, to an extent, in physics and biology, where pure mathematics seems to be of value. They've established better contacts with physicists on string theory. Whether that pans out I don't know, but certainly the geometers have made good progress, and that's beneficial. But in the past math wasn't very connected. So we had to change all of these things, and that's where the difficulty and the friction came.

Who ran the applied math committee?

Lin was the official head of the committee, and I was the secretary. He and I agreed on the philosophical basis of applied math, but he didn't want to run things. He was not confrontational, and we came from very different cultural backgrounds: I from Brooklyn, and he from the upper class in China. He did things by not getting hit.

I'll give you an example. We had a common council before this committee was set up, and we were going to argue for the necessity of appointing people in applied math who were not endowed with a mathematics degree. This met with a chorus of vituperation. Lin had brought it up, but I carried the whole debate very forcefully and in my own style [laughs]. Towards the end of the session, about 40 minutes later, someone turned to Lin, having forgotten that he brought it up, and he said, "C. C., what do you think about all this?"

And C. C. said, "I'm indifferent."

Well, you had to have a sense of humor, knowing his background. There was this tumult going on: even though he had started it, as he was supposed to, he then wanted to get back so he could negotiate in some other way, or at some other time. So you had to have a laugh about that, and I did, most of the time.

That's more a traditional Chinese value, right? No direct confrontation.

Yes. It wasn't a Brooklyn style. Lin could have waited ten, twenty years; I couldn't wait that long. That wasn't me, and I had a career to pursue as well. I would not commit my life to this if it couldn't be done. I already had an offer to go to Caltech, and we had considered that seriously. Caltech was building up in applied math, too, with Gerry Whitham there and now Philip Saffman from Cambridge: they had a very good fluid- and aerodynamics department, which was where it started. But I had made a decision that Boston wasn't a bad place to live. I wasn't too thrilled about living in Los Angeles, although the professional opportunities there were much better than at MIT. So progress at MIT had to come in a reasonable time scale.

Pretty much everything we wanted to do was objected to and rejected. We had to get around the departmental structure, which was then in the hands of Ted Martin and his deputy, Richard Schafer. But the organization of the department could be changed when we had the dean, Jerry Wiesner, on the committee. He was officially the head of this committee, although I did the work. We would meet, and Ted Martin would voice an objection, and Wiesner would overrule him, in effect. This was an untenable situation for Ted, and he resigned, though I do not know whether this was the only reason.

I'm surprised that if Norman Levinson supported this, Ted Martin would have such strong objections to it.

Levinson was certainly behind the appointment of quality people in pure math. Ted didn't do that, although he received much of the credit. Levinson was the main advisor on quality and the main judge. But Levinson was on the committee, too. Whether I took the thing faster than Levinson wanted, or certainly faster than Ted envisioned, that may be; I don't know. Ted's objections were always attributed to Reissner, who could always be counted on to say no if a negative answer were possible, and since he was the senior person in applied math, the implication was that his opinion should hold. Reissner decided not to be on the committee, which was for the best.

Applied Mathematics Appointments

How did the appointments go?

We appointed numbers of people in all of these areas—and they were really considered from all departments: meteorology, economics, computing. We participated in some of the early computing programs at MIT, mainly symbolic manipulation and numerical analysis, and had people in that area who have gone on here and elsewhere to fame and fortune.

We held to several standards that were put in place so that we didn't become too ingrown. For example, a new senior appointment had to either be exceptionally good—better than people we already had—or have a new capability. Not a capability when looked at by the specialists in that area, who always take a microscope and see vast distances between their activities that are not apparent to anybody else, but to others on the committee. And they had to really be strong candidates. Another rule was not to appoint anybody who could not also be appointed in another department. Those in meteorology had to merit consideration by meteorology if asked.

Did that work? It seems that the "pure" meteorologists could say, "Well, this guy has a lot of mathematics, but as far as real meteorology is concerned, he really isn't of high enough caliber."

Oh, I suppose every discipline of more than ten people is going to have that hierarchy. But it wasn't a problem. As I said, we had Jule Charney on our committee, and some of Charney's students were on our faculty. They have people who are very mathematical now and extremely good at constructing models. The advent of mathematics in other departments has been so rapid that every department at MIT is now but one facet of applied math.

Conversely, I can imagine a theoretical physicist saying, "We all do a lot of math. We don't need to be in a separate department."

Ah! All right! That's true, and two of our staff are physicists and could easily have been in the physics department—and yet they seem comfortably placed. I'm sure they collaborate with their colleagues; we all made such connections with other groups. Joint courses were encouraged. What's attractive about what we do? Because we're taking a general concept, with a general structure that may apply to economics as well. There are sound waves; there are waves in the economy; there are waves in traffic. Wave phenomena are quite common. The general study of shock waves finds application in all situations.

You said that all your appointees had to be of a caliber that would be appointed in another department as well, but of course you also made applied math appointments from within the math department itself.

For math we had to make that exception: we had to visualize math as a separate department to justify appointments that we were making in combinatorics. People like Dan Kleitman and Gian-Carlo Rota were in the department, but they weren't being treated as full members of pure math, and they wanted different placement. Kleitman really was a combinatorist who was applied; he had done—still does—a lot of work with government agencies, the organization of oil pipelines, etc. He was the person I thought would be most valuable to guide the discrete side, and I asked if he would join the committee.

So most of the MIT applied math people were on the continuum side in the early days, while discrete came later?

Yes. It happened only because we in continuum theory decided to do this and divide our resources to develop this unified concept of continuous and discrete applied mathematics. We were the only ones to attempt a union of applicable mathematics, both in the study of the mathematics involved and in its application to science and technology.

At one time, Marvin Minsky indicated interest in having artificial intelligence in applied math. But this meant absorbing an entire group, larger than our own, which was beyond our capabilities. We did make a few appointments, including Seymour Papert, who was an associate of Minsky's. Papert left to pursue his interest in computers for grade school education but we did eventually have a number of other people.

Teaching

How did the separation and growth of applied mathematics affect teaching in the department?

We introduced many courses. I think I counted ten that I introduced myself in applied math, three or four in advanced methods and an equal number for undergraduates, including calculus. Calculus was fiercely opposed in the department, because the enrollment here was really the strength on which the pure math faculty grew, although they didn't always do terribly well with it. George Thomas and a few other people did the service work for the department. Huge lectures. George at least got wealthy from his book, a calculus book which sold a million or something copies.[8]

But we introduced our own version nevertheless, as part of a complete professional curriculum. We were not competing in terms of quantity—we didn't have the staff for that—but we did offer high quality and a different approach: "Here's how scientists use mathematics." This was very much the approach that Lin embarked on with your father [Lee Segel] at the junior-level courses, which became two books.[9] Many courses wound up as books. Dave Benney and I wrote a calculus book,[10] and former students of ours also wrote books inspired, perhaps, from our notes. Many courses in various disciplines were added to the graduate curriculum.

Lin by this time had gone to work in astrophysics, so we had that as a component; new faculty in theoretical physicists added their own interests to the growing list of offerings. We divided the curriculum into continuum and discrete programs: discrete being statistics, combinatorics, and all of computing science, and continuous being solid mechanics, fluid mechanics, and whatever else—bio-fluids, chaos, etc. We were under a lot of pressure because of all this activity—we did a lot of teaching, and we didn't have the resources to hire enough instructors. At our request, the administration gave us a yearly supplement of money to hire new instructors, applied math instructorships, and funds for five premier student fellowships to award yearly, which we did. Wiesner also provided money to open up a laboratory

---

[8] Thomas, G. B. (1952). *Calculus and analytical geometry*. Reading, MA: Addison-Wesley; now in its eleventh edition as *Thomas' calculus*.

[9] Lin, C. C. & Segel, L. A. (1974). *Mathematics applied to deterministic problems in the natural sciences*. New York: Macmillan); Segel, L. A. (1977). *Mathematics applied to continuum mechanics*. New York: Macmillan; both reissued as part of the SIAM Classics in Applied Mathematics series.

[10] Greenspan, H. P. & Benney, D. J. (1973). *Calculus: An introduction to applied mathematics*. New York: McGraw-Hill.

for the study of genuine physical phenomena. And we hired Willem Malkus, an experimental and theoretical physicist from UCLA, who was then working on geophysical problems. He built a laboratory in the basement with funds from the Institute and government grants; he did fine research and had some excellent students during his tenure.

So it developed that way. Early specialization was not encouraged in this program. We tried to really show students what mathematics is, how it's used, and what its unifying concepts were. We taught mathematics that scientists and engineers would also need; our students came from, and could go to, many departments. We brought into the calculus not only its great historical successes but also current techniques that would receive greater attention later on: perturbation theory, asymptotics, numerical analysis. Our motivation was to give the best students the kind of course we would have liked to have taken. It would introduce concepts I had to learn by myself in higher courses that should have been presented earlier, but weren't.

Harvey Greenspan (left) receiving an honorary degree from the Royal Institute of Technology (Kungl Tekniska Högskolan), Stockholm, Sweden.

That's the kind of book we wrote, and that's why we didn't make a million dollars: too hard for the average student. The successful book is written for students who will forget everything they ever learned in five years and never use it. Ours was a structured book, starting with chapter one and advancing on from there.

It wasn't primarily the case method: "Let's look at a particular problem, and then we'll see what we can understand from there."

Well, no. There's a lot of technique to learn before you can play the violin. So there's a certain amount of basic mathematical technique you must have before you can play anything.

You yourself never took on many PhD students.

No, I never wanted to take on a lot of average students. I asked that they at least express a definite interest in some subject I was researching before I would give them something to do myself. This happened infrequently. And I never guaranteed their success. That's a big, big consideration for students, and I didn't fit their bill. Of the PhDs that I managed, three came over specifically from Sweden to complete their requirements, two from Israel, and a few from MIT. I mainly worked by myself, and also with post-doc instructors and research associates. I may have had seven to ten over the years. Some of them were really very good.

Chairman of Applied Mathematics

When did you become chairman of applied, and how did things develop at that point?

I took over the chairmanship of the committee, effectively in '64, but officially in '65. I moved fairly quickly to have the third floor of Building 2 consolidated as a location for applied mathematicians where they could congregate and communicate. I wanted a secretarial staff dedicated to our objectives and interests. So a number of such things were done.

After Martin, Levinson became department head. He didn't want to do this job, and I talked to him about it. I told him that he wouldn't have any more busy work than he did before, and he didn't. So Ken Hoffman, who was chairman of pure math, and I would often go to the science council meetings of the dean. Norman was an excellent head of the department. He didn't waste his time micromanaging. Most heads tend to fill their time with useless work, but he delegated all the routine jobs to other people. He wasn't someone the administration counted as on their "team" and would sometimes bring up faculty issues that don't ordinarily occur to the administration. And so when he wanted to resign after only two or two and a half years, they gladly accommodated him [laughs].

One of the interesting things about the applied math committee was that it elected its own chairman every two years. Which, again, was anathema to the administration, where an election was unknown. They never got used to that, and we always had to keep pressing to keep it that way.

We had about one-third the share of the department at that time, in faculty and in students, and we argued for one-third of the share of resources, a one-third/two-thirds ratio. We got that approved during Levinson's term, and that never changed, as far as I know. Applied math had more than paid its own way with new funding and more teaching—very little besides salary was from the department's ongoing budget. Unfortunately, most of the extra funds brought in have long since been absorbed into the general budget, ignoring the original reasons and continuing need for this support.

The structure of the department was reworked. The chairmen of pure and applied math were on the budget committee with the department head and resolved financial issues. Appointments from applied math and pure math were reviewed by a departmental committee, but if the departmental committee didn't approve these appointments, we could override them and go straight to the dean.

But we were functioning reasonably, and somewhere in the early seventies I thought it was time that we try for a separate department. I wrote a proposal for the creation of a separate department, showing that we were functioning that way, and that we *could* function better that way. It didn't succeed, mainly because Lin didn't support this move for personal reasons. And Levinson probably didn't support it either. So I was speaking through the wrong end of a megaphone, and Lin was transmitting even that at reduced volume.

Lin was afraid that we wouldn't get a large student base. He didn't want to be embarrassed by having a separate department with few students. I wasn't afraid of that because we were already teaching a high level of calculus with very good lecturers. It would not take much to make the course competitive to attract many more MIT students and educate them in the best way possible. The headquarters staff had never supported our efforts anyway, never really saw this course as worthwhile, never recommended it, never pushed it, never advertised it. I didn't care, but as a separate department, they would have found serious competition.

The students from the United States—aside from MIT, and Harvard, and a few other places—had a perception of applied math that wasn't exactly ours. As part of the mathematics department we found it difficult to attract first-rate students from other universities who didn't just want to play with the mathematics but were also interested in science. We never changed the popular notion of what applied math meant—nor could we do it, placed as we were.

When Ken Hoffman became head of the department, there was a great deal of trouble, because he kept trying to whittle away at what had been built up—and I kept blocking these actions. In '75, when I could see that the high point had been reached, I took myself out of contention. I stopped being chairman of applied math and left all these concerns to my successors. From that point on, like all sandcastles, the structure eroded. Attrition, lack of attention and will.

And that's when it stopped being an Institute committee?

Pretty much. Our status as an Institute committee, which had served us so well, soon disappeared. The argument given was that we were all one family now. Very few people have a taste for confrontation, even when it is necessary. That's often acceptable, I guess, but it is not a platform to build upon or even to maintain what already exists.

On the other hand, Kleitman became head of the department, as did Dave Benney. Not that having applied mathematicians as the heads of the department meant anything good for us. To keep the peace, they often bent over backwards, to our disadvantage.

And yet how successful would you say the applied math effort at MIT has been, as it stands today?

I think the people—the outstanding faculty we appointed—were a most successful effort.

The research environment was improved immensely: teaching was reduced from four to three courses per year at my initiative; salaries came from so-called "hard money" from the Institute and did not depend on securing grant funds, as was the practice in most other departments. Faculty had as much freedom to pursue their interests as I could give them, but they could not buy their way out of our teaching responsibility using outside support.

Our graduate students were required to diversify their courses somewhat, one or two courses of eight total, so that they experienced applied math outside their specific area of interest. The objective was to train students more broadly than the faculty had been educated; a written qualifying examination was given to test this requirement. This was difficult to administer and found little support from most students and faculty. It withered away for the most part after I ceased being chairman. I don't know what the status is at present.

But applied math in a mathematics department is an unstable adventure. Those on the continuum side are difficult to attract because there are many better professional opportunities with less related difficulties attached, and those on the discrete side are in overabundance. The same can be said about the students who applied for admission. The program veers naturally towards pure mathematics without constant attention and effort, like bicycle riding.

Another serious deficit was the lack of a charismatic leader who could inspire loyalty, unity, and willing commitment from the staff to a common purpose. The truth of the matter is that we didn't really have a father figure, like Courant or Goldstein—people, who, by virtue of charisma or a parental concern for everyone, sort of brought a feeling of family unity. Perhaps a separate department could have helped, but I am not sure of this. I could alone administer, lead, and protect the group from many things and treat people honestly and fairly, but I had no desire, time, or talent to give much in the way of an emotional component, TLC. And no one else could provide it either. I did my best with problems that affect most large groups: mental illness, alcoholism, marital difficulties, conflicts of interest, dishonesty, unacceptable behavior, personal and professional antagonisms, etc. There are few givers and many takers in every joint undertaking, and much time is taken by a few who think that any resource not spent on themselves is a waste.

Maybe the subject was just too broad for anyone to grasp in its entirety. We couldn't create professional contact or much interaction among the faculty, and the seminars were very specialized: aimed at experts only, as they must be. We started joint courses, principles of applied math, and taught it by team-teaching, two faculty members per semester per course. Two of us would take each course on continuum and discrete principles and try to unify and connect the presentation.

The attempt to develop statistics was not successful despite a serious and costly effort for several years. There was an Institute committee looking into this. I went to one meeting, which went like this: "Anyone have any ideas? No? Let's eat." I volunteered to find a statistician. So we looked over the whole list of people that were available. We made an offer to George Box from Wisconsin, who decided not to come. We wanted somebody from Stanford. But you know, good people don't have to leave.

Finally, Herman Chernoff accepted, and he brought statistics. We gave him a lot of support, in the sense that we approved many appointments that we wouldn't ordinarily have made. So we spent a lot of capital and reputation. Up to that point, no appointment that we ever recommended had failed. But with statistics, it was another matter. The candidates were just not impressive enough.

Herman did get a statistics lab set up, with Institute funds, in a room near the Sloan school, so that statisticians from economics, etc., could meet. And he too was moving in the direction of a separate department. I told him, "That's fine with me, but we can't support it with money, because we don't have it, and it's not appropriate! You can get more money on your own to do that, and I won't object." But it didn't succeed, and Herman finally went to Harvard.

And we had a number of efforts in various directions that were not complete successes. We had taken over a rather poor journal, *Journal of Math and Physics,* and we formed it into *Studies in Applied Mathematics,* which was supposed to highlight the activities of applied math at MIT. In this regard, it was not a success. It still exists, but it did not serve its purpose. Very few people contributed their best papers to this, because their best papers would then go unread. Rota even started his own journal.

Some efforts *were* complete successes. Combinatorics was a marked success, but it brought in a lot of people who really weren't applied: they were more towards the pure side, and they must be one of the bigger groups now. Physics was a good group. Computing science brought in new people since I left. Some of them are very good—and applied. So computing science I think was a success, and it probably will remain a success. I don't know about the new development of a biological group. Presumably the subject is sufficiently mathematized at the present time so that theoreticians can be of assistance.

The Computing Laboratory

I found a letter that Jerry Wiesner wrote to you saying that "the Administrators of the Sloan Basic Research Fund have approved allocation of $30,000 for partial support for a computing laboratory in the Department of Mathematics. As I agreed with Dean Alberty, the remainder of the $50,000 you requested in your proposal will be provided by him and *the Department of Mathematics." The letter was dated June 30, 1969. What do you remember about getting a computing presence off the ground in the early days?*

The early days are sort of interesting. We had a room where there were old electro-mechanical devices that we used from the forties onward. We looked over the situation and decided that the very least we could do was to bring in a couple of machines that could do long division or square roots. So as a trial, we bought only a couple of Monroe 620 calculators, because I had a feeling things were changing very quickly. The Monroe had four transistors. And within about four months Hewlett Packard came out with their HP-35, fully capable hand calculators, and we bought one for every faculty member.

At about this time, people in the department were also doing calculations on the older IBM machines, which required a deck of cards, and equipment was purchased to facilitate such use of the Institute's main computer.

The punch cards: "Do not fold, bend, spindle, or mutilate."

That's right. They would write a program using FORTRAN at first, then punch and compile a deck of cards. But we decided to upgrade our machines further and set up a laboratory, where it would be useful for both computation and Macsyma, a symbolic language that was being created in electrical engineering.

Space was grabbed, as much as I could have. The former head, Ted Martin, had given away the entire basement area, which had been a student model shop, etc. The chemistry department inherited most of it. So when I took over, I asked for rooms down there for computer science, computer equipment, and a machine shop for Malkus. Wiesner and other people generally supported it. Chemistry was in the process of constructing a new building, so it was easier for them to release enough for us. We had a place for another computing lab.

A lot of our new PhDs began to show that the problems that were being treated in electrical engineering and thought to be of extreme use to everybody were really missing the point somewhat. Exact integration that produced answers five pages long was not of really great use. We'd rather have a short formula—less exact, but very much more informative. That was called an asymptotic expression, because you looked at extreme values to get it.

Some of the new PhDs in the math department began to program this on the computer and it filtered back into electrical engineering, making Macsyma a more useful tool. Now, of course, symbolic manipulation is very highly developed. But for a long time I would try some iconic problems that machines were not able to do, just to see how it was going. I haven't made this test recently.

We set up a large, well-equipped computer laboratory, a task undertaken by Steve Orszag, who was a first-rate talent working in fluids and in numerical solution of physical problems. We received five or ten thousand dollars extra to establish this center—though very much more was actually spent: Orszag got a contract with money to set up things as well, and we used part of our budget to pull together whatever it cost. And we got another five or ten thousand dollars to set up a machine shop, which we still have at MIT to this day.

Consulting

You mentioned Kleitman's consulting work with government agencies. He said that you also did some consulting about fuel tanks?

Most applied mathematicians did some consulting. It was Kleitman's associate who approached me to do a study of liquid natural gas containers: how safe were they, and what modes of collapse were possible. The actual facility had a tank with a dike around it for safety. I said, "That's a fairly simple experiment. I'll show you." So I took a foam cup full of water and placed it upside down in a lab or Petri dish of similar volume. The amount of water in the cup was equal to the amount that the dish could contain, just as it was for the LNG facilities. I just picked the cup up, and much of the water overflowed because of the impact of the water bore, or shock wave, against the containing wall.

"Oh yeah, that's good. Now we have to have it so that the congressman can appreciate it."

So I said, "What do you want? Do you want a model with little people around?"

No, a deeper study was required. So I did a deeper study for them in the lab. It was mainly a model of the interaction of a shock with a barrier. The results were basically the same.

I was also a long-time consultant in Sweden, for *the* Swedish manufacturer of centrifuges. I had written a book on rotating fluids[11] and I was asked by people I knew to come to Alfa Laval in 1980 and see whether I could help establish a research group. I did, and it became a long-term and very enjoyable association with exceptional young researchers.

---

[11] Greenspan, H. P. (1968, 1990). *The theory of rotating fluids.* New York: Cambridge University Press; revised edition, Brookline, MA: Breukelen.

A new centrifuge designed and patented by Harvey Greenspan (patent no. 4,842,738; date 6/27/89), assembled and tested in the applied mathematics laboratory.

The Final Arbiter

The very term "pure mathematics"—math for its own sake—seems to imply a hierarchy, with applied math as something of a second-class hybrid. Yet you don't seem to feel that at all.

Well, the development of mathematics, for the sciences and for everybody else, does not often come from pure math. It came from the physicists, engineers, and applied mathematicians. The physicists were on to many ideas which couldn't be proved, but which they knew to be right, long before the pure mathematicians sanctified it with their seal of approval. Fourier series, Laplace transforms, and delta functions are a few examples where waiting for a rigorous proof of procedure would have stifled progress for a hundred years. The quest for rigor too often meant rigor mortis. The physicists used delta functions early on, but this wasn't really part of mathematics until the theory of distributions was invoked to make it all rigorous and pure. That was a century later! Scientists and engineers don't wait for that: they develop what they need when they need it. Of necessity, they invent all sorts of approximate, *ad hoc* methods: perturbation theory, singular perturbation theory, renormalization, numerical calculations and methods, Fourier analysis, etc. The mathematics that went into this all came from the applied side, from the scientists who wanted to understand physical phenomena. You most likely couldn't even learn these things in a math department. That's where our curriculum came from. It was called Methods of Applied Math at the graduate level.

So much of mathematics originates from applications and scientific phenomena. But we have nature as the final arbiter. Does a result agree with experiment? If it doesn't agree with experiment, something is wrong. You haven't put the right facts into the model; you haven't analyzed the model correctly. So there's formulation of a problem, whether it's about tissue formation, biological growth, whatever, followed by its mathematical statement. And then there has to be a solution of that formal problem.

Well, these are two different tasks. The solution of the problem may involve you inventing mathematics, or doing irregular things to get to the answer. That's what physicists do. But then you've got to say, What does it mean? Is the result or conclusion right? If theory says that such-and-such will happen or be seen, or that the structure of a star is such-and-such, then experimental corroboration or verification is necessary. And so nature is the final arbiter. If it doesn't agree, then the model is not terribly good even if the mathematics is elegant and correct. Something is wrong or at least not quite right or complete. Another tack might be tried that you feel is justified but can't be proved. All right. In applied math, we don't write in a theorem-lemma format.

But rather?

State the physical problem in mathematical terms; give its approximate solution and describe its relevance to the original problem and the data or experiments. You see? That's *science.* Scientific study has, as its vindication, that the results are correct when compared to nature and the available data.

~ December 20, 2006

Bertram Kostant

Chemistry First

J.S.: *Several of the faculty members represented in this project were initially more attracted to another science, most commonly to physics. I read that you were fascinated with chemistry as a boy. What was your early school experience like?*

B.K.: Well, I went to Stuyvesant High School [in New York City] during the war, and I essentially did very little school work. I spent my time either shooting pool or doing chemical experiments in the basement of my home, where I had a very well-stocked laboratory. I would normally stay up until 2 o'clock in the morning and wake up—or rather my mother woke me up—at 6 o'clock to take a combination bus/subway from Brooklyn to high school. Stuyvesant was on 15th Street between First and Second Avenues in Manhattan, and it took about an hour to get there from Borough Park. My household consisted of my mother, an ailing father for whom I and everyone I knew had tremendous respect, and two older sisters. My older brother was already a medical doctor, and the husband of one sister and the fiancé of my other sister were in the armed forces. At that time I had the reputation in the family, unlike that of my brother, of being a rather wild and undisciplined kid.

I *loved* chemistry at Stuyvesant High School. One teacher, Mr. Lieberman, sticks out in my memory. His superb lectures, although no doubt unsophisticated from the perspective of college chemistry, were devoted mainly to teaching us the formulas of chemical reactions. There were no complications involving energies of reaction, equilibrium constants, etc. Be that as it may, here I am more than 60 years later, and I still remember most of those formulas. Of course, without such things as equilibrium constants, one is not really doing science. In fact, I was not doing science in my home laboratory: I was more interested in producing exotic reactions. I was fascinated with the magic and *power* of chemistry. While making gun powder I accidentally blew my hands up and had to be rushed to a hospital. In experimenting with white phosphorus I accidentally set fire to the house. Fortunately for me, my father was vacationing in Florida at the time.

There was a place called Eimer and Amend near my high school where I used to buy a lot of chemical equipment. There were no restrictions on what one could buy. For example, I had ammonium nitrate, the chemical which blew up the Murrow Building in Oklahoma City. It is interesting, because if you heat ammonium nitrate gently, laughing gas, nitrous oxide, is produced. During the war I put on exhibits of thermite. We were afraid the Germans would drop thermite bombs. Thermite has a very striking chemistry: You take iron oxide, Fe_3O_4, and you add aluminum powder, and it is ignited with magnesium strips. The aluminum replaces the iron and you get this intense heat. It is a terrible weapon since water does not douse the flames. I set off the thermite in a sand bucket in front of a neighborhood audience to show how this reaction works.

Bertram Kostant climbing a tree.

I wasn't unusual in having a chemistry set: it was common. Kids didn't have computers; they had Erector sets and chemistry sets. In fact, there's an interesting story about that. Once at a meeting of the National Academy [of Sciences], we were bemoaning the fact that so few young people were going into science. There was a crowd of maybe over a hundred people discussing this issue, and one guy got up: "Who in this room had Erector sets when they were kids?" *Every* hand was raised. It is a very constructive kind of toy, where you had to create and build. I loved my Erector set. It was interesting that having an Erector set was such a uniform thing for all these scientists.

Undergrad at Purdue

You went to Purdue as an undergraduate. What took you there?

I applied to many places, but my applications were rejected. I hardly ever opened a book in high school. I got straight As in math at Stuyvesant even though I did no work, and I got 100s in my mathematics Regents tests. But I also had many Ds in nonscience subjects. And I was competing against returning vets because I graduated in '45, and the returning vets were given first choices. I was very lucky to get into Purdue: I was initially rejected, but then I got a telegram one day. It said that if I answered within 24 hours, I would be accepted, and that changed my whole life. I was later, in 1997, to receive an honorary degree from there.

Purdue was teeming with all these guys wearing bomber jackets, and I was an outsider from New York. I was housed in a huge dormitory full of returning pilots from World War II. I believe Gus Grissom and other future astronauts were my fellow students. [Neil] Armstrong later went to Purdue.[1]

I did very well in math, but I didn't like many nonscience required courses. I was ready to quit Purdue after my sophomore year, mainly because I hated having to take French. But also I didn't do well in English. As a matter of fact I wasn't reading books yet, though later I did in fact learn to read and enjoy reading at Purdue. They had this wonderful reading room in the student union; it was very cozy and wood paneled, and I would sit in an easy chair and start pulling books off the shelves, and that's where I first learned to read, truly.

---

1 Over twenty Purdue alumni have participated in space flight over the past 45 years.

The dean was Bill Ayres,[2] who was a topologist, and somehow he empathized with my problems. I never met him, but I heard that he also had flunked French in college. He gave me special permission to take three graduate courses—and only these courses—during my junior year as an undergraduate. The courses were wonderful, and I *loved* them. They opened a whole new world for me. I worked hard and did extremely well. It was a life-altering experience.

One of the courses was taught by a refugee from Germany who survived by managing to keep one step ahead of the Nazis. His name was Arthur Rosenthal. He was chairman of the department of mathematics at the University of Munich during the early 1920s. One of his students was Werner Heisenberg. I later met Heisenberg on a visit of his to MIT, and he told me he remembered Rosenthal very well. Rosenthal was also mentioned in [David C.] Cassidy's biography of Heisenberg, *Uncertainty: The Life and Science of Werner Heisenberg.*[3]

Rosenthal's course in measure theory was absolutely beautiful. Some of the material appears in a well-known book, *Set Functions,* by Hahn and Rosenthal.[4] I was very happy and honored when he presented a copy of his book to me in which he had inscribed, "To my brilliant student, Mr. Bertram Kostant." There were many nights when I had dinner with Rosenthal. Occasionally other students joined us. He was not married. Here he was, this kind and gentle European intellectual, in Lafayette, Indiana. I imagined he must have been lonely and sad here. But perhaps he wasn't. It was such a rare privilege to be able to have a faculty member as a dinner companion. He had coauthored some articles for a well-known German mathematical encyclopedia,[5] and he would tell me stories about European mathematicians. For example, he told me the story of when [C. L. Ferdinand] Lindemann, who was his teacher in Munich, proved that one couldn't square the circle—that π was not an algebraic number. This was a problem going back to ancient times. Rosenthal told me that Lindemann was vacationing in the Bavarian Alps and took a walk in the woods and came back from the walk with a big smile on his face. Someone joked, "You probably solved the problem of not squaring the circle," and he says, "Yes, I have!" Stories about Heisenberg; stories

[2] William Leake Ayres was dean of the School of Science Education and Humanities, which included the Department of Mathematics and Statistics.

[3] W. H. Freeman, 1991.

[4] Hahn, H. & Rosenthal, A. (1948). *Set functions.* Albuquerque: University of New Mexico Press.

[5] *Encyclopädie der mathematischen Wissenschaften* (Leipzig: Teubner).

about the many European mathematicians he had known. From Rosenthal I learned about the long cultural history of mathematics, that it existed and still exists as an area of thought where one could contribute new ideas to the many structures that it encompasses.

After my junior year I decided I would "do" mathematics. I of course knew that mathematics was important in science, but it was the beauty of the subject, not the applications, which interested me. There was also a more mundane reason for pursuing mathematics. Arguments back home in Brooklyn were generally won by people with fast-talking, persuasive personalities and little else. But here something existed in the world where there was no debate as to the truth of the statements made! It wasn't a question of personality; it was a question of actually *proving* something! That was so appealing. It was marvelous to see that something like that existed.

My whole world changed at Purdue. The mathematics had the unexpected effect of helping me grow intellectually in other directions. I also learned a great deal from friends I made there. One was a graduate mathematics stu-

Bert Kostant in his late 20s.

dent, Jimmy McKnight: [Paul] Halmos writes about him in one his books. Jimmy was from Alabama. Although he and I came from totally different backgrounds, we became very close friends. I also learned a lot of mathematics from him. Then there was David Caplan, who was a physics graduate student and a marvelous pianist. He played Beethoven's *Appassionata* as well as any professional I have ever heard. We and a few others—intellectuals, non-Purdue types—formed a little society. I forget the name: Formal Logic or something to that effect. We used to meet once a week, have an invited lecturer, maybe go out to a good restaurant, and then engage in an all-night bull session.

I remember a lesson I learned from one of the members of our little group. He was a son of the Finnish ambassador. One of the things he taught me was logical empiricism. I would argue the Newtonian point of view, that if you knew the position and velocity of every particle in the universe at one time you would know everything from then on. He says, "No, that's not true. Everything is probabilistic. You can't even prove that if you drop something, it's going to fall." This was such a *radical* idea for me! That mathematics was not to be identified with physics; it was only a model. I'd argue with him day in and day out, until finally I was convinced that he was right. What he taught me has stayed with me all my life. This was also true of other new ideas I had picked up as an undergraduate at Purdue.

At Chicago with Irving Segal

You went to the University of Chicago on a fellowship, which seems to have been another high point. In your obituary of Irving Segal in the Notices of the AMS, *you wrote, "The intellectual atmosphere was such that one was made to feel that doing mathematics was the most important thing one could do with one's life."* [6]

Chicago was fabulous in the early part of the fifties. Robert Hutchins had put together this great university.[7] Chicago still carried the aura of where the Atomic Age was born. Walking down 57th Street in Chicago, one sensed that this was one of the great intellectual centers of the world. Marshall Stone was the one who built the math department. Happily for me I was there at the high peak of its fame, with André Weil, S. S. Chern, Saunders Mac Lane, Antoni Zygmund, and Adrian Albert. Albert arranged for me to get my first government fellowship, an Atomic Energy Commission fellowship. Among the younger faculty

6 Kostant, B. (1999). Irving Ezra Segel (1918–1998). *Notices of the AMS, 46,* 661.

7 Robert Hutchins was president of the University of Chicago from 1929 to 1945 and chancellor from 1945 to 1951.

there were Segal, Halmos, Irving Kaplansky and Ed Spanier. In physics there was Enrico Fermi. In chemistry there was Harold Urey. Coming to Chicago, I was amazed at how much people *knew*. They knew much more mathematics than I did. Here again I learned much from my fellow students.

Do you have any memories of Marshall Stone?

Marshall Stone was the son of Harlan Fiske Stone, who was a chief justice of the Supreme Court and a real Yankee.[8] He had a lot of his father's sense of right and wrong, a guy with tremendous integrity, but he also had a terrible temper. He was also the world's worst teacher. One of the stars at Chicago was the great André Weil, who also had a terrible temper. Stone was the only one with the courage to talk back to him. So I had the following incident.

As a new graduate student, I was late getting some papers signed. I knew Stone would be angry with me, but I had to get these things done. So there I was, outside his office, hesitating to knock and have him do what was supposed to have been done at some earlier time. At that moment André Weil came rushing by, *threw* me aside, opened the door, and went over to Stone's desk, pounded the desk, and screamed about something that was bothering him. Stone just sat there and looked at him, then saw me at the door. He got up and said, "OUT. That man was here before you." So I fell in love with Stone. I mean, a graduate student being treated this way—this, to me, was American democracy at its finest.

How did Irving Segal become your thesis advisor?

Segal was not reputed to be a good teacher. But he taught the 260 Analysis sequence, which was required for potential PhD students. Segal worked very hard on one course I had to take. He wrote up the notes before every lecture, and they were very precise. It was much better than I expected. But he also had a forbidding style that seemed to discourage having a personal interaction with him. I had never spoken to him.

One day I noticed that there was a mistake in one of his lemmas, and a friend of mine, Bob Heyneman, later at North Carolina, told me, "Look, why don't you go to Segal's office and explain this error to him?"

I said, "Well, I don't want to start in with him." But Bob convinced me otherwise. So I went and knocked on his door. He let me in, and I proceeded to point out his error with a counterexample. It turned out it wasn't very seri-

[8] Harlan Fiske Stone (1872–1946) served as dean of Columbia Law School and U.S. Attorney General. He was appointed Associate Justice of the United States Supreme Court in 1925, then served as Chief Justice from 1941 until his death.

ous. He had simply forgotten to put in some natural assumption. The matter being settled, I was ready to walk out. As I was leaving he turned to me and said, "What do you know about Lie groups?"

It so happens I was taking my first course in Lie groups at the time, and I told him that. He said, "Okay, hold it." He went back to his desk, and he started writing and writing and writing.

"Okay. [Peremptory tone.] Here's your PhD problem."

I was stunned. I wasn't even asked to consider whether or not this is what I wanted to do. But I felt it would be impolite to say no. It was amazing. I walked in to discuss some minor point, and then here I was walking out with not only a thesis advisor but also a thesis problem!

As a thesis advisor Segal was very kind to me. He wrote numerous letters on my behalf. Another nice thing he would do was to have open house every Friday at his apartment. It was great. He lived right off the Midway, and the open house was the only chance for graduate students to meet both resident and visiting faculty on a social basis. I looked forward to those Friday evening get-togethers at Irving's apartment.

Princeton

You were also at Princeton before coming to MIT.

Yes. Irving Segal made it possible for me to get all sorts of fellowships. Initially I had an Atomic Energy Commission fellowship. At some point the AEC stopped supporting graduate students. Instead the NSF came on line, and I received one of the first NSF fellowships. Segal's support got me into the Institute for Advanced Study, and this was actually before I had my doctorate. I received my doctorate in '54, but I was admitted as a visiting member [to the IAS] in '53. I was there until '55, though I went back to Chicago in the summer of '54 for my thesis defense.

Some of the world's foremost scientists were at the IAS at that time: Einstein, John von Neumann, J. Robert Oppenheimer, Hermann Weyl. Whom do you remember from that period?

I remember all of them. I was very fortunate in that I seemed to be at the right place at the right time: this would have been the only opportunity for me to meet not only Einstein but von Neumann and Hermann Weyl, since all three were dead within a few years.

I met Einstein on Good Friday in 1955, about a week or so before he died. He was pacing around Fuld Hall looking for Jack, the driver of the Institute

Bert Kostant, March 1953.

bus. But Jack was off that day, apparently because of the Good Friday holiday. Einstein often walked home with Gödel, but I think Gödel was not around. I was the only one sitting in the Fuld Hall common room, and I said, "I'll be happy to drive you. It will be an honor."

"Ohhh." [Waving off the compliment.] So I drove him to 112 Mercer Street. I was very nervous. God forbid I should have an accident! Einstein was a hero. Even my sisters, who didn't care about science one bit, regarded Einstein as a near deity. Once, when I invited my sisters to the IAS, I took them into Einstein's office. Office doors were generally unlocked when I was there. My older sister, who was nine years older than I, went out of her mind. [Imitates high, excited voice:] "Einstein! This is Einstein's office!" She was running around in a delirious state looking for souvenirs. The younger of my two sisters was very proper and was screaming at my older sister for behaving this way. We were both running after her, taking things she wanted out of her hand. It was like a scene from Charlie Chaplin.

What did you talk about when you gave him a ride?

He was extremely friendly. He didn't want to get out of the car. We had a very pleasant talk. At one point he asked me what I was working on. I said Lie groups. He said, "Oh, that will be very important." I was so pleased with that comment and rather surprised that he knew who Lie was. He talked about why he never learned much mathematics. Among a lot of physicists, he said, mathematics is a diversion away from physics.

And we talked about the Oppenheimer affair.[9] I sensed he didn't like Oppenheimer. Besides a personality conflict it was known that Oppenheimer didn't take Einstein seriously any more because of Einstein's opposition to quantum mechanics. Einstein wrote this very famous paper with Podolsky and Rosen,[10] pointing out that if quantum mechanics were true, you'd get this obviously impossible situation, where two photons a billion miles apart would know properties of each other: in effect the paper was saying how ridiculous this is. But as it turned out, the photons do have this mysterious property. According to [Armand] Borel, Einstein regarded Oppenheimer as a man without a heart—there's some sort of German expression along these lines. Atle Selberg also confirmed to me the mutual antipathy between the two.

Einstein's rejection of quantum mechanics may be understandable for the following reasons. Germany had great schools in many areas of mathematics and physics, but I think there was very little communication among the various schools. Algebra was highly developed, for example, but I believe that this was an area in which Einstein was not strong. Initially, say around 1920, I think this was also true of Heisenberg. When Heisenberg went into physics—moving away from Lindemann's influence—he was getting these numbers for atomic spectra. It took him a number of years and the influence of Max Born at Göttingen to realize these were the eigenvalues of a matrix. Rosenthal knew measure theory, but he didn't know very much about linear algebra—there was no linear algebra in his classes. So I got the impression that there were many German mathematicians who, while first-rate in one field, knew little about other fields.

You mentioned meeting Hermann Weyl and John von Neumann as well.

I told the following story in a lecture when I received the Steele Prize. Hermann Weyl was one of the great developers of Lie theory. One of his accomplishments, in the 1920s, was in the area of the finite-dimensional representation theory of semisimple Lie groups. Later, in the early 1950s,

9 After serving as scientific director of the Manhattan Project, which built the atom bomb, Oppenheimer was stripped of his security clearance by the Atomic Energy Commission in 1953, ostensibly because of his association with Communist organizations during the 1930s. His defenders claimed that the true reason was his opposition to the hydrogen bomb.

10 Einstein, A., Podolsky, B., & Rosen, N. (1935). Can quantum-mechanical description of physical reality be considered complete? *Physical Review, 47,* 777.

the infinite-dimensional representation theory was being developed by Harish-Chandra, who was visiting the Institute. Harish-Chandra had been a physicist in England. He came to America and seemed to have learned Lie theory from Chevalley at Columbia. In the process he wrote a beautiful set of notes, which I read and found to be marvelous. I was quite friendly with Weyl, and I told Weyl about Harish-Chandra's efforts to develop an infinite-dimensional representation theory.

Weyl was *very* skeptical. He didn't think there was much of a theory there. But then one day I found out more, maybe from Borel, about what Harish-Chandra was doing. One of his great theorems had to do with what we now call a Harish-Chandra module. I explained this to Weyl. When he heard this, Weyl *completely* reversed himself. He now believed that there was indeed an infinite-dimensional representation theory to be developed. Most of the early subsequent approaches to the infinite-dimensional representation theory were via analysis, but Harish-Chandra's results opened the door to an algebraic approach. This is the direction I took and it led to my Steele Prize paper.

I was fortunate to get to know [John] von Neumann as well. I went into his office one day and told him a theorem I had proved, now known as the linear Kostant convexity theorem. I thought this might interest him, since it was related to one his results. In one direction it wasn't all that hard, and he was quickly able to see a proof. But the other direction was fairly subtle and was quite new to him. I was pleasantly surprised that he gave me, I think, an hour of his time, inasmuch as he was a member of the Atomic Energy Commission and very busy with that kind of activity. I was also pleased that he turned out to be warm and friendly.

I have very bad sleeping habits from high school, and going to college I thought all mathematicians were a little bit wild and would stay up to all hours of the night and work. I was kind of shocked when I went to the Institute to find out they lived 9-to-5 lives. One night I was the only one at 3 o'clock in the morning in the library in Fuld Hall, and it was very dark. I couldn't even see my way; I remember walking down the hall and coming near von Neumann's office, and there's this guy with a *gun,* right in front of von Neumann's office. I almost got a heart attack! It was an FBI agent. He was guarding von Neumann's office, because von Neumann had all these secrets in his office. But he didn't say a word until I almost banged into him.

I also had some experience with Oppenheimer. He was very bright, but his personality oscillated between friendliness and arrogance. The mathematicians disliked him. When I had been there for about two years, I needed to find a future teaching position. I was happy to receive three offers, but I needed advice as to which offer I should take. I imagined that Oppenheimer could help me, so I went to him and asked, "Can you help me make a selection here?"

His only comment was, "Well, when *I* was your age, I had 28 offers!"

From 1955 to '56 I was a Higgins lecturer at Princeton. This gave me the opportunity to interact with a brand new selection of very prominent faculty. I met Emil Artin—that's Mike Artin's father; Solomon Lefschetz, who submitted some of my early papers to the *Proceedings of the National Academy of Sciences*; and Don Spencer. I was very close to Don Spencer, and I would get private lectures from Emil Artin.

From Berkeley to MIT

Then you went to Berkeley, where you moved very quickly, academically speaking.

Yes, I went to Berkeley as an assistant professor and was there from 1956 to 1961–1962. During that time Berkeley was experiencing a period of tremendous growth, and it was nice to be part of it. As a consequence of a number of well-received papers, I was awarded Sloan, Guggenheim, and Miller Fellowships and was given very accelerated promotions. I had one PhD student at Berkeley, Jim Simons, who is known for his contributions to Chern-Simons theory. He was chairman of the Stony Brook mathematics department in the 1970s, and at present he has achieved considerable prominence as a financier and a patron of mathematics.

What—and who—brought you to MIT?

While at Berkeley, I received an offer of a full professorship at Chicago, mainly due, I suspect, to Irving Segal, my PhD advisor. But one doesn't *leave* Berkeley easily. I had been very treated well there; we had a lovely house up on Creston Road, and I really hated to leave Gerhard Hochschild, who was my closest friend. I took a trip to Chicago to look it over, but I decided, no, even though Chicago was where I received my doctorate.

But then Segal moved to MIT, and I suspect that he and [I. M.] Singer and [Warren] Ambrose arranged for me to get a full professorship offer here. Irving Segal is a *very* persuasive guy, and for me it's very hard to say no to him. He took me around all over Cambridge and Belmont: "all these great places you can live." He was a very good salesman. And in a certain sense I was getting tired of Berkeley. When I first went there, I thought I was in heaven. But the scene never changes. You look at the bay, it's beautiful, but there's a certain stagnancy built in. People talked about their gardens, and I was getting a little bored with it. I have a lot of Brooklyn in me [laughs] and longed for a place with a higher energy level. So I accepted the offer from MIT. But the decision wasn't easy.[11]

When I got the MIT offer, Berkeley counter-offered with a full professorship. This kind of promotion, at that time, was quite unusual. The norm, if I remember correctly, was six years as assistant professor, four years associate professor, and then full professor. Another reason to stay, aside from the fact that my research was going well, was that Berkeley had been experiencing an enormous growth over the previous five years. I don't want to exaggerate my role in this, but I was part of a group of young Turks who agitated for change in the chairmanship. A change indeed was made with John Kelley becoming chairman.[12] Kelley brought in some very good people, mainly from the Midwest, like Hochschild, Chern, Spanier and Rosenlicht. Sadly, I was leaving Berkeley during this growth period.

Who were the personalities who stood out when you got to MIT? Whom were you close to?

Well, first of all, Ted Martin was chairman of the department. He was very much of a hands-on guy. He took his responsibilities very seriously and yet he maintained a contagious sense of humor. He was from the start very friendly to me, and I was appointed to the department hiring committee—the split

---

[11] Calvin C. Moore, former vice chair of the Berkeley mathematics department, portrays this in the context of Berkeley's unrealistic assessment of what was required to keep such an able mathematician, given the competitive environment of the time. "It was argued that [Kostant's] appointment would strengthen algebra at Berkeley, which it did, but it achieved a whole lot more. His work on Lie algebras and Lie groups expanded to work in differential geometry and work on the geometry and algebraic structure of homogeneous spaces, representation theory, and cohomology theory, areas that he pioneered. He won very rapid advancement, being promoted to associate professor in 1959, after only three years, and then to full professor in 1961. He was much sought after, and many universities came calling ….

"Kostant's case well illustrated the red-hot job market in mathematics and the reluctance of the Berkeley administration to respond to it at the time. There was a raging battle between the department and the administration over the rate of Kostant's advancement. The end result was that Kostant went on leave without salary at MIT for the year 1961–1962, and then finally resigned from Berkeley effective July 1, 1962…. The loss of Kostant was a serious blow indeed for Berkeley, and perhaps the only positive outcome of his departure was a more realistic attitude on the part of the administration to the realities of the mathematics job market." Moore, C. C. (2007) *Mathematics at Berkeley: A history* (pp. 165–166). Wellesley, MA: A K Peters.

[12] John Kelley arrived at UC Berkeley in 1947 and served two terms, 1957–1960 and 1975–1978, as chair of the mathematics department.

between pure and applied was to come later. I believe I was the youngest member of that committee. With a small committee, decision-making at MIT seemed much more efficient than at Berkeley, where everybody with tenure met about hiring decisions, people made long speeches, and often little got done. On the other hand Martin would occasionally make decisions without consulting the committee. I remember that Warren Ambrose, in particular, was very angry with Martin for doing this.

It was always interesting to talk to Ambrose. He was free-thinking and very unconventional. Although he and I had very little mathematics in common, his personality resonated with me. There wasn't an ounce of deceptiveness about him: what you saw and heard from him was what he was.

Much later on I became close to Norman Levinson. Socially, not scientifically: Norman had been a Communist in the thirties, but in the sixties he totally reversed himself. He was very unsympathetic to the student rebellions of that period. He would often come into my office, and we would have heart-to-heart conversations about the way the academic world was changing. Like Ambrose he would always say exactly what was on his mind, but his opinions were diametrically opposite to those of Ambrose. Norman was very self-confident, but he was never pushy. He listened carefully to other people's opinions and thought about them. He was also an *extremely* powerful figure at MIT. He could go into the president's office and get what he wanted. He knew all these administrative bigshots—president, provost, and whomever. They were all friends of his. I was very deeply saddened when he died. In some sense, I feel that the math department has never really ever recovered from his departure.

Did you have any contact with Norbert Wiener?

When I first came here, Wiener came into my office one day and said [low, conspiratorial voice], "Am I still the greatest mathematician in the world?"

[Stammers incoherent syllables] "Well, I, guh, bl—…." He told me that he was descended from Maimonides. How would he know anything like that, a thousand years ago? He was very insecure about his reputation. His two attempts at novels emphasize being cheated or not getting proper credit for things. There are so many Wiener stories. Like that time at a dinner party he went upstairs to get something, but having gone upstairs, he thought he was going to go to sleep. So he got undressed and went right to sleep, in the middle of a dinner party.

I would have liked to have had a serious conversation with him. Sadly, he died not long after I arrived at MIT.

Did you remain close to Irving Segal?

To some extent initially, but I eventually broke with Irving, as many of his former students did. I finally was able to say no to him. He was very insistent on what you should do with your time—he wanted you to work on things that interested *him*. He was a very passionate person about math and science: that's what attracted so many people to him. He made it exciting. You *sensed* his love of mathematics, particularly as it applied to physics. But you had to see things his way. As a teacher he was very kind to his students. But once you became a colleague of his, it was a different matter.

And I was going off in my own direction. My courses in Lie groups attracted large audiences. I supervised nineteen PhD students while at MIT. Originally they were students in my classes. Some of the nineteen are now quite famous mathematicians. I still keep meeting people who say, "I was in your class," and tell me how the classes affected their mathematical perspectives. Research-wise things were going very well: for example, I formulated the ideas of geometric quantization a few years after I arrived at MIT. At the end of the sixties, I traveled to Budapest, where I met I. M. Gelfand, one of the great figures of twentieth-century mathematics. It was the first time he was permitted to leave the Soviet Union. I am happy to say that he and I have been close friends ever since that meeting.

Representation theory eventually became one of the strongest groups in the department.

Well, I think I made a good case for Lie theory as having a central role in mathematics. It had relations with so many other areas. It tied up with topology. It tied up with both differential and algebraic geometry. It tied up with physics. It also tied up with algebra and harmonic analysis. People might very well have felt that they could relate Lie theory to their own fields.

I think it is fair to say that at MIT people don't get appointments because they are close to someone or work with someone in the department. One really has to make a strong case for an appointment. At many weaker departments, prizes or membership in well-known academies play a major role in selecting people for appointments. One of reasons I respect MIT is that here we are sufficiently self-confident to make our own judgments, rather than being influenced by the fact that some committee somewhere decided to give this or that person some honor. I am very grateful to my colleagues at MIT for accepting my recommendations for appointments in Lie theory: Michèle Vergne, Victor Kac, George Lusztig, and David Vogan, who was also a student of mine. Aside from Michèle Vergne, who later left, our group became the nucleus for further appointments, making MIT a world center in Lie theory.

In these days of collapsed distances—telephones, e-mail, jet travel—is it still valuable to have colleagues in the office next door or down the hall who are working in your field?

One can give a surprising number of examples where a critical mass of people in close proximity produces an explosion of results. I am thinking, for example, of the school in algebraic geometry centered around Oscar Zariski at Harvard, or topology centered around Lefschetz at Princeton, or Bourbaki in Paris after the war, or analysis at NYU. Other classic examples are the school centered around Gelfand in Moscow, where they have six-hour seminars, or Göttingen in the 1920s, where quantum mechanics was born. I think when you get a group of talented people together, a spontaneous combustion can possibly happen. It's mysterious. Many great cultures seem to arise when there's a whole collection of people in communication with each other and doing similar things: Italian painters in the Renaissance or music in Germany in the late 1700s and early 1800s; Greece in the third/fourth/fifth century, B.C. I don't quite understand why, but I think that with a high density of talented people, the whole is greater than the sum of its parts. There is some sort of a strange, accelerated interchange of inspiration and ideas.

Have you had any fruitful dialogue with people in the department outside your immediate area?

I have profited from conversations with analysts, probabilists, topologists, algebraic geometers, and combinatorists in our department. I especially spoke a great deal with Gian-Carlo Rota. The topics included Hopf algebras and combinatorics. In the sixties and seventies Victor Guillemin and I had many conversations on symplectic matters and geometric quantization. I am cited in a number of his works with Shlomo Sternberg.

Teaching

You mentioned that early on in your career you had taught a very large calculus course at Berkeley. Did you teach undergraduates here as well?

Yes. Initially I was teaching only graduate courses: I enjoyed that and, from what the students reported back to me over the years, I think my enthusiasm was contagious. Eventually I had to fulfill obligations to undergraduate students, and quite frankly that was a less pleasant experience. I taught calculus and related courses. It was clear that students were mainly interested in get-

Bert Kostant at his desk, c. 1960s.

ting high grades, and to do so they had to absorb an enormous amount of material. This created a time pressure that prevented me from being able to elaborate on some of the interesting ideas in these courses. When I lectured on two centerpieces of Western scientific culture, namely the works of Kepler and Newton, the general reaction was, "Is this going to be on the test?"

When I said, "No. It's not going to be on the test," then, except for a very few people, I lost their interest. Of course I understand the dynamics behind this attitude. Their tuition costs are huge, and they have to make this money back. Thus they need good jobs, and that requires good grades. As a consequence, a laid-back intellectual environment, which should be part of the undergraduate experience, is diminished.

Lie Groups and Beauty

How did you first become attracted to Lie groups?

One of the aspects of Lie groups that drew me to the subject is that there is an interplay of a number of different areas of mathematics, like a musical symphony. Differential geometry, analysis, and algebra all play major roles in the structure of these groups. Lie theory was born in the 1800s.

Major contributors were [Sophus] Lie, [Wilhelm] Killing, and Élie Cartan, who wrote his thesis in 1899. But the subject was somewhat murky until Chevalley came out with a clean, logically consistent book in the early fifties.[13] I was a graduate student in Chicago at the time, and the book had all the earmarks of postwar Bourbaki precision. I became acquainted with the book since it was used by Spanier in a course I was taking on Lie groups. It also fit in with an epoch-making course, which I was also taking, by Chern in differential geometry. Chevalley's book became a bible for me. I poured over every page. If you looked at my copy of the book today, you would see that almost all the pages are black with commentary notes on the material.

Unification of different areas of mathematics has always been important to me, and Lie theory was a theory which did just that.

A write-up for a conference at Leipzig University entitled "100 Years After Sophus Lie" said that his work is still not completely accessible, owing to its depth and complexity. Where does Lie theory stand now?

Today there are many good introductory books to Lie groups and Lie algebras, but it's true that students have to learn lots of different structures and to see how they fit together. I would say that in general it takes at least two years of learning before reaching the point where something original can be done.

There has been a great development in Lie theory over the past 50 years. The PhD problems of interest have changed accordingly. Also, any current PhD student will have selected an advisor, and the problem no doubt will reflect the interests of the advisor. A sampling, then, of current problems could be in the area of unitary representations of real or p-adic groups, Hecke algebras and Kazhdan-Lusztig theory, Kac-Moody theory, the Langlands program, Poisson and symplectic geometry, quantum and super groups.

Gian-Carlo Rota, in distinguishing between proofs that merely verify a given conjecture and those that give a satisfying reason for it, cited your work in Lie groups as an example of the latter. He talked about your "leap of faith to find a reason for the existence of the five exceptional Lie groups in an outer automorphism of the orthogonal group in dimension eight, finding a reason for that by a tour de force that remains to this day a jewel of

---

13 Chevalley, C. (1946). *Theory of Lie Groups, I.* Princeton, NJ: Princeton University Press.

mathematical reasoning." [14] *Can you give a nonspecialist explanation of what requires a reason here, not to mention a leap of faith?*

The classification of the simple Lie groups goes back to the 1800s. There are four infinite families, known as the classical groups, and then there are five marvelous exceptions, each with its own "personality" (related to symmetries of some exotic structures in relatively low dimension), G(2), F(4), E(6), E(7), and E(8). Where do they come from? Well, something strange happens in eight dimensions: the triality principle; the existence of an outer automorphism of order 3 of the rotation group in eight dimensions. No other simple Lie group has such an automorphism. One can show that the triality principal gives birth to these five exceptional Lie groups.

These groups have fascinated me ever since I first became aware of them more than fifty years ago. In a sense which I can't explain here, the classical groups, other than the general linear groups, can be associated with the number 2 and the groups G(2), F(4), E(6), and E(7) with the numbers 2 and 3. The most marvelous of all is E(8). It is associated with 2, 3 and 5. Dealing with E(8) is like looking at a diamond. Its intricacies are awesome. From one direction one sees 2s all over the place. From another direction one sees 3s, and from a third direction one sees 5s. In my opinion E(8) is the most magnificent object in all of mathematics. It is a symphony in the numbers 2, 3, and 5. One feels that it is too wonderful not to play a role in our understanding of nature. In fact, the exceptional Lie groups have attracted the attention of many physicists, particularly since the emergence of string theory. Indeed, there is a huge number of papers by physicists having to do with the exceptional Lie groups. However, as far as I am aware, this has not, as yet, led to a breakthrough in our understanding of the natural world.

Do mathematicians ask themselves why? If you ask scientists, they'll say, "That's teleology; we don't deal with why." But in mathematics it seems like one should deal with that question. Why on earth does this happen in dimension 8?

I am sure I can't answer that question in a satisfying way. What mathematicians can do, and what for me is very satisfying, is to show that phenomenon A and phenomenon B, though seemingly unrelated, are indeed both manifestations of a more general phenomenon C. Unifying, seeking a common source, is intellectually very pleasing to me.

---

14 Rota, G.-C. (1997). The Phenomenology of mathematical proof. *Synthese, 111*(2), 183–196.

I think so. In a way, I'm surprised that it hasn't taken over faster. One drawback is that the process of programming the computer doesn't help your understanding of the mathematics; it's a separate process. So that's the trade-off: you have to either have something that's very easy to use, or build into your planning the time that it takes to program the computer to do what you want.

Or even just running the program: One of our teams had a student who was using a spreadsheet to make some experiments. And it's so tempting to make one more computation. But then your brain isn't really operating. It's the computer that's doing things. He wasn't thinking about what he was finding. Of course, that may have been an extreme case. But he didn't have to do any *programming,* essentially; he just had this spreadsheet with all this data on it, and he noticed some interesting things happening. But did he add it up? He was amassing data, and processing it in ways that didn't necessarily make sense, because he was too involved in the computer and always having it make one more calculation. I've done that, too!

How do you choose problems to work on, and how would you advise a student seeking guidance on choosing problems?

I don't know if I can answer that. It's not at all conscious, choosing a thing to look at. I've always just done what seemed like a good thing to work on. I've been fairly successful at that, but I can't really say where it comes from. I just sort of follow my nose.

Michael Artin in his gorilla suit with his wife, Jean, 1985.

Ann and Bert Kostant, c. 2000.

word *beautiful* as almost a negative statement. I suppose from the perspective of this mathematician we should restrict our search to people who can solve down-to-earth, hard, technical problems: problems which have a clear-cut utility. The description of a candidate's work as being beautiful carries no indication that the candidate can do this. The ensuing discussion certainly illuminated, for me, a definite divergence of opinion as to what constitutes high quality mathematics. Fortunately, the way the department is now set up allows for both perspectives to flourish.

What does a mathematician mean by beautiful?

It's easier to give examples than to define beautiful mathematics. Mathematicians will almost uniformly agree that the work of Jean-Pierre Serre is beautiful. He has been called the Mozart of mathematics. I think beautiful mathematics leaves one not just with a problem solved but with pictures and ideas in one's head. Mathematics, I think, sits between art and science. This allows for an aesthetic judgment. Formulas may be very useful, but unless the formula illuminates a concept, it is not likely to be beautiful. An obvious example is Einstein's $E = mc^2$. The formula is simple, but the concept behind it, the interchangeability of mass and energy, is magnificent!

~ June 5 and October 11, 2006

Michael Artin

Family Background

J.S.: *Your father, Emil Artin, was one of the most prominent mathematicians on the American scene during the first half of the twentieth century, and the general outlines of his biography are a matter of record. Your mother, Natascha Artin Brunswick, originally studied math with your father, then went on to become a mathematics editor for many years at Communications on Pure and Applied Mathematics, is that right?*

M.A.: She did. She taught math, too. During the Second World War she also taught Russian to American soldiers, presumably training them to be spies [laughs]. This was in Bloomington, Indiana, at Indiana University, as part of the ASTP, Army Special Training Program. She became very friendly with them: two of those people were her friends for the rest of their lives. One of them is still alive. He is not well, but he lives in Manhattan. He worked after the war for Radio Free Europe or for one of the CIA outfits that was beaming radio programs. He was in Munich for quite a few years. The other one became an English professor at Columbia, but he died.

I taped my mother about twenty years ago: about three cassettes, just reminiscences. She was born in St. Petersburg. Her mother was a dentist and lived in Moscow, at least before she was married, and her father came from a well-to-do Jewish family in St. Petersburg. He was an active Menshevik.[1] When the Bolsheviks came to power, they had to flee Russia. When the revolution started in St. Petersburg, my mother was in school, and her mother came to get her. They had to go home while there were machine guns on the roofs: they ran from doorway to doorway when the machine guns were reloading. That's the way my mother put it.

---

[1] The Mensheviks and Bolsheviks were rival factions of the Russian revolutionary movement in the early 1900s.

The Artin family c. 1956; from left to right: Emil, Karin, Thomas, Michael, Natascha, with their dog Frisky.

They spent some time in Kharkov. One story is that they were hiding guns in the apartment for some group of partisans: Mensheviks, no doubt. My grandfather and my mother both attested to this. They got word that the soldiers were coming, so the two of them carried the guns out and put them in the alley beside the house. The soldiers came, searched the apartment, didn't find any guns, and left. My mother and my grandfather went down to the alley and brought the guns back in. In my grandfather's memoirs, he writes, "There are many such stories, but I won't tell them, because they wouldn't be believed."

They had to leave Kharkov when the Bolsheviks came. They went to Tbilisi in Georgia—Tiflis was what my mother called it—for a year or so, and then took a boat to Constantinople and went to Europe. My mother went to high school and university in Hamburg and met my father there.

So your mother was half Jewish, and you children were a quarter Jewish?

Right. *Mischling zweiten Grades*—that means a quarter Jew. "A mixture of second degree" is more or less the translation. The Third Reich forbade anybody from having a government position who was married to somebody who was even half-Jewish. So in 1937 my father was fired from the university in Hamburg—they didn't call it fired; they called it *in Ruhestand gesetzt*. I think you could say "retired"—retired in the transitive sense of the verb. That's when my family came to America. I was three years old.

That's too young to remember much.

Yes. I have some rehearsed memories from home movies and things like that, but I don't actually remember Hamburg from that time. I've been back to see the house. It was in the suburbs and not destroyed. It was in Langenhorn, which is a suburb of Hamburg that's very close to the airport. I think [the Allies] didn't bomb that because they wanted to land there, but I don't know. The center of the city was pretty much destroyed by bombing.

I understand that you were met at the docks not only by your maternal grandfather but by Richard Courant and Hermann Weyl as well.[2]

I think that's correct, but I don't remember being met at the dock. I was three years old. I've seen photos of the family on the boat, so I know that we were on it. It was called the *New York*, I think, a boat on the Hamburg-America Line.[3]

Today that scene seems emblematic: Here are Richard Courant, who as a Jew had been expelled from Göttingen in 1933 and established what later became the Courant Institute at New York University; and Hermann Weyl, whose own wife was Jewish and who had left Germany for Princeton in 1933, meeting Emil Artin and his family, including his three-year-old son, Michael, later a prominent mathematician in his own right. That one snapshot seems to capture the future of world science departing the shores of Europe and coming to the United States.

Yes, a lot of it. And both Courant and Weyl were quite active in trying to help refugees from Germany get positions in America. They also established a fund during the War—a bank account. We have the bank book. It had maybe $500 in it. They gave people who just arrived $50 or something, which was in those days enough to live on for a while. So it's a pretty inter-

---

[2] Yandell, B. (2001) *The Honors Class* (p. 231). Natick, MA: A K Peters.

[3] The steamship *New York* was built by the Blohm & Voss shipyard in 1926. It capsized during an air attack on Kiel on April 3, 1945, was refloated in 1947, and was finally scrapped in 1949. See the Norway Heritage project site at http://www.norwayheritage.com/p_ship.asp?sh=newyk.

esting document. I saw it after my mother died. She was one of the people who worked on that, so she had the bank book. We are planning to send it to Hermann Weyl's papers, but I'm waiting for my brother to find a letter from Weyl.

Your brother, Tom, had a career in academia but now makes his living as a jazz trombonist?

That's right. He also does a lot of photography.

It seems there were at least two ways that an Artin could have gone: into math, of course, but there was also a lot of music in your house.

There was. My father played several instruments—keyboard and flute mainly. My parents had a harpsichord and clavichord that came with their things from Europe, and then they got a piano, and they also had something called a *Tafelklavier* that was more like a table and some old-type piano.[4] In Indiana, my father sold that in order to buy an electric organ, because he wanted to play Bach on the organ. But the organ didn't have enough pedals, so he spent quite a lot of time adding pedals in the base of the organ and then had to go in there and rewire. This would have been around 1940: it was all done with tubes in those days, and it was fairly complicated. They had tubes that emitted a sine wave, basically; then they tried to mix those to get overtones. But of course the overtones aren't supposed to be pure, because organ pipes aren't perfect, so they deviate from the true [tempered] harmonics. It was not a very good thing from the point of view of fidelity.

Arthur Jaffe, the quantum field theorist, remembered that even as a visitor at the IHÉS[5] you had a piano.

That was for my wife. It was one of those beautiful French upright pianos—a beautiful-looking instrument, but not very good quality. It was old.

And you played several instruments yourself?

---

4 The *Tafelklavier*, or square piano, was an eighteenth-century effort to produce a keyboard that could play at different volumes, unlike the harpsichord. They were sturdy, had an excellent sound, and were enormously popular.

5 The Institut des Hautes Études Scientifiques, located just south of Paris in Bures-sur-Yvette, is a center for advanced research in mathematics and theoretical physics, similar to the Institute of Advanced Study in Princeton.

I played and still play the violin. And for a few years, I played classical guitar and lute. This was when I was in college and graduate school: I stopped playing the lute when I got married and had no more use for it [laughs]. Serge Lang[6] was a student of my father, and I converted him to the lute also. Now you can get lute music, but in those days there was no printed music. So we used to get microfilm of old manuscripts from Elizabethan England from the British Museum or the Cambridge Library and then print it out and play from that. It was pretty exciting.

Formative Influences

In your Steele Prize response you wrote, "I never made a conscious decision to become a mathematician."[7] Growing up as you did, you must have had the worry, whenever you did think of going into math, that "I might never be as good as my dad at this."

Yes. That was, of course, clear from the start. It's still clear, but I think I just had a good defense mechanism against it. So I didn't plan, really, to do mathematics research, back when I was a graduate student. What I was hoping to do was to get a job at a place like Bowdoin or Williams, in a small town, and become a math professor. That was my idea. I didn't really like living in cities.

But it didn't work out that way. I got too involved in the research to do that.

Let's back up a minute. You went to Princeton as an undergraduate.

Yes. I got free tuition there: it saved $500 a year [laughs]. That was the tuition back then. It sounds crazy.

I originally thought I might do topology. When I was an undergraduate, my advisor was Ralph Fox, who was a topologist, and he got me doing things. In Princeton—in those days, at least—you had to write a junior paper, and also a senior paper. I did what was thought to be original research in my senior thesis, but I never published it. I think Ralph mentions the result in one of his books, but I didn't think it was good enough.

[Biology was my other] field when I was an undergraduate. But in those days, mathematical biology was . . . zero. People studied the pecking order

[6] Serge Lang (1927–2005), member of Bourbaki, had wide-ranging mathematical interests, especially algebraic geometry and number theory. He shared the AMS Cole Prize in Algebra in 1960 and won the Steele Prize for Mathematical Exposition in 1999.

[7] 2002 Steele Prizes. (2002) *Notices of the AMS*, *49(4)*, 470.

in broods of chickens by drawing a graph—it was *really* Mickey Mouse. But now it's a fascinating subject. Maybe I chose the wrong field.

Why didn't you ever publish your thesis?

Well, one thing I learned from my parents was high standards. Not that they did anything actively to teach me high standards, it's just that *they* had very high standards, and it rubbed off. That's probably why I thought I'd go teach at Bowdoin or something, and also probably why I never thought about publishing my undergraduate thesis—or my PhD thesis.

One really nice story that my sister [Karin Tate] told me about my father was that when we first went to Indiana, she came home one day and said at the dinner table, "People in school ask me what church we go to, and I don't know what to tell them." In Indiana at that time, what church you went to was a big thing.

My father said, "Well, let me think about it." So he thought about it for a day or two, and then he said, "You should say, 'We don't go to church, but we read the Bible at home.'" And in order that that should be true, he began to read the Bible to us in the evenings.

He read it in English or in German?

I remember him reading German to us quite a lot. That was one of his ways to keep our German there. We also had to speak German at the dinner table. The rest of the time it didn't matter, but at the dinner table, German was spoken. And then after dinner he would read to us.

You came over in 1937, so there must have been a fair amount of anti-German feeling during your first years in the country.

It's interesting that we never experienced that. Sometimes if we were playing war—which was most of the time, right?—then maybe I would be the German. But otherwise I didn't have that sense. It's possible that it was just an accident, where we were living. But our neighbors, they weren't university people.

At what point did you commit to a career in mathematics?

Well, I graduated in '55 from college, and '56, my first year as a graduate student, was a rocky year. I was trying to decide what I wanted to do in life and having an unhappy love life at the same time, and that didn't go too well. But the following year things started to work. I just got sucked in to mathematics, I guess, and I didn't ever emerge.

Do you remember when you first met Oscar Zariski, your thesis advisor at Harvard?

I first met him when my father and I came to the International Congress in Cambridge in 1950. We had lunch together, I think, at a dump in Harvard Square, and then I left. I wasn't going to the meeting: I was going to Cape Cod to be with some friends. But he was certainly a strong, attractive personality. I decided to work with him after taking his class in commutative algebra. Just because of his personality, maybe.

Is it possible to give a sense of how his personality came across?

I'm not sure I can. I think Mumford described him coming in to teach the class completely unprepared and getting sort of nervous in the middle of the proof, not quite remembering how it was supposed to go, but then, as it went on and he understood it, picking up speed and enthusiasm. When he was at Johns Hopkins, he taught twelve hours a week and then taught night school. So he was used to teaching without preparation, because you couldn't prepare so many hours well. And that was when he was doing his best work.

Oscar Zariski, with the inscription "To Mike and Jean with affection."

That's astonishing. How did he manage that?

I don't know. Not much of a life outside of mathematics, I assume.

Is it fair to say that, great as Zariski's contribution was, algebraic geometry had moved beyond him in his later years at Harvard?

Well, he lived to be eighty-four, and he tried to continue working until the end, but he wasn't up to it mentally. There's no question that by the time he was in his seventies it had moved on, but he was quite remarkable. He was born in 1899, and in the mid-fifties, when Serre published his work on sheaf theory in algebraic geometry, the NSF organized a summer program, and Zariski wrote the notes for that. So he took that very seriously at that time.

He did his best work at age fifty. There's absolutely no doubt. He developed a lot of tools. It's just that he didn't use the language of schemes or sheaves very freely. But he had laid really strong foundations. He was very forward-thinking in algebraic geometry up to the age of fifty and Serre's sheaf theory. And he kept on, but obviously, he had to slow down. I think after that he still wrote one of the four volumes of his collected works. It's not at the level of his great works....

But still fairly respectable.

Oh, yes. Oh, yes.

You have mentioned John Tate as another influence.

Tate was at Harvard, and he's another one of my father's students. He was married to my sister, so I knew him socially as well as mathematically, and he helped me a lot. He got me started on the first sort of work I did that attracted much notice, which was on étale cohomology. That was based in part on some work that Tate had done with Kawada; there's a paper, Kawada and Tate.[8] Also, he was very much involved with the arithmetic of elliptic curves, and there the étale cohomology was a very natural tool, so I started to absorb that. And then Grothendieck came—what year was that?

He came to Harvard for the first time in 1958.

Yes. At that time, I still didn't have any idea what a scheme was. But the second time, in '61, I had heard that he had this idea for étale cohomology, and when he arrived I asked him if it was all right if I thought about it, and

---

[8] Kawada, Y. & Tate, J. (1955). On the Galois cohomology of unramified extensions of function fields in one variable. *American Journal of Mathematics, 77,* 197–217.

he said yes. And so that was the beginning. He wasn't working on it then—he had the idea but had put it aside. He didn't work on it until I proved the first theorem. He was extremely active, but this may have been the only thing in those years that he really didn't do right away, and it's not clear why. I thought about it that fall, and we argued about what the definition should be [laughs]. And then I gave a seminar, fall of '62.

You later went to the IHÉS to work with Grothendieck.

Yes. That was in the fall of '63, for a year, and that was just after I had proved the first theorem in that subject. And I went several times after that.

The IHÉS was modeled on the Institute of Advanced Study by founder Léon Motchane, the Swiss industrialist. Was Grothendieck the central pillar there?

Well, he was such a towering figure that he dominated everything. But no, René Thom was there in mathematics, and there were people in theoretical physics as well. But Grothendieck had a seminar every week. That was a big event. Many people came from Paris.

Grothendieck was a very interesting character, by all reports, intellectual achievements aside.

Yes. I think his parents were anarchists or something, and he expressed strange political beliefs. And he had strange rules about eating; he would only eat certain things. He liked cabbage. He was vegetarian at least part of the time, but you didn't have the feeling that he thought cabbage was particularly tasty: it was more a moral position on cabbage [laughs].[9] I found that very easy to relate to, because my father also took strange positions for no apparent reason. Like Grothendieck, my father never did this without having thought about it. But, for instance, pepper. He was down on pepper. It's not that he didn't like pepper, but he thought that people who put pepper on everything, that was not good. You shouldn't just sprinkle pepper indiscriminately.

---

9 Michael Artin's sister has similar memories of Grothendieck. "Grothendieck had 'an incredible idealistic streak,' Karin Tate remembered. For example, he refused to have any rugs in his house because he believed that rugs were merely a decorative luxury. She also remembered him wearing sandals made out of tires. 'He thought these were fantastic,' she said. 'They were a symbol of the kind of thing he respected—you take what you have, and you make do.'" Jackson, A. (2004). *Comme appelé du néant*—As if summoned from the void: The life of Alexander Grothendieck. *Notices of the AMS*, *51*(9), 1047.

Grothendieck wasn't publishing in journals anymore at that point, is that right?

Yes, that's right. He had three places where he published. One was his EGA, the *Elements,* and then also there were the seminars that he had weekly at IHÉS, and typically there were notes to those, part of which he wrote. And then also he gave lectures at the Bourbaki Seminar and wrote those up. They're some of the best reading, actually, because they're very concise. EGA is so incredibly big, it's hard to find your way in it.

What would be a typical agenda for Grothendieck's seminars?

Well, there would just be a program where one would start to rewrite the theory [laughs]. One was on étale cohomology, and that was just being developed. But then another one was on algebraic groups, and there he thought, well, everyone always assumes that you have a field, and that maybe the field is algebraically closed. He wanted to write everything down carefully, with minimal hypotheses. And that's what that seminar did, I think very well.

Did you also give seminars?

Yes. I was active in seminars when I was there.

I've read that a bit of a personality cult grew up around Grothendieck, and that some of his students' work went out on kind of an arid limb, but that you and David Mumford were more resistant to that.

Well, I don't know. It is perhaps true that we were resistant to excess abstraction. But I was working on étale cohomology: in those days it was considered very abstract, but I didn't worry about that. Grothendieck had students that wrote their doctor's theses under his direction, and some of them were *very* abstract: I would agree with the word *arid* applied to some of them. But there's a reason why they were doing very abstract mathematics; it's what Grothendieck assigned them as a thesis. Of course, these people were also active participants in the seminars.

MIT

You were already based at MIT at that time?

I had a post-doc at Harvard. I got my degree in '60, then was an instructor there for three years, and then came to MIT, though I was on leave the first year [at the IHÉS].

Princeton classmates Fred Almgren, Mike Artin, and Hyman Bass, all of whom gave lectures at the ICM in Moscow, at the Kremlin, 1966.

So you came to MIT in 1963. How did you come here?

Arthur Mattuck was here. It was his influence. By the way, he was also a student of my father, but I didn't know him well then. He worked more by himself, and didn't become a close family friend the way Tate and Lang did. I got to know him pretty much here. We started playing chamber music—and we still play chamber music.[10]

So Mattuck brought you over here; you were the second algebraic geometer?

I think so. Steve Kleiman came in six years later. So there were just three of us, directly in algebraic geometry, for a long time. I can't think of anyone else. The bulk of activity was at Harvard. For instance, [Masayoshi] Nagata visited Harvard several times—I don't remember the years—and [Jun-ichi] Igusa, also, and [Wei-Liang] Chow.

---

[10] Michael Artin (first violin) and Arthur Mattuck (cello) are currently joined by MIT alumnus and University of Massachusetts professor Dana Fine (second violin), and Dick Gross (viola), former dean of Harvard College, in a chamber music group that has existed for fifty years, the voices slowly rotating among mathematics graduate students and professors with an occasional physicist, biologist, or psychologist.

It seemed like Oscar Zariski's students had turned Boston into the algebraic geometry capital of the world by the mid-1960s, with you (and later Steven Kleiman) at MIT, David Mumford at Harvard, and Heisuke Hironaka at Brandeis.

There was a lot of activity here—and Grothendieck was visiting. Mumford and Hironaka and I were students at the same time, but the three of us never actually worked together, in the sense of publishing a paper together. Well, Mumford and I published a paper together, but that was much later.

In 1966, the year you turned 32, you were one of fifteen mathematicians worldwide to be asked to deliver a one-hour talk at the quadrennial International Congress of Mathematicians, which that year was being held in Moscow. Our official relationship with the USSR was hardly a friendly one at that time. Leonid Brezhnev had just taken power, the Cuban missile crisis was a recent memory—how was the trip handled?

Well, there was a slight window in the Iron Curtain. In those days there was something called "Intourist," the Soviet government tourist agency. Visas and airline tickets were done through the Math Society. But you had to decide what level of tourist you wanted to be, and then they gave you meal tickets—too many to use. There was essentially nothing to be had in Moscow. You could stay in your hotel and eat fairly reasonably, but a lot of the food was terrible. Going out of the hotel to a restaurant was difficult: there would probably be a line, and because you were a foreigner, you would be going with some budding spy who was assigned as your guide. He would say, "Go to the front of the line" and get you in, but it wasn't very comfortable, and the food wouldn't be good. So we ate almost entirely in the hotels, big hotels. Or the meeting was in the university, and lunch would be served there. I found that the best thing to have for supper was red wine, Mukuzani, from Georgia; borscht; and caviar. So I used to always order that. I've almost never eaten caviar, except on that occasion.

They also had a banquet in the Kremlin, in some modern building. It was an enormous room lined with tables, big enough to feed all of these two thousand people or however many there were at the meeting, and the tables were filled with food and vodka and wine. We have a photo of me with [Yuri] Manin, and I'm explaining to him which is the good vodka [laughs]. This was the middle of the Vietnam War, and there was a Vietnam petition that was being passed around. It was kind of crazy to wander around in the Kremlin asking your friends to sign the Vietnam petition, but that's what we did.

We also had time to take a trip. I wanted to go to Tashkent and Samarkand, because I remembered those names from when I was a child, and they sounded so romantic. But that tour was full, so I chose Yerevan. You know, my name is Armenian.[11] So I went down there, and that was interesting. At that time, many people, including me, thought of the Soviet Union as a monolithic state. So I was quite astonished to find in Armenia a completely different society, with its own language and its own architecture. When our guide went around the table, towards the end of our stay, to check on our return flights, everyone had to spell their names, even when that name was Smith. But she had no trouble with mine.

Noncommutative Algebra

Some years ago you switched to a new field, noncommutative algebra. What was the attraction there?

Well, there's a tie-in between that and algebraic geometry, because of something called the Brauer group, which classifies noncommutative division rings over a field. It also has to do with one fundamental question: if you have, for instance, an algebraic surface, which cohomology classes are represented by algebraic curves? Some are called transcendental: that means algebraic geometry doesn't see them. Others are represented by curves. It's a difficult question, and that got me started. Already in the sixties I wanted to find some new rings, so I decided what I should look for, and I ended up showing that there weren't any—that there were just the ones that we already knew. I got, in other words, the opposite result from what I was looking for.

That sounds like a significant one, though.

It was a fairly important result, actually. That was my first foray into noncommutative, and it was interesting, but that was all I did. Then in the seventies, Ron Irving, who got his degree at MIT—I was his thesis advisor—came to me and said he wanted to work in noncommutative rings.

I said, "All right, but I don't know anything." But I gave him an assignment, which was to write down all the rings, and also all the classes of

---

[11] "The name Artin comes from my great-grandfather, an Armenian rug merchant who moved to Vienna in the nineteenth century." 2002 Steele Prizes. *Notices of the AMS, 49*(4), 469.

rings, and then find all the inclusions between these things. So he did that for one semester and at the end of that he was on his way. It worked pretty well.

You said you didn't know anything about this field, which makes your suggestion for how to proceed all the more interesting. It seems like such a simple idea, something one would think would have been done already.

I think that this had not been done before. Now, probably in many fields that's not the way people worked. In ring theory, at that time, people worked very abstractly with a property. They assume the ring has some properties, and then they start to talk about it. Coming from algebraic geometry, I was much more interested in, you know, what are they? That was my motivation.

Where did Ron Irving go from there?

He wrote a thesis on representations of certain algebras. It wasn't directly related to his assignment, but he was at home then.

Anyway, during the time Irving was here, I started thinking about noncommutative algebra again, and then a few years after that I got to know Bill Schelter and worked with him on a computer project related to noncommutative algebra, and then I just went into that full-time. But the transition took about ten years.

Bill Schelter was the Lisp guru, father of Affine.[12] *How did you come to work together on that?*

We found a great experiment. What happened was that I was going down to spend a semester in Austin, and when I got there he had been sucked into the computer. He became an incredibly good programmer—he was working for the computer companies to fix their programs. One day he told me, "Yes, I made it run ten times faster. They were pretty happy."

I said, "How did you do it?"

"Well," he said, "it's easy. I just asked the computer where it was spending time. Then I looked at those places."

Anyway, so he was sucked into that, and I thought, God, this is going to be a disaster. But there was a problem that I was thinking about. It was in

[12] Affine was part of Maxima, an open-source descendant—developed and maintained by Schelter—of Macsyma, an MIT-based symbolic computation program from the late 1960s, predecessor of Maple and Mathematica.

just three variables, so the computer could handle it, and we started doing experiments and finding really interesting things. So it was a big success. I wanted to prove something: as usual, I found the opposite.

You may have been looking for a little peace and quiet in the noncommutative field, but then the theoretical physicists started getting interested in it.

Yes. I once joked when I was giving a talk that I thought this was a nice quiet pond where I could spend my declining years, but the water was getting a little turbulent. It's become quite fashionable.

Administration

You are one of relatively few people in the department who has served the math community at the national level. For example, Arthur Mattuck said you narrowly missed being chosen department head by virtue of being chosen president-elect of the AMS.

Yes, I erroneously thought that president would be less work [laughs]. My reasoning was that as department head, you're here in the department every day. Whereas, the AMS business, there are meetings, but the rest of the time.... What I failed to take into account was that the AMS is very complicated. It takes a lot of brain just to figure out what's going on, and none of it is things that you really care about. So it turned out I was wrong: department head would have been a daily agony, but less consuming of the mind.

When I started, I called up someone in the Sloan School and asked for some reading. Whoever it was said, "Ah, so you're the transient perturbation that hits the Society every few years!" He recommended a book, *Designing Complex Organizations,* by a man named Jay R. Galbraith. A brilliantly written book, just a model of how you should write. He had two examples that he carried through. One was Boeing, and the other was a restaurant. A restaurant is complicated, because you don't know what's going to happen today.

What did the complexities of the AMS emerge from?

Well, there's the volunteer side, the president and the council and committees; and then there's the executive director's side, which runs the Providence and Washington offices and does things like arrange meetings and publish journals. They're both very complicated. And often at the council meeting, there are political issues brought up, which concern the Math Society only very peripherally, and which are very divisive. For instance: I first was on the council, then I was a vice president, and then I became president. These were separated by *years,* right? I went to the council meetings during these periods, and I had to chair them as president. Well, during each of these three periods, the same issue was brought up, and that was: should Israel close Birzeit University?[13] And *never* was there a consensus on making a declaration. Never. That was the most amazing thing, to see that same issue be brought up three times.

When you agreed to be president, did you have certain things that you wanted to do?

I had absolutely no idea. But then, after six months of immersion, suddenly you do have an agenda—and very likely you can't do most of the things you want to do.

There was one big change made in the organization during the time I was president, which was to establish subcommittees that reported to the council on various parts of the structure. I felt that the organizational chart was too primitive. There was just the council, and then there were committees that chose people to give talks at the meeting, for instance, or nominated people to run for positions. So they established some sort of larger oversight committee, a task force on meetings on national educational policy, which is supposed to talk about meetings in general, to take a longer view than just who's going to talk at the next meeting.

You were the first editor-in-chief of the Journal of the American Mathematical Society. The overview of the Transactions of the AMS says it is "devoted to research articles in all areas of pure and applied mathematics." The overview of the Journal, on the other hand, says it is "devoted to research articles of the highest quality in all areas of pure and applied mathematics." What's the story behind that difference in wording?

Back in the thirties, the *Transactions* was an excellent journal. It was almost on the level of the *Annals*. But it got watered down over the years. Not that the papers weren't good, but just by publishing too much. What happens to these Society journals is that they get bigger and bigger. If you're an editor, you're constantly under pressure to accept one more paper. And there's never

[13] Birzeit, located near Ramallah in the West Bank, is one of the top two Palestinian universities (the other is Bethlehem) and the site of much political activity.

really a large difference in quality, at the borderline, between this paper and the next one. So that tends to blow the journal up: there's nothing that controls their size, except dollars.

That's happened with the *Transactions.* That meant that if somebody had a really good paper, they'd probably want to send it to the *Annals* or *Inventiones [Mathematicae],* if it is an article in pure mathematics, or some other journal, not necessarily to *Transactions.* So there was really no flagship journal for the Society, and so I agreed to do it. I tried to get the Society to decrease the *Transactions* page allowance by the number of pages they were putting into the *Journal.* I did not succeed.

So this is the life cycle of journals, as it were. A journal starts out; it's very good, very selective, but it gradually gets a little bigger, a little less selective; people start paying less attention. Then someone starts a new and better journal, and the cycle repeats?

I'm afraid there's some truth in that. I mean, the *Mathematische Annalen* in Germany went through a similar evolution, and then they started *Inventiones* to be the good journal, right? *Mathematische Annalen* is still a good journal, but *Inventiones* gets the best papers.

You were in administrative positions in the department, including pure chairman. Danny Kleitman said you were part of the effort to bring Is Singer back to MIT, which obviously was an accomplishment.

Right. He did grow up here, in some sense. He had left to go to Berkeley, and I think he just didn't like it in Berkeley. But I was reasonably friendly with Is at that time; he sometimes stayed in our house. So I think that probably helped, to have a rapport with him.

John Deutch was the dean at that time, and I went with Singer to talk to him, and it was clear that the dean was in good hands [laughs]. So I left. And then Singer decided to come, and I saw Deutch in the hall, and I shook his hand and congratulated him. And he said, "Don't come back to me with anything else this year."

And I said, "I won't come back with anything you can refuse."

That's a good line. Did you go back to him with any more offers he couldn't refuse?

Nope. But I didn't want to rule it out.

Michael Artin giving a lecture, Montreal, 1970.

Teaching

You taught a year-long undergraduate algebra course for about thirty years, which ultimately became an algebra textbook.[14] *The user reviews on Amazon are very interesting. People either don't get what you're trying to do, or it's the book they would take to a desert island. There's no middle ground.*

Yes, that's right. It's also not easy for the traditional algebraists to use it as a textbook, because it has other stuff in it.

Did you set out to write a textbook that was not just "business as usual"?

I didn't start out to write a textbook at all. I just wanted to teach a class, and to do some topics that weren't traditional. So I handed out notes for those, and eventually I started using them instead of the textbook. And then I revised them every year.

I'm in the process of thinking about a second edition. Every time I teach it, I have new ideas about how something should be done. And also I collect

---

[14] Artin, M. (1991). *Algebra.* Upper Saddle River, NJ: Prentice Hall.

exercises. It's very hard to find really good exercises. You want an exercise that teaches the students something different. You want a hook; you want it to inspire them and, if possible, to teach them something useful, maybe from a different field. And you don't want to give it all away, you know? "Prove this" is okay—but then you've already told them what to do. So the ideal problem would start with the hook, right? Something that catches their interest, and then maybe asks some questions. Then they're supposed to do something where I don't exactly tell them what, or I don't tell them how. They have to be inventive—but they should be able to do it. That's the ideal problem.

So what would be a hook to start off with?

I have a problem on patterns, Escher pictures. Maybe they're supposed to find a way to compute the area of one of those geese that's flying: that's the hook. And the answer is that you can find probably a little polygon, some nice, easy geometric shape that has the same area. Because the geese cover the plane, right? Then it goes on to ask them to analyze something called the fundamental domain—that piece that you move around, that covers the plane—and to relate it to the symmetry group. So by the end they're asked to do something that's more serious.

Being able to keep coming up with new ways to teach this course, even after so many years, is astonishing, in a way.

Maybe it's a corollary of having written a book, that I don't get tired of it. It's interesting to me that writing a calculus book, for instance, is not typically considered an honorable thing to do. It's sort of looked down on. Whereas I would think that it would be a fantastic thing to do. We don't *need* another calculus book, and that may be one of the reasons why people feel that it's not a worthwhile thing. But from my point of view, the devotion to the pedagogy that is required to write a book is paid off to the university over and over again in the teaching.

Encouraging the students themselves to formulate what needs to be done sounds like the same principle at work in the Project Laboratory in Mathematics (18.821), a fairly new course, which you taught with Haynes Miller.

Right, though that's much less directive. They are given a bare-bones topic, and they're supposed to figure out what to do. But then they're a team, and they have a mentor, an on-staff person who meets with the team, who can give them a hint if they're floundering.

Do you think we're going to see an increase in the use of computers as teaching tools?

I think so. In a way, I'm surprised that it hasn't taken over faster. One drawback is that the process of programming the computer doesn't help your understanding of the mathematics; it's a separate process. So that's the trade-off: you have to either have something that's very easy to use, or build into your planning the time that it takes to program the computer to do what you want.

Or even just running the program: One of our teams had a student who was using a spreadsheet to make some experiments. And it's so tempting to make one more computation. But then your brain isn't really operating. It's the computer that's doing things. He wasn't thinking about what he was finding. Of course, that may have been an extreme case. But he didn't have to do any *programming,* essentially; he just had this spreadsheet with all this data on it, and he noticed some interesting things happening. But did he add it up? He was amassing data, and processing it in ways that didn't necessarily make sense, because he was too involved in the computer and always having it make one more calculation. I've done that, too!

How do you choose problems to work on, and how would you advise a student seeking guidance on choosing problems?

I don't know if I can answer that. It's not at all conscious, choosing a thing to look at. I've always just done what seemed like a good thing to work on. I've been fairly successful at that, but I can't really say where it comes from. I just sort of follow my nose.

Michael Artin in his gorilla suit with his wife, Jean, 1985.

I would tell students to trust their instinct. If it's not good, then they're not going to be very successful, I guess! Of course, when you have a student working on a thesis, maybe your biggest job as a teacher is to set them straight when they're going off in a direction that's not going to be fruitful. But after they get their degree, then it's more that they have to follow their instinct.

And you can't always predict which student is going to do the best work. Some are very smart, and you know they're going to do good work, but then they might dry up. On the other hand, one of my fairly recent students studied Japanese while he was in graduate school, and he got a fellowship to go to Japan for a year. So there he was in Kyoto, and he learned about Mori's Program for algebraic geometry.[15] He was able to apply that to a problem that he was thinking about, and it turned into a gold mine for him.

You have said that your father loved teaching as much as you do.[16] Is it possible to say what you learned from him about teaching?

Very hard, because I never took a course from him. I only on very rare occasions—I can think of only one at the moment—heard him give a class. But on the other hand, he spent a lot of time with me when I was a child, teaching me things. We used to go on walks in Indiana in the woods. It was a family outing, and I guess my sister and I would always ask, "What's this flower?" Of course he had no idea—he'd just come to America. But he learned them, and then we learned them.

That seems like the mark of a true intellectual, a potential for interest in just about anything.

Right. You know, when I was an undergraduate, I had friends who had not come from intellectual families, who discovered for themselves the joy of intellectual activity. I was sort of jealous of them, because for me, I'd never been able to do that. It was always there.

Speaking of the joys of intellectual activity, rumor has it that you have occasionally lectured in a gorilla suit?

---

15 Mori's Program or the Minimal Model Program, initiated by Fields medalist Shigefumi Mori, relates to classification problems of algebraic varieties of dimension three. See for example Hironaka, H. (1990). On the work of Shigefumi Mori. *Proceedings of the International Congress of Mathematicians, Kyoto, 1990 I*, 19–25.

16 2002 Steele Prizes. (2002). *Notices of the AMS*, *49*(4), 470.

I got a gorilla suit for my fortieth birthday. I asked for it. There was this movie, *Morgan.* It's about a guy who's kind of crazy, he has broken up with his girlfriend, and in the end he comes to her wedding in the gorilla suit. So that was one of my inspirations. I kept asking for it, and finally I actually got one. I only once wore it in a class. I had practiced drawing a banana, the almost parallel lines, so I came in and drew, very carefully, this banana. And then I drew a stick figure. And the class was silent [laughs]. So I took the suit off. It's pretty hard to talk. Not only your voice is muffled, but even breathing is a noisy affair. You have to take deep breaths.

It was a great thing. Unfortunately, once on Halloween I was answering the door in the gorilla suit, and there was a girl about this high [indicates height] who ran screaming off into the darkness. That took some of the kick out of it.

But I had a number of adventures in it. One was, I think Egbert Brieskorn and his wife were over for dinner. After dinner we took a walk around the block, and I happened to have on my gorilla suit. I was slightly ahead of them, lurking behind trees, and a police car went by, slowed down, made a U-turn, came back. The officer in the passenger seat held the flashlight out, and I went up and introduced myself. By this time my wife had come up, and she said, "It's all right. He's with me."

And the police officer said, "I said to my buddy, 'It's got to be a guy in a gorilla suit.' My buddy said, 'I sure hope so.'"

Then he said, "Now, don't get shot," which was good advice, and they left.

Arthur Mattuck said that as undergraduate chairman you taught some of the big lecture courses yourself in order to experience them firsthand. What was your own development like as a teacher?

When I came, I don't think we were teaching calculus in big lectures; at least I didn't do it. I taught calculus in small classes for years, starting when I was a graduate student at Harvard. But the first time that I tried to lecture, one of the freshman calculus classes, it didn't work very well. I was too formal, probably. You're in front of 250 people or so, it's a sort of show game, and I had a hard time. The last time I did it, it was fine, but I'd learned a lot. Before trying it again, I attended Arthur Mattuck's lectures for an entire semester, and I learned a lot from him.

Another thing that really helped me was being videotaped. The first time I was videotaped it wasn't in the calculus, though I've been taped teaching those courses. It was an upper-class course. I looked at it, and I saw that I was doing something terrible: I avoided looking at the class. Even when I turned around, my gaze was down. It just stared you in the face when you watched me. I decided that I was going to correct this, so I had to figure out: why was I doing this? And I decided the reason was that looking at the class would break my concentration on what I was saying. I'm pretty sure that's what it was. But it was *extremely* difficult to stop doing it. I probably still do it, but not as much. One of my brothers-in-law suggested just programming looking at the class into my preparation. It was a good idea. He's a doctor. You know, treat the symptom [laughs].

There was one other problem that also I can't do anything about. When Arthur sees my tapes, he is very critical, because I sometimes skip steps. The reason I can't do anything about it is because it's really a brain-function thing—my brain has skipped the step. But that's one reason that I prepare very carefully. I've written everything out exactly the way it's going to go on the board, for instance, so my board has gotten pretty good. I still have this skipping steps, but it's probably not as bad as not turning around often enough to look at the class. I get reasonably good ratings, so these aren't terrible. It's just that you want to be perfect.

~ September 28, 2006

Daniel J. Kleitman

From Physics to Mathematics

J.S.: *You started as a physicist but arrived here a combinatorialist, is that right?*

D.J.K.: Yes, that is correct. How did that happen? Well, you know, when I grew up I never actually got the impression that it was possible to make a living as a mathematician. My older brother went into physics, and I was a few years behind him, so it was the natural thing to me to study physics. I loved mathematics in college, but I went to graduate school at Harvard and got my degree in physics in '58. Then I had an NSF post-doc, and I went to the University of Copenhagen's Niels Bohr Institute for a year, and then the second year I spent at Harvard. And then I got an assistant professorship at Brandeis.

While I was there, I encountered a former classmate in college, David Lubell. He was a Benjamin Peirce Instructor at Harvard. He invited me to dinner and told me about a problem that he was working on, which he had gotten from a book by [Stanislaw] Ulam. In Poland there was a tradition, a book in which people would write down unsolved problems that they were working on or interested in. And Ulam published a problem book based upon that idea, being that he was Polish.[1] Lubell had seen this problem in that book, and I went home, and I thought about it, and I came up with a solution to it, which I communicated to him. And then I borrowed the book, and I was able to solve another problem in the book! And actually a third, but it was not a very hard one. So I decided to try to find Ulam, to ask him whether anyone really was interested in this problem, or [whether

1 Ulam, S. M. (Ed.). (1957). *The Scottish book: A collection of problems* (Translated from a notebook kept at the Scottish Coffee House for the use of the Lwów Section, Polish Mathematical Society). Los Alamos, NM: Los Alamos Scientific Laboratory.

Daniel Kleitman, 1950s.

he knew] anything about it. I wrote to him and got a response that *he* had gotten these problems from [Paul] Erdős, and he suggested I contact Erdős. But there was a problem in contacting Erdős, because Erdős had no fixed address, which made it difficult to locate him. You know, "Paul Erdős, The World," and hope it gets to him!

Erdős lived in the United States during and after the Second World War, I believe, but the Hungarian government (which I think had a mathematician among its Politburo) wanted him to come back to Hungary. They offered him a deal, in which he could be a Hungarian citizen and visit his mother, who lived in Budapest, as much as he wanted, but could also maintain a bank account abroad and travel freely. Most Hungarian nationals at the time could do none of these things, and only those politically most favored could travel at all. This was arranged by having him a citizen of Hungary but a resident of Israel. In this way he was, as far as Hungary was concerned, a nonresident, with all the travel restrictions that Hungary imposes on nonresidents, which is none: unless they stay too long in Hungary, in which case they may try to change his status to that of a resident. Thus, while he liked to spend time in Budapest, with his mother and with Hungarian mathematicians, he never wanted to stay too long there. So for at least nine months of the year he traveled, all over the world. Wherever he went he created mathematical interest, got mathematicians who were not doing much to become excited over research, and was a great positive influence. The honoraria he got were enough for him to live on, since he spent very little money on himself and had few possessions that he did not carry in his suitcase. He lived like this until he was into his eighties, only stopping on his death.

Anyway, these problems were much more exciting to me than the things I was doing as a physicist. I mean, they appealed to me in a different way, and *more*. So, at one stage I decided to write them up. Writing things in those days was quite different from today. Because there was no such thing as a word processor: the things had to be typed. And they had to be typed by a secretary, because they had to be perfect. If you made a mistake, you had to throw away the page and start all over again. Equations had to be written in by hand, because the typewriters did not *have* equipment for putting in equations.

Anyway, so I wrote them up, but I wrote them up perhaps in a way that I was used to as a physicist, and it was somehow different. So I got back, fairly quickly, an answer from the journal that I sent them to, dismissing them. Which annoyed me some, so I threw it in a drawer and I didn't think of it much.

But then about a year or two later, I said, "Well, wait a second, this really is good stuff." I went to a fellow in the math department at Brandeis, Edgar Brown, and said, "Help me write this up." So he helped me write it up in the way a mathematician would. I submitted them to two different journals, and maybe a month or so later, I got a letter from Erdős! It seems they were both sent to Erdős to referee, and Erdős liked them! A letter from Erdős in those days was on this very thin, air mail–type, folding stationary—things were supposed to be light to go in air mail, so it was those little blue things that you folded up and it was an envelope. The first paragraph was personal things, something about he got these papers and he liked them, and then the next paragraph was a problem! "Let A_1 through A_N, being a collection of sets obeying the condition, etc., etc. … How large can N be?" I was very pleased. I remember that the following day was a faculty meeting in the physics department, which is probably the most boring thing known to man. So in the middle of it, I took it out—I had it in my pocket—and thought about the problem, and by gosh, I figured out how to solve it. It was a total miracle. So I went home and wrote it out and sent it to him. And a few days later I got another letter from him—not in response to that; it crossed in the mail. And it had *another* problem! So this time I thought *very* hard about this problem, because I had so much luck on the previous ones, and by gosh, I figured out how to do that one, too. So I sent it back. So we were sending back and forth, and I think I solved four of the problems. I wrote papers on these things!

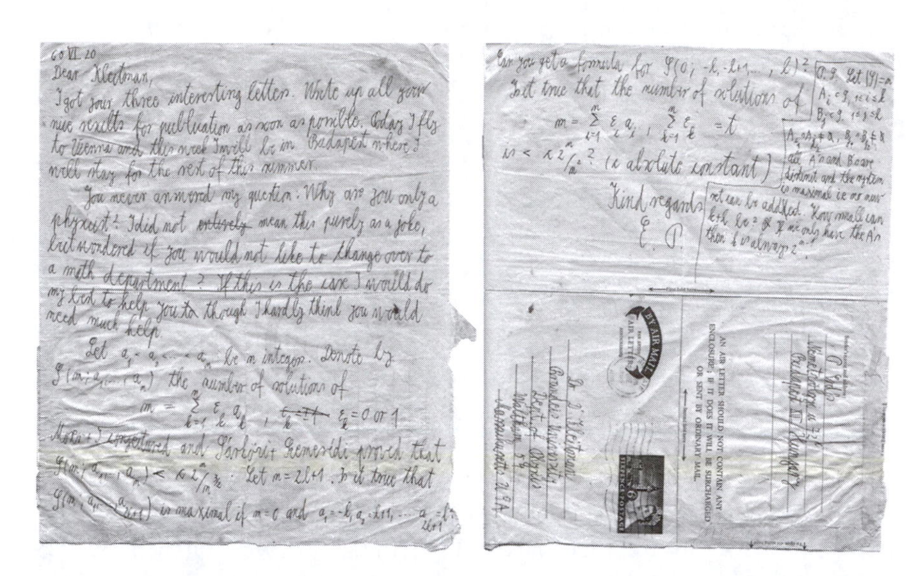

60 VI 20
Dear Kleitman,
I got your three interesting letters. Write up all your nice results for publication as soon as possible. Today I fly to Vienna and this week I will be in Budapest where I will stay for the rest of this summer.

You never answered my question: Why are you only a physicist? I did not ~~entirely~~ mean this purely as a joke, but wondered if you would not like to change over to a math department? If this is the case I would do my best to help you to though I hardly think you would need much help.

Let $a_1 < a_2 < \dots < a_n$ be n integers. Denote by $g(m; a_1, \dots, a_n)$ the number of solutions of

$$m = \sum_{k=1}^{n} \varepsilon_k a_k, \quad \varepsilon_k = 0 \text{ or } 1$$

~~Moser conjectured~~ and Sárközi + Szemerédi proved that $g(m; a_1, \dots, a_n) < c\,2^n / n^{3/2}$. Let $n = 2l+1$. Is it true that $g(m; a_1, \dots, a_{2l+1})$ is maximal if $m = 0$ and $a_1 = -l, a_2 = -l+1, \dots a_{2l+1} = l$?

Can you get a formula for $g(0; -l, -l+1, \dots, l)$? Is it true that the number of solutions of

$$m = \sum_{i=1}^{n} \varepsilon_i a_i, \quad \sum_{k=1}^{n} \varepsilon_k = t$$

is $< c\,2^n / n^2$? (c absolute constant)

Kind regards
E. P.

Let $|S| = n$, $A_i \subset S$, $1 \le i \le k$, $B_j \subset S$, $1 \le j \le k$, $A_{i_1} \cap A_{i_2} \ne \emptyset$, $B_{j_1} \cap B_{j_2} \ne \emptyset$. The A's and B's are distinct and the system is maximal, i.e. no new set can be added. How small can $k+l$ be? If we only have the A's then k is always 2^{n-1}.

BY AIR MAIL
AIR LETTER
AN AIR LETTER SHOULD NOT CONTAIN ANY ENCLOSURE; IF IT DOES IT WILL BE SURCHARGED OR SENT BY ORDINARY MAIL.

Letter from Paul Erdős to Kleitman, dated June 20, 1960.

Now, you got to understand he had an amazing knack of knowing what people could do. It was more him than it was me. Anyway, after about four or five of these, he said, "By the way, why are you only a physicist? You should be a mathematician." I ignored it, and there was more back and forth, and finally he wrote—this letter I have—"You never answered my question, why are you only a physicist?" So I wrote him [splutters], "Well—I mean—you know—what?" And the next thing you know, he wrote to people he knew, and I got a call from Gian-Carlo Rota, who was teaching here at the time. He said, "Come, let's talk," and I went, and I talked to him, and a little while later I got two job offers, one from MIT and one from City University in New York. But I'm lazy, I wanted to stay here [in Boston]—and I've been here ever since!

Norman Levinson told me that when my appointment was brought up, somebody said, "This guy doesn't have a degree in mathematics. How can you make an associate professor appointment to someone who has no degree?"

So he said, "Well, you know we had somebody else with no degree in this department."

He said, "Who?"

"Norbert Wiener!" His degree was in philosophy. So that silenced that particular objection [laughs].

How did you end up in applied math?

Well, pure math and applied math were organized differently. Originally the applied math committee was conceived as an interdepartmental committee. The idea was to have people from all different departments and to cultivate the kind of mathematics that was of interest to other departments. I think it was in reaction to the fact that the traditional math department was relatively pure and wasn't interested in anything like this.

I'm not sure who were the ringleaders, the ones who started it. There was Norman Levinson, C. C. Lin, Harvey Greenspan, Lou Howard, Willem Malkus, Dave Benney—these were the people who were there when I became sentient. Eric Reissner: he was an older guy. I'm leaving out one or two. But as more and more people were appointed in applied mathematics, it came within the mathematics department. In other words, there were no members who were not members of the department.

The committee's main function is appointments. Everybody in applied math with a high enough rank is on the applied math committee and takes part in decisions about young people and appointments and whatever. In pure math, it was not so. Pure math was somewhat larger, and there was a pure math committee. If you were on that committee, then you worried about appointments and things. Otherwise, you had no connection with such matters.

Now, I came as an associate professor in pure math. And in due course I was promoted, but I was not on any committees, so I knew nothing about appointments. But once, it must have been around 1971 or '72, Harvey Greenspan—whom I had known; he had been an instructor or a T.A., I think, when I was a graduate student at Harvard—said, "Would you like to be in applied mathematics?"

I said, "Well, uh, what do you mean? What does that mean? Why?"

And he said, "Well, if you have somebody in mind, you can hire an instructor and have influence on who gets appointed. Your name will say Professor of Applied Mathematics instead of Professor of Mathematics."

And, you know [laughs], no one had ever asked my advice on anything at all in pure math! So it seemed like a clear winner: I could appoint an instructor. I said, "Oh, okay!" And so I moved upstairs, and generally there's been one junior position that I've had a lot of influence in appointing.

The year that I came, 1966, Rota left. He went to the Rockefeller Institute in New York. But after a few years, he came back. And, after a while, when Gian-Carlo was back, he said, "Can I be in applied mathematics?" Because pure mathematics has tended to favor the traditional areas of pure mathematics, and combinatorics was not quite a traditional area. So it had not so much influence downstairs, in pure mathematics. There was some question whether what he was doing could be considered applied math, but then again the same question existed about what I was doing. So he went to applied mathematics and got involved with appointments.

I was asked to be on the applied math committee in an attempt [on the committee's part] to branch out in the discrete direction. They foresaw that computers and computation were something of central importance in the future, in mathematics and everywhere else. Their first effort in that direction did not work well. They had Seymour Papert, who was a good guy. He was one of the early figures in computer science. But we never saw him. He existed in some other realm. He did teach a course; it was a pretty good course, but he was nonexistent as a member of the department. You wouldn't see him from one year to the next.

But they figured I knew something about science, my degree being in physics—that maybe I could help them develop in a discrete direction. And I did a little bit, eventually, in operations research, that kind of thing. And I actually helped them when it came time for appointments.

Colleagues and Social Life

What do you remember from those early days? Whom were you close to?

Well, mathematically, Rota was the closest. He was a wonderful, very charming guy. He was interested in philosophy, he was very lucid, and he had a tremendous wit. He could write the most beautiful recommendation letter you've ever seen. You would read it and it was *hypnotic*; you couldn't not give the person the job, just by reading his letter. He had a beautiful hand.

When I came here, I was downstairs on the second floor. It's now a secretary's office, on the hallway that goes along the river. I remember Gerry Sacks was in the next office. I had known him as an undergraduate; we were contemporaries. Eventually, I got to meet everybody. Ted Martin was a very nice guy and amazingly well organized. On the other hand, when you spoke to him, after a few minutes, you could see that his mind had moved elsewhere [laughs]. He was already thinking of something else. That was a little disconcerting. But he would have parties, and so I got to meet most people.

Gian-Carlo Rota.

I remember people who played bridge and ping-pong. Frank Peterson was one of those. There was an intramural table-tennis MIT competition. It was a league, and any organization at MIT could field a team. Most of the teams were with various fraternities or clubs—in fact, I'm not sure the other departments had teams at all, but we had a team and we played. We did pretty well, but there was Chinese A and Chinese B, I think even Chinese C. I think one of them was more Taiwanese and the other was more mainland—who knows. They tended to beat everybody. I remember this one guy whom *nobody* could beat. He had won national championships. I think I actually met him years later; I can't remember where. Interesting guy. So they generally won. But we would beat everybody else.

What was the social scene like in the department? I gather that socializing at people's homes used to be more common than it is now.

Well, the department in the old days used to have a party or several parties and invite the faculty. That was more or less typical. I don't know whether that continues. There *are* parties, but they tend to be at school. There's an

Endicott House party, which is kind of fun. That's a tradition that's gone on for a long time. I know back in the eighties when I was department head, it was already standard. I changed the food a little bit, but that was it. So maybe that was sort of a substitute. And now there are little parties at the beginning of the term and at the end of it, and now and again during the year.

But parties at people's houses—over the years, I think society itself has had less of these. In the old days social life often revolved around dinner parties that people would have, small groups of people. You'd go to various of these parties, and then you'd build up obligations to reciprocate. It was kind of fun and a way to socialize. That's rare today, partially because women are working more and just don't have the time to devote to being hostesses and setting up such things! And life is at a slightly faster pace these days. People are busy with this and that. I don't think that's only at MIT, but in general. Also, we've gotten older, my wife and I. So her willingness to have a party that's not actually required is less.

But back when you were department head this was something you felt you should do.

Yeah. You wanted to do something, particularly for the newer and the younger people. But, of course, you knew the older people, so one tended to include those. Now I think it's more of a burden to a department head to do that at home. If you're going to do it in small groups, you have to have a lot of them. And if you're going to do it in a large group, that's really a big job.

So what would happen in the old days when people would gather at these parties? Was there a lot of math talk?

No, no. It was a social event, social thing.

And what about within the department itself? Steve Kleiman said that they used to serve tea every day, with samovars.

There is still tea every day. In fact, there are two of them, one on the second floor [pure] and one on the third floor [applied]. Although anybody can go to anything, and people do. I do remember one thing from—gosh, it must be close to thirty years ago. In those days they would set out a plate of little cakes, a big piece of cake cut into small squares. I think that the tea would begin at 3:30. And people started complaining that they came in at 3:45, and all the cake was gone! I happened to have a discussion about this, and someone said, "Yes, there's somebody who comes, I think from the chemistry department, 3:30, with a bag, and comes down the hall, this way and this way"—and described how to me—"and fills her bag with cake. And then goes down the stairs, and then goes down the hall, and then *up* the other stairs, and down there." And then he looked down the hall, and *there she was* going up the stairs!

So I raced after her; I grabbed her bag of cakes. I said, "Don't you ever do that again."

She said, "Can I have my bag back?"

I said, "No." That solved that particular cake problem.

Work

How does your work in combinatorics differ from Richard Stanley's?

Well, he does more algebraic combinatorics. It's closer to pure mathematics. You know, all mathematical problems have combinatorial aspects. In fact, what are called combinatorics used to be the things that—you got a certain amount by wonderful, general arguments, and you were left with some details. So the combinatorics was sort of the ugly details: that was the attitude of mathematicians. But Richard was able to find beauty in these, and wonderful things, about the details. Like for example, in algebra, the aesthetic approach is to do things invariantly, without introducing coordinates. But relations among different coordinate schemes turn out to have lots of wonderful properties. So that's sort of a combinatorial direction.

My original things that I worked on were extremal problems. You're given a set of conditions; how big can some structure or some parameter be, given those conditions? The Erdős problems were of that flavor. But I also worked on things in operations research and odd things in all sorts of areas. I do anything that seems interesting [laughs]. A hodge-podge.

Were there problems that emerged because of collaboration with people in other departments?

Yes, some of that, and some from consulting. You hear about problems from other people. I even worked a little in biology.

Now, that came about in the strangest way. A colleague, a physicist, whose wife was a biologist, once came to me with a problem that involved RNA structure. See, RNA is a single strand of nucleotides, and it forms some sort of shape, because the parts interact. There's certain rules that this obeys, and so trying to model their interactions to determine the shape—that was the problem. So a student and I figured out an algorithm for doing this, many, many years ago, back in the seventies.

A long time later, for reasons I don't know, I got a call: would I take part in some symposium, which was involved in trying to get computer science–type people interested in biological problems? It was a big meeting at the National Academy in Washington, and various biologists made presentations, and I was supposed to give comments on these presentations. I said, "Okay, why not?" My assumption was that they would send me something so I'd know what they were going to say—but no one ever sent me anything! So, the meeting came, and I went down to Washington the night before. Well, I was supposed to talk—about what?? So I tried to imagine what they would say, which was, I had not the vaguest idea! And you had to have slides and all of this, comments on something that you never heard. But I tried, and I cobbled something together. Well, when I get to the meeting, there's like a thousand people there. And again, by some sort of miracle, what I prepared wasn't that far from what was appropriate.

Anyway, I was impressed: I learned a tremendous amount by attending this meeting. And Bonnie Berger, who had just received her degree from Course VI,[2] in computer science, came as a post-doc to work with me. She was looking for a new field. So we had discussions, and I said, how about applications to biology?

Now, there are two kinds of applied mathematics. One kind involves working on problems that arise out of science. You hear about a problem and you work on it, without paying attention to what's going on on the scientific side. Eventually you may solve that problem, but you would look at it as a mathematical problem. So it's really mathematics, but mathematics motivated by problems from science. There's a whole school of people at NYU, at the Courant Institute, who mainly do this sort of thing. The other is to actually do things *in* the science using mathematics—using mathematics to make progress in the science. You look at the science and try to develop what you need to handle the science. C. C. Lin and [Harvey] Greenspan were big advocates of this latter business.

So I encouraged her. I said, the problem of understanding the production of proteins from the DNA sequence is a great problem, and I heard [other] interesting ideas from people. This is a good thing to work on, I said, but you have to really understand what the problems that the biologists are facing are, because these things change. They are now confronting real

---

[2] Course VI at MIT is the Department of Electrical Engineering and Computer Science (EECS).

mathematical problems, difficult mathematical problems, that they're not equipped to answer. But these change! Eventually there will be answers to it, and then they'll go on and be interested in other things. So you can't just say, "Here's an area of problems, and I'll study it for the next twenty years." Stay close to the people who are in the trenches, and find out what it is that they want and need, and work on those.

So she went, and she took courses in biology and she got to know various biologists, and they had all sorts of problems that they wanted to do. She worked on them, and she solved a few, and she's working here on the faculty and has a whole group of students working with her. So she has been very successful in doing that.

Do the continuous people, who work on systems of differential equations, have a different mind-set than those who do discrete math? Do you have people floating between the two groups?

No, I don't think so. There is an area of overlap, but it's not really well developed. The differential equation–type people, they're different: there are real problems that they want to apply these differential equations to. The new methods are computational: how to handle types of problems that were intractable before. These are discrete mathematical questions of interest combinatorially, as well, in terms of the problems that they address.

The difference is, in order to succeed in the more continuous areas you really have to embed yourself in the area of the problem. It's not enough to know mathematics. John Bush has been working with students on the way bugs move on the surfaces of liquids, how they propel themselves on a lake, the surface tension, how do they go uphill, and understanding the mechanisms or the physical laws that they use to move.[3] To do something new and interesting there, you have to acquaint yourself with the phenomenon, the actual bugs whose behavior you're trying to understand, and make models of the phenomenon. *Then* comes the mathematics, which is where the differential equations come in and may involve some numerical computations as well.

And so to succeed in the traditional applied area nowadays, you have to be able to span that whole spectrum. You have to really embed yourself in science as well as mathematics. So it becomes harder. I do the easy stuff. There's plenty to do in combinatorics or computer science, traditional problems, where the models are well established, so you can afford to just look at

---

[3] For a description on John W. M. Bush's work on the hydrodynamic propulsion of water striders, see, for example, Dickinson, M. (2003). Animal locomotion: How to walk on water, *Nature. 224* (6949), 621.

the more mathematical aspects. Operations research: you're trying to optimize, minimize the length of time, or the cost of something. What's there to model?

Consulting

You mentioned that interesting problems have emerged from consulting work. What were some of the experiences you had working with the world outside of MIT and academia?

Well, when I was teaching at Brandeis, there was a student there whom I knew named David Rosenbaum. He actually was working for someone else, but he got his PhD, and I was on his committee. He got a job I think at B.U., and then later he went to work for the government. And he remembered me and hired me as a consultant. This was in the 1960s.

He worked for the Office of Emergency Preparedness. And the guy who was running it, who was a pretty bright guy, somehow had a mandate for hiring mathematicians to try to analyze problems. So he had hired this fellow, and he worked on various things, some of which were classified, which I knew nothing about. But they also used their skills to work on problems brought to them by other departments, and somebody at some other department had come to them with a problem.

The problem was, they were developing offshore pipelines in the Gulf of Mexico. They would make a platform and drill. And when they got the gas, they would build deep pipelines to bring the gas to the shore. There were lots and lots of these platforms, lots and lots of these wells. Dozens. It would be silly to have a single pipeline for each separate platform, so you could merge pipelines together at different platforms. The question is, what pipeline should be built, because you wanted it built in the network of pipelines. And these were huge pipelines to connect. There were seven different sizes of pipe at different platforms, and the pipes have different costs depending upon how thick they are. There were different locations, with a certain pressure at each location. You have a given amount that you want to flow from each place. So the question is how you design it so that the stuff will flow in and do the right thing.

And this organization was responsible for approving the pipelines. What do you do? What should be built? Are the proposals that they got good ones? What should they approve? They wanted some criteria. They wanted a mechanism for designing an optimal pipeline.

Okay. So he hired a group of people to consult, sort of a summer thing. I stayed here, but I visited a few times during the summer. And we developed an algorithm for a computer. This was around 1970, I'd say: in those days, computers were ridiculous, simple things. But we did it: I was fortunate enough to come up with one of the key ideas. It was an optimization program. It could design wonderfully. It was dramatically successful—did exactly what they wanted. And it was fun. You know, it was just the summer, half a dozen guys, pretty good guys.

At the conclusion of that, they held a meeting. They invited all kinds of people from different government agencies to explain some of the things they did, and a woman from the General Services Organization said, "The government has this problem with their telephone lines." See, people rented dedicated lines from the telephone company, private lines. So you'd pick up the phone and you were connected to your person in Texas automatically or through your own switchboard. Lots of companies had lots of dedicated lines, and the government had thousands, tens of thousands of dedicated lines. And they were complicated lines. The Weather Bureau had lines that were actually not just two terminal lines, but a thousand terminal lines.

Now, it had become possible for companies to build their own networks. The government had approved a tariff in those days called a TELPAC. The TELPAC cost a certain amount per mile, but you could have a lot of lines go through. The bigger-sized TELPAC could handle 240 lines, and the mileage charge of this TELPAC was like the cost of 20 lines, so it was a twelfth of the cost. So this is bringing in a lot of sources here, and a lot of syncs there, and you would put lines into the TELPAC here and out there. On the other hand, you had to deviate to *get* to the TELPAC, so the actual mileage was longer. So it was a tradeoff between how much you want to deviate and how much you save, because once you already have the TELPAC, it doesn't cost you anything extra to put another line in it, until you reach its capacity.

This was for billing purposes only. It was a virtual object, a billing device. In other words, as an order would come into the telephone company, it said, "You can build a network, so we will bill you on the basis of the costing of that network," but it was strictly a paper network.

So the government had the problem that there were several thousand nodes and lots of TELPACs all here and there. Not only that, you could change them: there were a hundred changes every day. There was a big place outside of St. Louis with a lot of people working there, and every time a change would be ordered, the telephone company would implement it and send it to this place, and somebody would decide where to route it through these TELPACs, and then they would enter it in the file for each TELPAC.

This was all done by hand. So, clearly, the problem was to create an algorithm for solving it efficiently. You had to create a system for handling the data and an algorithm for designing the TELPAC. They said this could not be done! At the time they thought the problem was too big to be solved on any computer.

Now, there were all kinds of problems, which we learned subsequently. You see, what would happen was a TELPAC would fill. Well, then something that they would normally route *through* the TELPAC, they would route *around* it, which would be more costly. And then, at some later stage, maybe space would open up inside—but they had no record of what was routed around it: they only had a record of what was *in* it. Maybe they'd order a second TELPAC because there's a big increase in volume. If they did that, then what was routed around it they'd certainly want to put in there—but they wouldn't even know what that was. So what was done was very inefficient. Because these were *huge* files. It was a gigantic enterprise, with all the government lines, tens or hundreds of thousands of lines. So they had no idea.

Worse than that was that they actually shared the system with the military! The TELPACs were ordered jointly, and they each had their share of the TELPAC. It was senseless to have them done separately. But the question of how this should be arranged involved political decisions as to who should be in charge: you had to pay attention to how you did things and how you interpreted things. They couldn't just change this and that: it had to be coordinated. If you wanted to change TELPACs, you'd have to call up some other place where they had a similar thing.

What do you do? The cost function for individual lines was something which depended on distance. The longer it was, the cheaper it was per mile. But then there was the TELPAC. You could put TELPACs anywhere; you had a lot of cities and a lot of other things. So you had this strange cost function, very nonlinear. It could be phrased as a multicommodity flow problem.

I was concerned with the question of designing algorithms for solving this optimization problem. And this young guy came to my office by the name of Armin Claus. Actually, he was sort of my student in a way, but he wasn't an MIT student. He taught operations research in Canada and later at Montpelier in France. And he said he had been working in Canada on a problem similar to the pipeline problem. So I told him about the telephone problem, and he came up with a good idea for doing it, which I communicated to the people down there. But he was not a citizen of the United States, so they couldn't hire him. However, what we did instead, since they were the General Services Organization—the General Services Organization is responsible for purchasing everything for the government, so they were actually paying for all the computers at MIT [laughs]. So they let them use the computers at MIT free.

He conceived the idea that since he couldn't work for them, why doesn't he do consulting? He said, "Other people must have the same problem. Let's offer to do the same for other companies." His English wasn't perfect, so I helped him, and we sent a letter to the president of all sorts of companies saying, "Hey, you've got this problem, and we can help you."

And we got response! You see, nobody believed this tariff would last forever, and they were perfectly right. So they didn't want to spend a lot of money to buy a system that would work for this tariff. However, what we offered them was, "We will look at your network. We will make suggestions, which will save money on your bill. You just have to communicate them to the telephone company. And we will charge you two months' savings."

We dealt with a lot of big companies. We went to Westinghouse. The General Electric people were rough customers. They wanted to find out what we knew, and they weren't interested actually in purchasing anything, but we went and visited them. It was all very ridiculous. Then there was one guy who did this for his company in Connecticut, and he thought he was a wiz. So he said, "Okay. Here, look at it and find something." I didn't even use a computer: I just looked at it, because it was a small thing compared to these others, which you couldn't dream of looking at. So we found [savings of] $2,500 a month. But I didn't want to insult him, so I wrote this nice letter saying, "Boy, we expected to find much more. We're terribly disappointed. Your guy is really doing a great job. *But* we did find this small amount."

So they sent us the money. And the guy liked the letter so much that his brother-in-law, who had the same thing at another company, called us and wanted to do the same thing because he wanted a letter like that! So we saved him a thousand dollars or something. This is early seventies: it's like $10,000 now. So we got a little money.

One response was from an airline was, "Very interesting. But we have this company, ARINC, Aeronautical Radio, which is located in Maryland, that does this stuff for us." And so we said, "Well, gee, why don't we contact them?" So we got a contract for a couple hundred thousand dollars to do this for *them*! We did it, although the vicissitudes you don't want to hear. It was too horrible. And the computers were pathetic—pathetic by Commodore 64 standards. The mainframes were, like, 30K memory. So it wasn't easy to do.

There were all kinds of problems that developed, which wised me a lot on the ways of the world, so it was a valuable experience. Ultimately, when they had this thing up and running for the government, it was quite successful. And the thing we did for ARINC was very successful.

What is MIT's policy regarding spending the time and getting paid to do consulting?

You are allowed to consult one day a week during the year, and then whatever time you have in the summer. It depends: if you're drawing additional summer salary, you're limited. You're not supposed to take two salaries at the same time. But you are allowed one day a week to consult. This wasn't salaried, and it wasn't taking anything *like* one day a week.

What would they call you in for?

Oh, occasionally I'd go down and talk to people, and see what they were doing, and make suggestions. And sometimes those suggestions helped. What you want to do is stimulate people: give them some idea that they didn't have before. You don't want to solve their problem for them—or you do, that would be wonderful, but you really want to energize them. You talk to them, and you try to mention a few ideas. And maybe one of them will rub off and help them; maybe it won't.

The same guy who hired me initially, David Rosenbaum, got another position with the government. One was, he worked for the Nuclear Regulatory Commission: its name has changed over the years. I consulted for that on safety of reactors, at first, and then on a number of other issues. Because there were statistical issues, but also I was trained as a physicist, so I knew something about what it was all about. Then he worked for the General Accounting Office, and they did studies on various things including the safety of LNG facilities, liquid natural gas facilities. I did consulting on that, and various people in the department worked on it.

There were interesting problems. During the Second World War, there was a gas facility in Cleveland where there was a big disaster. There was a tank that ruptured, and this liquid gas flowed out, and it ruptured the adjacent tank. This stuff flowed as a liquid into basements of buildings, and then it started to evaporate. And when it evaporated, it hits the pilots and BOOM! the apartment houses all blew up.[4] And so one issue was, if there

[4] In October 1944, a tank belonging to the East Ohio Gas Company allowed a vapor cloud of LNG to escape to the surrounding streets and sewer system. The natural gas in the

was a rupture—if somebody flies an airplane into it, this kind of thing—will this stuff flow out? They build the tanks with containing walls around them. These are designed so that the stuff, if it came, would not flow over. But, dynamically, if it ruptures, there will be a shockwave as it comes out. It's not a static situation, it's a dynamical one: it's a fluid *flow* problem. Fluid dynamics is the model.

Harvey Greenspan designed a little scale model with water, a little thing with a tank. It scales: water is a liquid, and, sure enough, they showed it would just flow right over. On the other hand, you could put baffles in which would prevent that. I'm not sure whether it ever changed the design of any gas facility, but it should have! Because it was stupid not to have these baffles. There was no big deal about them, and it would render the thing much safer.

Do you ever get papers out of consulting projects?

Yeah! Oh, sure. And I've had students who have had trouble, who weren't progressing very fast in their research. Normally, we support people for four years. If there's a fifth year, we've said, "No, go get a job. Do something." They've gotten a job and got interested in a problem that they encountered during their job, and that's what they wrote their thesis on. That's happened more than once. That's why I encourage them to do it: because it's useful for one's career and for one's research to get involved with a specific problem. In mathematics there's *so* many potential problems; there are so many potential areas. It's tough to focus on one! Why do this one and not these eight zillion others? Well, if this guy's paying you to do it—okay, I'll focus on it. And then, once you're focused on something, then finding something related that interests you, and you can contribute. It happens.

Committee Work

You've also been involved in administration, starting from around 1969. That was the time of the student uprisings at MIT.

Ohhhhh, yes. When I was promoted, which I guess was in '68 or '69, I felt an obligation (which I probably should have felt from the beginning) to do service to the Institute. So I got a call saying, would I serve on an Institute committee. In those days there were representations on various Institute

---

cloud ignited, causing the deaths of 128 people. Investigations concluded that World War II rationing of stainless steel alloys had compromised the tank's structure.

committees that were elected from the faculty, and there was something called a Nominating Committee, which was charged with getting victims who would be willing to be elected. This election was something of a farce, because it was hard enough to get *one* candidate for a position, let alone two, so that any competition was strictly nonexistent. So I said, "Look, I don't want to do this"—of course—"but if you can't find anybody, I would." Which, I did not realize, was the most positive response that they ever got. That was acceptance. Slamming down the phone, that's one step downward [laughs]. Ripping your phone out from its roots and never answering again—that's negative.

So I guess that summer or that spring I had a sabbatical. It was 1970. When I came back they said: You were elected to be a member of this committee, the Committee for Education Policy (CEP), and each member of this committee is supposed to join some *other* Institute committee as sort of a liaison, so this central committee would know what's going on with the other committee. Since you were away, everybody else chose theirs, and you're left with the Discipline Committee.

Now, this was something terrifying, whatever it was. See, the previous year there had been some problems, I guess it was an occupation of Wiesner's office, and it was a discipline matter, and this one committee had met constantly for days, weeks, months. But what was I going to do? Tell them, "No, take it all back; un-elect me, have a special election?" And they said, if you'll do this, we'll cut your teaching load one course; we'll do this, we'll do that. So I agreed. And everything worked out in my favor because there was almost *nothing* going on during the two-year period when I was on this committee.

The Discipline Committee was a fun sort of thing. You don't want to punish people—you don't want to ruin their lives. On the other hand, you want them to do the right thing, so the question is, how you do that? Well, the standard thing is when people transgressed and they were suitably apologetic, one wrote them a letter saying, "This is a very, very bad thing [adopts severe expression], and you must never do that again, and if anything happens in the next year, there will be *serious* consequences." That was it! So every case ended like that. Except one: I think it might even have been a child of a faculty member. This person had taken a course at Harvard and had submitted a paper that clearly was directly copied from some journal or something. It even had the illustration, the graphs. The Harvard instructor reported the incident to someone at MIT, who referred it to the Dean for Student Affairs, who was on the committee. But this individual—I think it was a she—never showed up for her hearing. And if you do *that*—you get tossed out. What I learned is, *any* authority, even the most feeble [laughs], if you disrespect it; if you don't show up, you're in trouble. Do anything, but make sure to show up. So that was a good lesson.

And then there was this other incident at the faculty club, which was really a lot of fun. I mean, it was entertaining, but in a gruesome sort of way. And there, I actually was useful. The incident was this: The faculty club employees belonged to a union. There were several kinds of employees, one of which was people called banquet set-up men. The members of that group were largely black—in fact, all black, I think. Now, they had special privileges, because they were around for the banquets: they got to eat the banquet food. As a result, the union, which represented all the others as well, insisted that they be paid slightly less, to cover the value of this perk that they got. This was the union! It had nothing to do with the faculty dining room. The dining room couldn't care less about whether they got fifty cents or two dollars more.

An organization called SDS[5] got wind of this and decided that this was an unfair discrimination against blacks, so it made a big stink. This was a time when universities, MIT included, had made a big effort to get more black students. Today it makes similar efforts, but it does so in a better way; in those days, it simply admitted more, which had pernicious effect. Because these included people who you could say *could* be talented and it was just that their backgrounds were bad. But because their backgrounds were bad, they didn't have the work habits that their colleagues had, and they didn't have the [academic] background; and lacking those two things, their chances of success were very small. So they tended to self-segregate themselves and to be very unhappy. Those who had the qualifications and the work habits were fine. But those that didn't would eventually fail. The people who had done this were proud of themselves for all the good they were doing, but it was in effect ruining these people's lives. Because they probably could survive, and thrive even, at schools which didn't assume the kind of background and work habits that were standard at MIT. Those were much higher than most schools. These students would have been average students elsewhere, but at MIT they were strictly at the bottom, and there were no provisions for them. This kind of thing happened all over the country, and there were student riots and all kinds of unpleasantness came from taking people and being very proud of putting them in a horrible position. They didn't like it, and they rebelled! And I can understand that.

---

5 Students for a Democratic Society, a radical student organization in the 1960s.

So this Black Student Organization supported this, and in response, they disrupted some sort of party, a faculty party, that took place at the faculty club. Wouldn't let people attend and so on. So this was a big crisis, because technically, disrupting a class, or obstructing an official university meeting, is very, very bad. Horrible!

Now, you have to understand something about the composition of the Discipline Committee. There was me; there were other faculty members that arrived at it from other means or directly agreed to be on; and there were student members. The student members, particularly at that time, were sort of a left fringe, SDS types, and they were on the committee in order to further the cause. Now, the interesting thing was, all the ordinary meetings, which weren't very many, were unanimous: the students' recommendations were identical to anybody else's. But *here* was a case that they felt that they could do something for the downtrodden. So they conceived of the idea of temporarily giving up their positions on the Discipline Committee and have themselves replaced by members of the Black Student Organization.

So there was a meeting of the Educational Policy Committee to consider, should they be permitted to do this? And there was big controversy in the student papers and all this hullabaloo. So I got up in a meeting and I said [speaking slowly and carefully], "Look, on the Discipline Committee…it would not be proper…for members of a group that's accused of something to be voting on the punishment for them. That's unethical. The committee would *welcome* having representatives of the Black Student Organization come and present their views and advise the committee. But in terms of voting, you can't do that; that would not be right. In fact," I said, "the Chairman of the Committee, Elias Gyftopoulos, happens to be on some governing committee for the Faculty Club, and he's disqualified himself because he's on that committee! So why don't the students stay on the committee, and we'd be *delighted* to have some advisors, who can come and tell us their point of view."

And so they did! And there never was a hearing. *Completely* disappeared, out of sight, just by my opening my mouth! So I was very pleased with myself. And it led to no Discipline Committee meetings at all. It was obvious there was going to be a letter. They had attorneys arranging the wording of this letter so it was pleasing to everybody. The letter said, [serious tone] this is very bad, you really can't do this, and don't do it again.

The incident that did lead to meetings happened I think in June, and my committee appointment ended July 1, but I had to attend the meetings anyway. There was some sort of trouble somewhere, maybe at the ROTC offices, but I don't really remember. By this time the administration had a clever strategy for handling riot-like events. It diluted them. If a group of people attempted a riot, a huge number of administrators and assistants and friends would descend on the area until they represented a majority of those present, and they would then dominate the conversation. This would take the fire out of the event and tend to render it innocuous and even boring to the participants.

But somehow this one got out of hand, and I think some participants were arrested. The Discipline Committee had nothing to do with that and essentially had nothing to do in the entire matter. However, there was an extreme left-wing faculty member on the committee who wanted to use this opportunity to issue a manifesto denouncing the government for its behavior in the Vietnam War. There were enough sympathizers on the committee to make issuance of this manifesto a real possibility. I didn't think that issuing political manifestos was within the mandate of the committee, which was after all an official MIT committee. Nor was it in the interests of MIT, which might be held responsible for it, after the inevitable publicity. I was asked to stay on the committee all summer long while this was being debated, until the matter was resolved.

What about the Committee for Education Policy itself? Who ran that, and how did that go?

The chairman of the CEP was William Ted Martin—this was after he was department head. And, boy, he ran a committee something beautiful to behold. What I remember about the CEP meetings was their great efficiency. Much of this was due to Ted, but also to the members of the committee, all of whom ardently desired to get the meeting over with as quickly as possible, so they could get back to doing what they wanted to be doing.

First they began with a very nice sandwich lunch. This meant that when the business part of the meeting began there were no straggling latecomers: everyone was there to get the lunch. Ted would arrange a full agenda and send information to the committee members about each item. Thus there were no boring presentations by advocates to put us to sleep. On each item Ted would go around the table and solicit comments from everyone in turn. He would then entertain motions or, if people showed particular interest in an item, would ask them to draft a motion on it to be voted on at the next meeting. In this way everyone got in his or her two cents' worth of discussion, and few people gave more than a nickel's worth. Once I made the error of remarking that I heard about a problem with one of the educational experiments then under discussion. I got appointed to a committee of one to figure out how to assess that problem. I managed to come up with something, and my recommendation was actually followed and accomplished.

There was generally lively discussion, and the committee decisions were generally very reasonable, though the committee consisted of faculty members, students, and a few others. Much of the committee activity consisted of drafting motions on the issues under consideration to be voted on by the entire faculty. As far as I could tell it did an excellent job, and morale was very good. The only trouble I saw was that, in part because the committee did such a good job in drafting motions that they were quickly approved, faculty meetings became quite dull.

Ted made it seem like the committee members themselves came up with all the ideas. He seemed to leave everything up to everyone else. Yet in fact he got the committee to do exactly what he wanted it to do, by some sort of magic. I can't really remember much of his technique, but it was surely brilliant.

Department Head and Chairman of Applied

You were department head for five years, 1978 to '83, and then chairman of applied math after that?

Yes, twice after that and once before that, I think.

You came on as department head after Ken Hoffman. I gather that his tenure as head had been somewhat controversial.

Well, see, there had been some conflict. Now, the conflict, I think, was silly, and could have been avoided. But it infuriated the people in applied mathematics. That is, the people in applied mathematics, Benny and Greenspan and Lin and so on, wanted to be able to make their own decisions about things in applied mathematics. Now, there was no question in terms of appointments that they would do that. That's the important thing, and that really wasn't a bone of contention. The issues of contention were in trivial matters, which I think Hoffman assigned to his assistants and didn't pay the slightest attention to. And the assistants did things that annoyed people upstairs. They would assign offices to new people, and sometimes they wouldn't be appropriate. Little things like that would cause friction. People would get annoyed if it wasn't done right. And instead of listening to the people in applied math, Hoffman was gruff, and said no, and simply backed up his staff, whose judgment wasn't terribly good in these matters. That caused friction.

What were your goals when you came on as head? Did you have basic things or principles in mind that you wanted to put into effect?

Well, the department at that time was ranked among the top very few departments in the country. So my first goal was to see that it stayed there and that I not destroy the department [laughs]. See, department head is the king of the department—can do anything: make the rules; change the rules, such as they are; and do exactly what he or she pleases. On the other hand, normally one doesn't do very much unless one sees a need to do it: just to change the department for the purpose of changing it makes little or no sense.

The important thing was hiring and trying not to lose people. But one thing I found [as department head] was that the mathematicians were more or less happy. They wanted more of everything, of course, but, heck, they didn't waste their lives moaning that they couldn't get it. They just lived their lives, and did their thing, and more or less accepted their fate. And they didn't pay a great deal of interest in trying to see who could make hires.

There were committees that were responsible for the hiring process, and they did a very good job of hiring young people. But in terms of recruiting older people, there wasn't much interest. The actual *doing* it is done by the committees, but I tried to drum up interest, to get people. I remember going to Mike Artin, who I think was pure chairman at that time, and saying, "Let's *do* something. Let's hire someone. We need people." And that actually led to hiring Isadore Singer. He had been at MIT and had left, and we got him back.

Upstairs nobody was terribly interested in doing that. We *grew* people in computer science, like Mike Sipser and Tom Leighton. But the people in the continuous areas—it was a very distinguished group, and they just didn't see or couldn't recognize young people who were at the same level. That's why they didn't hire anybody [laughs].

The trouble is, applied math changed so rapidly, and most older people did their work without reference to computers. It was more in the mind rather than computational, whereas today it is computational. And so the older experts don't relate as well to what people do and what's being done today. So you write to all the leading lights in the world, and it's not their kind of thing, which makes it hard to get the kind of enthusiastic letters you need. Now that's changing. I think the new generation of leading lights is stabilizing, more or less, and is familiar with the newest generation of young people. But there was a time when they weren't, and there was a mismatch. At that time, I made some effort to get the applied people to think of doing something, but I didn't succeed at all. We have not had much success hiring older people in applied, except for Peter Shor.

Although their attitude—and I think it was a remarkably good attitude, and a very rare one—was that they did *not* try to just hire in their area. So instead of hiring more people to do what they did, they encouraged hiring people in computer science, or in any new area that came up. They wanted to keep up with the modern world and do what would be the right thing for the future rather than simply do what *they* did.

I knew a number of people in computer science, and I relied a lot on Albert Meyer for advice. He's in Course VI, in theoretical computer science. I would go to Albert and say, "Who are the good guys? Whom should we hire?" He would give me wonderful advice, which I followed rigorously. That was his field, and he knew what was going on: he followed it completely. In theoretical computer science, there were two big meetings: the STOC meeting and the FOCS meeting, and everybody submits papers to give talks at those meetings.[6] It is very competitive: there were three times as many papers as talks, and it wasn't parallel sessions like most meetings; it was one session. So it was a field in which, if you went to these meetings and paid attention, you knew what was hot and what was good, and who was doing the best things. You knew who were the good prospects. Of course, Albert was involved with hiring over there [in Course VI]. But there's a competition from all different areas, from the more practical, machine-oriented people, and a limited amount of money.

So we started from scratch, but we started with people over in Course VI, who were doing similar things. When we hired somebody, we arranged so that they had an office here *and* an office over there, and that they taught classes jointly. So they were in close contact. There were also people in the Sloan School who had the same interests. So it was really one group, which was interdepartmental. We built on each other, so the whole group became the best in the country. Because the more you had, the more people wanted to come. If this was the center where lots of things were going on, this is where people wanted to come, so it made it possible to get people relatively easily. If nothing is going on, how do you get the first person to come?

[6] The annual ACM Symposium on the Theory of Computing (STOC) and the annual IEEE Symposium on Foundations of Computer Science (FOCS).

Because people want colleagues to talk to. If someone comes to a math department and nobody else is doing this: "Who do I talk to? And who will pay attention and know that I'm good and that I should be promoted?" So we were very fortunate that they exist, and it worked out well. We couldn't have built a quality, active group without that.

We hired young people. We kept hiring people at the beginning level, instructors. For example, I hired Adelman and Shamir, and we kind of teach courses jointly with Rivest, and a year or so later, they came up with this famous algorithm, the RSA. And there were others as well: Ravi Kannan, Gary Miller, Mike Fredman, Éva Tardos, David Shmoys. Those are the ones I remember best: my apologies to any others I left out.

The problem in hiring anybody is that they want to live a normal life, buy a house. You know, we bought this house for $43,000. You could do that on this salary. Now, for someone to buy an equivalent house—a million-and-a-half, two million dollars, something like that? Well, come on! It's very tricky.

So the real estate in the Boston area really has an effect on this, you think?

Oh, definitely. It makes it much tougher to get younger people. You know, people don't think of it that way. If you're a young person, you have your ambitions and what you want to do, and you're not thinking about the real-estate situation. But when push comes to shove, I think it definitely does matter.

What would tend to happen is, other schools would want people before we were ready to consider tenure. People would be here, and they'd do great stuff and could get all these offers from other places before we could do anything. That was a problem, but as long as there was a steady supply of new, wonderful people, it wasn't *that* big a problem. You're better off with that problem than with the problem of having people that didn't get tenure and are disgruntled. That's a worse problem.

What about the rest of the combinatorics group, like Richard Stanley?

Richard was a student of Gian-Carlo. He got his degree at Harvard, and he came as an instructor. And we hired other people in combinatorics as well: I remember we hired three terrific people in one year. We hired Stanley; we also hired Curtis Green and Joel Spencer. Joel Spencer had worked at Rand, I think.

And, again, when Stanley came up for tenure, there was a question: is what he's doing applied, and in what sense? There was a little discussion of that, but we said, we care about quality.

I did feel that mathematics was being shortchanged in terms of resources, both by the NSF and by the Institute. I remember reading something that made my blood boil, saying that there were going to be relatively few retirements in mathematics, so the job market for mathematicians was going to go down. See, the retirement age when I entered had just been raised from 65 to 70. And I believe when I was department head not a single person in the School of Science retired. That's what I was told. So this was a recommendation that the government should stop supporting post-docs and graduate students because there wouldn't be any need for mathematicians!

This was at the *very beginning* of the computer revolution. The computer revolution had the effect of vastly increasing the demand for mathematicians. There was almost no such thing as a computer scientist in those days. And there developed a tremendous need to teach people about what we call computer science, and the obvious people to do that were mathematicians, so the demand for mathematicians rose. So this recommendation that the government stop supporting math graduate students and post-docs was *exactly* wrong.

There was also something called the David Report. The David Report pointed out that mathematics got the short shrift in every measure of support among the various sciences. Support for graduate students, support for post-docs—vastly more per faculty member in other departments. In those days, most faculty got two months summer support. And there was support for graduate students in research contracts, and even for post-docs from time to time. Gradually, all of that sort of slowly faded away. Slowly, slowly. But in every respect—its base, per faculty member, in just everything—mathematics got less than any other department.

Other departments had vastly more discretionary money, from this and that source. I had *no* discretionary money as department head. Other departments had all kinds. I tried to build up some money by saving it and getting money from here and there so that the department would have in the future. I built up maybe $100,000, but that doesn't go very far.

Then there are secretaries. You have this laboratory, so you need a secretary to handle whatever is associated with it. So there was typically one secretary for two faculty members everywhere else in the School of Science. In mathematics there were five or ten for one secretary. Now it's even more.

Mathematics had a higher teaching load. Why did it have a higher teaching load? There was a historical basis for it. Initially, the nominal teaching load at MIT was two courses a term per professor. This got reduced to nine hours, or three courses, which is two and one, at some stage, long before my time. Now, somebody who runs a big laboratory could say, "Look, you're spending a million dollars a year to support this laboratory. I need time to run this laboratory. Therefore, I should get some of my pay from the contract which is supporting this laboratory, and I should teach less because I have to do this." So all the other departments in the School of Science taught one course a term, and they bought out the other! That was the justification. There was one department—nutrition, I think it was—where I believe the teaching load was *one* course a *year*. But the standard in the math department was three, and the others were two.

Then, at some stage—I believe it was after I was head—MIT decided that this was bad, to be so dependent on government contracts. If, God forbid, the contracts were terminated, they couldn't pay the salaries. So they got everybody off soft money. But they didn't change the teaching load! In other words, MIT now pays the full academic-year salary for all the faculty in the School of Science, but all departments but math have a two-course teaching load. Mathematics has a three-course teaching load.

Now, there were a number of reasons for that. One reason is, mathematicians, their activities are in their mind. If a mathematician is frustrated and unhappy and can't do something, he can always close off the world and go into his world of mathematics and think about it, and derive satisfaction from accomplishing something, perhaps. Others tend to express their frustrations by reacting and pressing forward on their demands to the rest of society. So mathematicians tend to be a quieter, less demanding group. So whenever there's money to be cut, "Oh, we'll cut a little off mathematics," you see. That was one aspect.

Another aspect is in terms of influence and lobbying. Other scientists use equipment, lots of equipment, scientific equipment, which brings them allies: if they want more, all the makers of this equipment say, "*Yes. Yes.* Give them more money. They will use that money to buy our equipment!" So they have all sorts of allies. Whereas the pencil makers don't distinguish a mathematician from anyone else. Chemists use chalk just as much, you see. There's nobody out there lobbying for money for mathematics. All the other sciences, knowing that they have to raise money to exist, have perfected it. They have practice. They know how to do it. Mathematicians are diffident. "I'm great. Give me money"—it's not something that rolls off the tongue of your typical mathematician.

I became department head at a time when there were cuts in the budget every year. There would be a big meeting of all the bigwigs—the department heads, people in charge of the laboratories, the deans—all the administrative people. And we were told the budget figures for the year and how last year, we had *this* amount of money and spent *that* amount, and somehow, by taking money from this place and that place, we managed to pay everything, we were able to make ends meet. But this year we're going to run a deficit of so-and-so, so we're going to have to slash your budget.

Every year it was the same thing. After three or four or five years, you realized it was all bullshit. You know, as the endowment grew, when the stock market went way up, they kept the money. But when it went down—oh, they were all upset, and had to slash the budget [laughs].

Of course, our budgets were totally meaningless. What is a budget? You have a certain amount allocated for each item. But, as a matter of fact, the amounts allocated for these items made little or no sense. Of course, the major part of the budget was salaries, and that was quite inflexible. But a lot of the money came from people who went on leave, and we'd find that people were going on leave only long after the budget was prepared. We only found out about that at the very end—and that was where the money came for doing a lot of the things that we did, including—everything! So we want to cut the budget, we cut away the postage: *fsssst*—took away the postage budget. What does that mean? You're still going to pay the same amount in postage. But the money was coming from leaves. It had nothing to do with the budget.

The first year that I came there, we ran a deficit. So the dean yelled at me. "You can't do this!" But one year we were at $100,000 deficit; the next year we were at $100,000 excess. You see, things would appear on the list of expenses. And the timing was random, so some things got charged one year that should have been charged the next year. Like some contract from several years ago was overspent, so they decided to charge the department for it. Suddenly $20,000 is taken out [laughs]. What did that have to do with any budget? Nothing! So I did not lose a lot of sleep over the budget. I just ignored it. It was pointless to spend time preparing the budget because it had very little relation to reality.

The main chore in dealing with the administration was to restore the cuts that they made. It turned out that what I had to do was to try to invent new things to get a little money added to the budget. You asked for $10,000 for this year to hire somebody to do this or that, or you wanted to support some sort of meeting and you needed a few thousand dollars.

If you asked the dean, who was Bob Alberty at that time, he would tell you the rules. "No, you can't do that. It's impossible." But if you asked the provost, who was a jovial guy, Walter Rosenblith, for a small amount of money—under $10,000, for anything at all—he would just give it to you. "Sure. You want that? No problem. Just do it." See, at each level there was a certain amount of discretionary funds, so he had discretionary funds that the dean didn't have. So once or twice I didn't even bother to tell the dean that I was doing it. He said no, and I went ahead and did it. He said, "What are you doing?" I said, "Well, I asked the provost. He said it's okay."

So that was one fact. You could get anything at all in small quantity from the provost. And here was the wonderful thing. If it was a small amount temporarily, he would just give it to you. But then they'd forget about it! It would be of no interest to him. So it would stay in your budget forever! They'd never remember. A few years later, what, they'd take it away? "What? What was it? They gave him what?" Who cares? So if you added enough of those, you vitiated the decrease in your budget. We got the money back from postage by saying, well, we're going to do this great educational experiment, and we want money to give computers to a few people, and then we're going to do this important little thing and that. "All right. Here's $5,000." And that would be in the budget forever! So, they would take out $15,000, and you would just put it back, and you would be back where you started from.

So I made a big effort to try to pound away that we needed more money, more everything. Not with any great success in the long run. But, you know, you keep telling people, eventually something happens.

We *did* manage to get some space, in the basement of Building 2. Mathematics had some graduate offices that were miles away in some corner, and nobody ever saw the people involved. And there was a space in the basement from some chemist who had retired. Now, normally, the chemists would just take it. However, they were getting a whole new building built, and they were keeping most of their old space as well. There was a committee at MIT that was responsible for space, and the provost was the head of the space committee. I called the provost's office every day for three months. I would say, "What's happening with the space? We need the space." Eventually, they gave us the space!

Now, part of that space was a room in the basement that was used to see if people had exposure to radiation. To test that, you need a place where the person is absolutely shielded from all external sources of radiation. There are cosmic rays that come from the skies and all kinds of other things that would drown out the signals, so you have to be shielded from that. So they had built a chamber in that room. The chamber consisted of pieces taken from an old battleship: ten-inch-thick steel plates. Each one weighed several tons. It was a cube, about—oh, eight or ten feet high, and you went in there.

And, boy, *nothing* got in there through all that steel. But dismantling it took quite a while.

One thing people remember is that as department head you were the first person to put a computer on their desk.

Yes. That was a time where personal computers—extraordinarily primitive ones—became available. These were Commodore 64s. They were little things, but they were easy to program, easy to understand, because they were some pretty simple machines. BASIC was the language, and you could learn BASIC in minutes.

And you could do word processing. The printing was lousy, but it was acceptable; it was better than typing it. Before that time, you had to have a secretary type your things, and that was a terrible process. Because you would write something, and you'd be completely up on it, but then you'd give it to a secretary. But the secretary had other duties and would have to prioritize things, so a long-term project like doing a long paper couldn't be given first priority. Somebody comes and wants a letter that should go out tomorrow—you can't wait three weeks until this paper is done! You have to do the short-term things first, which stretches out the time on the longer ones. And so you'd get the paper three or four weeks later and by that time, you'd have totally forgotten what it was about.

And it would have *mistakes* in it. I remember I had a secretary once who, if the word *that* occurred several times in a paragraph, she'd maybe stop at one *that* and maybe start up from the other *that*. So if you read it without following the meaning, every pair of words or every triple of words made sense together. But the whole paragraph made no sense at all! And it was very difficult, unless you were following, to find it. So you would find more, and you would give it back to the secretary, and another three weeks would pass, when they retyped the whole thing. And you'd get it back *again*. With a bad secretary who continually made a certain proportion of mistakes, this was not a convergent process.

So the computers that you brought on were more for word processing, not so much for mathematical modeling?

Oh, no, they were. You could iterate; you could do things. The problem is that they had very limited memory: 64K. So you had to be very clever. And they were ridiculously slow, by modern standards. But various faculty members used them to perform mathematical experiments, like iterating functions and seeing what happens, and did so in a way that developed new mathematics. Nowadays even such simple things as spreadsheets are incredibly useful in mathematics. The fun thing is differentiating, integrating, solving differential equations. You can do all those things on a spreadsheet. Solving pairs of differential equations, solving higher order equations as well. And of course today there are readily available programs that perform all the standard mathematical tasks immediately, so that one can get from a model to consequences in very short order.

Anyway, I thought it was very important that people in the math department became aware of what computers could do and should use them. It was terrible to have people utterly ignorant—and we were utterly ignorant. We knew *nothing.* There was the odd person here and there, but the natural tendency is to keep on doing what you're doing. It was silly to have the last people in the world who would know anything about computers be the math department. The whole thing was to get people to realize the potential of these things. Then, as they improved, you could do more and more. Once you're aware of them, then you go on from there yourself. The point is to get people to the point where they were aware of these things.

Before that time the idea of a computer was this big machine. There were computers since the Second World War, but they were mainframes, these big, huge machines with vacuum tubes in special facilities, and it was a tremendous chore to get to use them. Nothing like what exists today. And you had to know how to use it. You had to make a special effort to get involved with it. You made your program on the IBM cards, and then you shoved them into a machine. But you had to reserve time in advance on these machines, and they were located in some special place somewhere. And God forbid there was some error on one of your cards; then you had to do everything all over again. To the people who did it, it wasn't so horrible, once you got used to it. But for the average person, it was something you heard about people doing, like you hear about people being pilots and flying planes. Those who at that time did computations—solved differential equations, this sort of thing—used them. But people in pure mathematics had no contact with them. I said, we've got to change that attitude, and we've got to get people to at least know what personal computers are! So I bought computers for everybody.

Also, I had another goal: to try to get the department air conditioned. They wouldn't do it for just plain people, but if you have computers *you need air conditioning,* so that they won't rot. They were more sensitive in the old

Daniel Kleitman lecturing.

days to fluctuations in temperature. So that was one reason. You know, one tends to think a decision is done for a purpose. But human decisions are threshold affairs. There's this aspect and that aspect, and when the planets are all in sync, *then* somebody makes the decision to do something. So each little thing helps.

Teaching

You used computer programs as aids for teaching as well. You developed OpenCourseWare (OCW) courses, right?

Yeah, I have several of those. One of the things faculty used to do to fill out our nine hours of teaching was to teach sections of calculus. I did that a number of times, and I got to see that there were some things that students had a great deal of difficulty with. Visualizing certain things: changes in variables, and multiple integrals, for example, and seeing how the limits changed. They're not really very exciting topics, but, boy, did students screw them up. They scored poorly on the exams.

So I thought that there should be computer aids to doing things like that, a little thing where students can play around and see: "Change variables; see what you've got." They'd do that for ten minutes and they'd understand completely. Whereas you could talk about it, and they didn't get it or, if they got it, they quickly forgot.

Now, my friend Armin Claus—his son, Jean-Michel, did this for him. He set up a program where you could fiddle around and do a linear program, do the simplex algorithm, and see things. I happened to be on the phone with him, and he told me about his son. So his son came over and I hired him to do things. The first thing he did is he took a course in calculus that was taught by David Jerison. He wrote notes, and we animated them and put little things that were visual. Animations. We had fun, and I thought it was a nice thing.

But these were demonstration-type things. I realized that it's much better to empower the students to actually *do* something rather than just look at it. So he built applets: I would think of things that I thought would be fun to have applets for, and then he would make them. You could sit back and watch, but you could also interact with it. You could put your own function in, whatever function you want, and look at it in three dimensions: see the tangent plane, see this or that. It wasn't just a demonstration. It was actually a tool which could be used.

Well, I wanted to test these, to try to use them in a course. I started it when Michael Brenner left. He went to Harvard. Brenner was in applied mathematics, and he was listed as teaching a course, 18.013A, which was nominally a course for people who'd had some calculus so they could take the first term accelerated. I said, "Let's use this in this course." So Jean-Michel made applets, and I wrote text.

Then I decided, let's do things differently. So I developed this crazy course that bore no relation to anything anybody had done at MIT before or since. You can see it on the Web.[7] First of all, it had all of these applets for illustrating things. Secondly, it started at the beginning, but it did everything once. Traditionally you did differentiation and integration; then you went back and did differentiation and integra-

---

[7] See http://ocw.mit.edu/ans7870/18/18.013a/textbook/HTMLindex.html

tion in several variables. Here, we just go through everything once. You differentiate one dimension, and then the trick is to use that—with a little geometric intuition—to apply the identical thing in any number of dimensions. Because the difference is not in technique or anything else: it's just applying the idea. Similarly, with integration, the idea of integration is the same in all dimensions. So you might as well do it all at once, rather than after half the people have forgotten what they knew about integration. And there were other things that were totally different from what was usually done.

There was a slight objection, but nobody really cared. And what I discovered was, with these applets people learned the concepts much faster than normal. To me, it's like the difference between somebody having a textbook where you read about how to ride a bicycle. You never touch a bicycle, but you read and take exams on riding a bicycle. Or somebody just—"Here. Here's a bicycle. Get out and try it." Well, maybe you fall down, but in about an hour you've got the idea. Whereas you could spend weeks at the text and still not have the slightest idea of what to do when you reach the bicycle. It's the same way with learning these things.

The course would generally number ten or twelve people, and they were not necessarily the ones who would want this strange version of calculus. We never advertised this course, so it got a very small audience. But I used doing it to make improvements. And then I said, well, wait a second. The real advantage of this stuff is it makes calculus itself accessible to people who are not mathematically inclined. They can learn this. Whereas, learning *about* it, in the bicycle sense, was beyond them.

So we made another course, Calculus for Beginners, and we put it on OCW, although it's not an MIT course. And it turns out that the applets were very popular with people: they were a first attempt at these things, but they are very useful educationally. And they do make a difference. You know, not too many people when they see something on the Web take the trouble to write to the authors to comment on it, especially if they don't know who you are. But I occasionally get things from all over the world about these things that are on OCW.

It's a first step: I'm sure people will be able to do it more classily and better. I think there's tremendous potential at the lower levels for doing this. Because at least in my day, the way mathematics was taught in high school was, you had some sort of lesson, and you learned that; then you went on and learned something else. But where it was all going—what you were aiming for—was a total mystery. There was no sense of what anything was for or leading to.

So, with calculus, people learn to differentiate. Here's an operation, something with limits, whatever. There's no exercises on the limit stuff, so nobody really feels comfortable about them. Then you have drilling on how to apply certain rules to algebraically differentiate things. And then you go on and do integration. What these things are for, why you're doing this? Nobody tells you. Half the class typically terminates at that stage, and all they remember was funny rules for differentiating. Not the vaguest idea of what calculus is about or what it's for.

One doesn't have to do it that way. You know, the fundamental idea of calculus is, if you want to understand change, you break time up into little pieces; you look at the change in each little piece of time; and that turns out to be much simpler. In a little tiny piece of time, you can assume all lines are straight, and then just use simple geometric statements. Different causes don't mix—they don't have time to mix. You can just add up the effects of different causes and make a simple model of something. So there are two problems: dealing with the small things, which is differentiation; and putting them together, which is integration. And you use that to model complicated situations. That simple fact, 99 percent of the people who take calculus, it never *occurs* to them. It's never taught to them. It's not in the curriculum.

Just to give you one example, I heard people say that they shouldn't teach Newton's method anymore because it's boring and students don't learn it very well. That was strictly because of the way it was taught: if you taught some principle that people rapidly forgot, and then either they had to do totally boring calculations or else you were just talking about a numerical method, then it was a useless thing. But when you see what the thing is, pictorially, and then can do it yourself with your own function and solve the equation you want, and you can see where it's good and what its problems are, it becomes very simple, very intuitive, and very powerful, and learnable in 15 minutes instead of being something that just confuses people and you should take out of the curriculum because it's boring.

It's totally different. People learned this so much faster that we had to enrich the subject matter. We went for much, much more material. We had differential equations, linear algebra, almost a whole curriculum that we covered. And people had not much trouble with it.

So others have been doing it. Their initial thought was similar to our initial one when Jean-Michel and I did that first course with Jerison, which was to have demonstrations. The initial prejudice in favor of demonstrations is that when *you* control the input, you can make the output beautiful; when you let people do anything.... But later we said demonstrations are not enough. You really have to have things that students can use and play with, and you can give them assignments where they're supposed to do something with them, which forces them to look at them. And when they look at them and play with them, they can't help but understand the concepts. I think it's even good to challenge people: see if you can find a function where this thing will screw up completely. I'm sure you can.

~ March 17 and May 9, 2006

Sigurdur Helgason

Early Years

J.S.: *You were born in Iceland?*

S.H.: That's right. I was born in northern Iceland, as a matter of fact. The town is called Akureyri. It was—as Iceland was at the time—rather primitive. It had around 3000 people, and yet it was the second largest town in Iceland: Iceland even now has only 300,000 people, total. And Akureyri was very isolated because there was in 1930 no passable road from Reykjavík, the capital, and the only communication was by boat or by horse. The farmers in that area brought their goods to Akureyri—milk and so on—and this was all by horse-drawn carts. So I remember very vividly horses in the town. At that time there were probably five or seven private cars in town. The owners were primarily medical doctors who needed them for their practice. There was a taxi station, too. So there were some decent roads in the town. The roads in the northern part of the country were mostly one-lane gravel roads.

Life in Akureyri was simple: no TV, of course, and the radio reception was rather poor. Sports and social life at all ages were quite active. The large fjord would be frozen over much of the winter, providing infinite space for skating. As children we would skate for hours until we were exhausted, then lie down on the ice, look up in the black evening sky, and admire the dazzling display of northern lights in various colors.

My father was born in 1892 and my mother in 1900. Both of them lived in Reykjavík in the early part of their years. My mother lived in Reykjavík because her father was postmaster general there. My parents' families go way back. You see, in Iceland it is very easy to trace families, partly because there are so few people but in addition because there is recently available software called Íslendingabók,[1] which can trace how an arbitrary person is related to you. My mother happened to be of a very large family called Briem, which went way back to 1750 or so. Iceland does not have family names, except from old times. Now you can use an existing family name, but you cannot introduce a new one. You have to use the patronymic system. That is, your last name comes from your father's first name: Helgason means son of Helgi, and so on.

My father became a medical doctor in Reykjavík, and he practiced for a couple of years in Iceland before he went abroad for specialization in ophthalmology in Germany, in Freiburg and Berlin. So he became an eye doctor and was offered the position in Akureyri by the surgeon general. At that time the number of eye doctors in Iceland was maybe two or three. He settled in Akureyri in 1926 and served until 1970 as the only eye doctor in northern Iceland. This meant that in the summer he traveled to the various villages in the north and set up consultations, two or three days in each place, depending on the size. There was very often a medical doctor in these areas where he would be housed. Thus people didn't have to travel to Akureyri, but they could see him in these villages. This took him, I guess, a little over a month every year.

My father's travels were rather primitive: first a little by car, but then by boat and finally by horse. So he had his equipment on two horses, and he set up consultations in schools or little buildings where he could house his equipment. I was with him on one of these trips, I remember. It was to a village called Raufarhöfn: there was a herring factory in the village, and I and several gymnasium students got jobs there for the summer, mid-June through August. This was convenient because the herring season coincided with the school vacation. My father happened to be going to that place on the boat with me at the time, so I saw his setup.

He actually produced the glasses too; he ground the lenses. After the summer he would be preparing dozens and dozens of glasses for the people he had seen during the summer, and the whole family was involved in packing these things. We would send them by mail to people, by boat.

My father found glaucoma to be abnormally frequent in northern Iceland and gave several talks on the radio about it. With no TV and only one radio station, this reached many people and may have made some difference.

---

[1] Named after *The Book of Icelanders*, originally written by a twelfth-century priest.

Sigurdur Helgason in Iceland.

The work at the herring factory in Raufarhöfn consisted of two six-hour shifts each day. It was a hard and messy job. I remember several hours of rowing a boat inside a huge herring oil tank while my partner stood in the boat spreading protective paint on the inside walls of the tank. We alternated positions occasionally. The smell was overpowering, and the boat bottom was slippery. Falling out meant certain death, since nobody could swim in that oil: he would simply sink. However, the pay was great, at least when the fishing was good. The last summer I worked there, the herring season lasted until mid-September. The bus that took us back to Akureyri got stuck in a snow bank during a blizzard, the motor died, and we spent the night in the bus. There was neither cell phone nor radio communication there. Help came the following morning from the village of Húsavík. I was 16 or 17 at the time.

In earlier summers I had done the usual thing for boys of my age, did some light work on a farm, haymaking, attending to animals, sheep, goats, cows, and horses. It was quite enjoyable.

A Scandinavian Education

You went to gymnasium in Akureyri, which started in 1930 as the second gymnasium in the country.

Yes. Gymnasium is about equivalent to high school, although it goes normally until the age of 19 or 20. So the last classes are a little bit more advanced than an American high school. At that time, there were just two: one in Reykjavík and one in Akureyri. Since then there are several others. There was a lot of time spent on languages, because Icelandic is only used by Icelanders: it resembles Norwegian but has more complicated grammar. So we had classes in Danish, English, German, and French. If you chose the "language line," as my brother did, you would also have Latin. Now this has changed, of course, because Latin has lost its usefulness. But the program was actually quite varied. This was a six-year school, so a bit longer than high school. I had chemistry, astronomy, physics, and mathematics, in addition to several other subjects.

I think the chemistry was somewhat inadequate because the laboratory facility was rather limited. Physics was pretty good. It was modeled after the Danish system: we used Danish textbooks. The mathematics program was fine. Our final exam included finding the radius of a sphere inscribed in a regular dodecahedron. So this was a pretty good school, I would say.

The teacher that I benefited most from was actually an astronomer, but he had some training in mathematics. He had a doctorate from Göttingen in Germany, which was a very prominent place, a center of mathematics in the world at the time. This would have been around 1930 or so. His name was Trausti Einarsson. He is no longer alive, but he was a highly respected astronomer and geologist. And he had this infectious respect for mathematics, which he conveyed to the students. He became a professor at the University of Iceland in Reykjavík.

You were in the gymnasium during World War II. How did the war affect Iceland?

Exactly the war years, yes. From beginning to end. It affected Iceland in several ways. Iceland of course did not have any army. But after Germany invaded Denmark, England believed that Iceland might be next, so England decided to occupy Iceland. That was in 1940. So they moved in there with a certain amount of army and their equipment. We were very conscious of the army there in Akureyri, because they had a fairly large group there.

Now, that meant that Germany no longer considered Iceland neutral, and that was very costly to Iceland because Iceland had this unlimited trade of fishing with Britain. So there were constant shipments of fish to England from Iceland, and this was a very significant part of Iceland's economy. But by the end of the war, Germany had sunk just about all that fleet. The war had this effect in Iceland that there was considerable loss of life, mainly because of the sinking of the boats. This included many boats in normal travel from one town to another, as well as ordinary fishing boats. These were completely unprotected, you see. My mother's sister and her husband were returning from the United States at the end of the war, in 1945. Two weeks before they left he had defended his PhD thesis in medicine at Harvard; she was a pediatrician. The boat was in a convoy coming from New York, and it was quite close to Iceland when it was shot down by a U-boat. The whole family was wiped out: both parents and three children. It was very tragic. As a matter of fact, in percent of population, Iceland lost more lives during the war than did the United States.

Iceland benefited from the war in some ways though. After America entered the war, the British were essentially replaced by Americans. Various construction projects, including the airport in Keflavík, caused considerable boost to the economy. At that time, Iceland became able economically to start on the geothermal energy project, first in Reykjavík, but gradually it spread over the whole country. This energy source, combined with plentiful hydroelectric power, has eliminated the use of coal and oil for heating.

In 1945 the war ended, and you went to the University of Iceland's school of engineering.

I did, yes. That was a temporary move, because I intended to go to Denmark. Copenhagen was the traditional place. Of course prior to that, around 1930, Göttingen was also popular. But in those days it was out of the question to go to Germany. Göttingen actually did not suffer bombing during the war, so the university escaped more than many others. But the Hitler regime had decimated the mathematics faculty.

But I was told that the University of Copenhagen hadn't quite gotten started during that first year. The teaching had become a bit fragmentary during the last years of the war. In mathematics, a couple of the professors had to settle in Sweden. I think Harald Bohr had to settle there, as did his brother Niels Bohr.[2] Werner Fenchel was another of Jewish descent; he also had to go to Sweden. So I was told it would be more practical for me simply to study at the engineering school in Reykjavík, because, as I said, the science education in Iceland was very much modeled after the Danish one, including the use of the same textbooks. So this was a practical move, just to stay in Reykjavík one year.

---

[2] The Bohrs' mother, Ellen Adler Bohr, was Jewish.

During that time, my fellow engineering students spent a lot of time on technical drawing. Actually the professor was the father of Vigdís Finnbogadóttir, who later became president. He had very high standards. Nowadays that drawing has been computerized, but at that time, much of the students' time went to that. But I could skip it. The courses I took were in mathematics, physics, and chemistry, and a small one in philosophy.

In 1951, you won the University of Copenhagen's Gold Medal for an answer to a Mathematical Prize problem, to establish a Nevanlinna-type value distribution theory for analytic almost-periodic functions.

Yah, the University of Copenhagen had a tradition that goes back—well, I don't know if it is centuries, but very far back, and they still have this system: every year the university posts certain prize questions, not just in mathematics, but in all fields. In literature it may be a literature study about some specific author. There is one in theology and one in physics and so on. So it would generally be an essay about a specific topic. In mathematics it is usually a real research problem, although sometimes it may be an exposition of a certain obscure subject.

Now, in Copenhagen, I had plenty of time. You see, the degree there was the so-called Cand. Mag.[3] This meant you studied mathematics, physics, chemistry, and astronomy to a certain level so that you could with ease teach these subjects in gymnasium throughout Denmark. Then after the two or three first years, you specialized—and I specialized in mathematics—and that would be like the beginning of graduate school. But the number of courses available was rather limited, which meant that I had lots of time on my own. I could have finished my degree much earlier than I did. But then I noticed one of these prize problems, because it was posted there in the university. So I decided to try that, simply because there was plenty of time for me.

The system is that you hand in your response after a year: you have the calendar year for this. The only condition is that you had to be under the age of thirty, so anybody in the world is allowed to try. And you do it anonymously: the author's name is not on the paper, but rather a certain symbol. I used the letter aleph. Along with the paper is a closed envelope where the author's name and the symbol are included. If there is no reward for the paper, the envelope is not opened. So you had nothing to lose if you didn't win! [Laughs]

---

3 *Candidatus* (m.) or *Candidata* (f.) *magisterii*.

How hard are these problems? How much did you work on it?

I worked on it for a whole year. Yah, it was a hard problem, at least for me, because I didn't really have the background needed, and since it was supposed to be handed in anonymously, I could not consult with anybody: I felt duty bound not to tell anyone. So I was on my own trying to get the background. It wasn't even clear what background I needed! So I think the first four months went on just finding out what the problem was about. But an indirect benefit, perhaps, was that I got used to working on my own without any guidance.

There were two subjects involved. One was a subject founded by a Finnish mathematician, Rolf Nevanlinna. There was a school of Finns that all went into that line, including one who later became a professor at Harvard, Lars Ahlfors, and who became a good friend later, after I got to MIT. Then there was the other side, the theory of Harald Bohr, who was a very prominent Danish mathematician, and he had started this almost-periodic function theory. The problem was to somehow combine these two theories. That is, to show that the Nevanlinna theory could be modified so that it fits the theory of almost-periodic functions and then prove the analogs of Nevanlinna's two main theorems, as well as the defect relations. So this was what I did.

The problem isn't completely solved yet because I had to make a certain restriction on the functions considered. I did actually revise the paper later with the intention of pushing further, getting rid of this assumption. This was after I got to the States. But then I somewhat lost interest. I have published an excerpt of the paper in the proceedings of a memorial conference to Harald Bohr. I have been approached lately by two Russians about what remains of the problem, namely certain subtleties concerning the so-called defect relations. But almost-periodic functions is a rather limited subject, although it was intensively studied in Denmark during the 1930s and 1940s. Harald Bohr developed the theory in the 1920s, and it was very prominent for a while. But it has dried up a bit.

The theory was a kind of a natural extension of the theory of Fourier series, which is a very prominent subject in mathematics. Later on, one can say it has been to some extent absorbed in the theory of Fourier series on compact groups, except for the part that I was involved with, which was more related to Dirichlet series. I would say that this part is very much alive. In fact, analytic almost-periodic functions do constitute a very interesting generalization of Dirichlet series, in that the frequencies are quite arbitrary.

Rolf Nevanlinna had a strong connection to Switzerland. There used to be a Nevanlinna colloquium in Switzerland regularly, and he enters a little bit in the story of André Weil. Weil was in Finland during the second half of 1939, and he was arrested by the Finns during the Russian-Finnish war and was in danger of being executed as a spy. But Nevanlinna managed to have him deported to Sweden instead. André Weil wrote his memoirs published by Birkhäuser[4]—not very large, but very interesting. There he relates the story. He had to go back to France and serve in the army, and he got into some difficulties, having been away when the war started.

Is there a distinct tradition of Scandinavian mathematics, in any sense, a tendency to work in certain areas?

In Finland, Sweden, Norway, and Denmark there is a tradition primarily in analysis. Norway stressed analytic number theory, and Norway has of course these prominent stars in the history of mathematics, Niels Abel and Sophus Lie, but they were in somewhat different fields. Abel was short-lived, and his work was related to algebraic geometry, although strongly to analysis as well. And Lie founded this theory of so-called Lie groups.

This is a relatively small group that we are talking about, so one might say it grew from the interests of very few people. In Finland, there were Lindelöf, the Nevanlinna brothers, and Ahlfors who continued in that path, so there was a certain school formed in this area of classical complex analysis. So it was a natural development. Now, in Sweden, the first mathematicians—for example, Phragmén—were also involved in analysis. Mittag-Leffler was very influential in this regard. Sweden has tremendous analysts today, like Carleson, Gårding, and Hörmander. They are getting up in years now, but their influence is enormous.

So that sort of explains Finland and Sweden. Norway had the number theorists Viggo Brun and Atle Selberg. Now, in mathematics in Denmark, there was for a long time a great emphasis on geometry. The principals were two people: Julius Petersen and Hieronymus Zeuthen. Zeuthen was of the Italian school in geometry. Petersen, a great geometer and a problem solver, later became an important figure in graph theory. But the analysis, that started I suppose with Niels Nielsen and mainly Harald Bohr and Börge Jessen. This, however, was somewhat different from the mathematics in Finland and in Sweden: there was not very much contact there. But this is I think a product

[4] Weil, A. (1992). *The apprenticeship of a mathematician*, translated from the French by J. Gage. Basel, Boston: Birkhäuser.

of those times, that certain special fields in mathematics would be tied to a very specific place or specific country, and that the communication between mathematicians in different countries was very limited. There would be an international congress and a Scandinavian congress every four years or so, and of course there would be journals. But this is nothing compared to the situation today, where mathematicians are constantly flying to other conferences, especially during the summers, and communicating electronically. So the atmosphere has changed completely.

You solved this research prize problem posed by the university. Did that start you thinking about being a research mathematician?

No, somehow at that time I wasn't thinking much about the future: I didn't have any particular sights on doing research in mathematics. I suppose I assumed that I would go back to Iceland and take a position as a gymnasium teacher, because that seemed a very pleasant job at the time. The teaching load was quite high, although nobody thought about that. But the teachers in the gymnasium in Akureyri, they were *extremely* respected people—oh, yah. In town, these were the people who were looked up to, and they seemed to enjoy their job. So I felt this was just an agreeable position to have.

At that time I discounted the possibility of having some kind of a position or a job at the University in Denmark. It just didn't seem possible: there were very few positions and relatively little research activity. As I pointed out to you, the number of math courses that were given was rather limited, and Denmark, in mathematics, took a while to recover after the war. Mathematics was only present at one university: the University of Copenhagen. There was another university, Aarhus, but there was no math department there at the time. The mathematics department in Aarhus came later, through the visionary initiative of Svend Bundgaard. That meant that there were two math departments in Denmark, and that turned out to be *very* beneficial to both places. There was not a real competition, but somehow they both got revitalized in the process. But it wasn't until the 1960s when that really took off. So now Danish mathematics is very versatile, very dynamic, but in those days it seems in retrospect to have been somewhat dormant, although the teaching was first rate.

Princeton

In 1952 you left Denmark and came to Princeton. What brought you here?

My parents told me about this possibility of a fellowship to go to the United States. So I applied to something called the Institute for International

Education, kind of a Fulbright.[5] So I filled out an application there, and I had the choice to go to either Harvard or Princeton, but my professor, Börge Jessen, recommended Princeton, and it was a very wise choice because Princeton did not have such a structured graduate program. There were no exams in the graduate courses. No homework. Complete academic freedom. And I didn't take many courses there! As a matter of fact, the courses in Princeton were, at that time, somewhat specialized, in the sense that it was nothing like the spectrum we have now at MIT—and maybe now in Princeton too. But the compensation for that was that there were many seminars. In particular, students would often organize seminars on topics they wanted to study. Then there was the Institute for Advanced Study, which had regular lectures. I went to some of them, particularly to some given by Arne Beurling. So I was on my own as a graduate student, which suited me fine. It was a great time!

I was only two years in Princeton, so when I finished my thesis there I really knew very little. There was one serious oral exam called the Generals[6]: on my exam committee I had Artin, Lefschetz, and Spencer. But I was still quite ignorant about several fields in mathematics. At the time I didn't feel that way, but later I realized that two years was a little short. So I could have stayed a little longer perhaps, but when you finish your thesis you are supposed to leave! [Laughs]

Salomon Bochner was your thesis advisor. How did that come about?

That was fairly natural because he was one of the major names in the theory of almost-periodic functions. He went to other things after he came to the States, but while he was in Germany, he was one of the founders of the theory. So I went to his lectures and somehow got acquainted with him. During the term when he was on leave, I came across a problem in almost-periodic function theory. After Bochner came back I mentioned this to him. He was at first skeptical: "This problem has already been considered by many people." But he quickly agreed that it was far from being solved. Actually, the solution was quite easy, but it required methods which were only known

---

5 The non-profit Institute for International Education, founded in 1919, administers some 250 programs, including the Fulbright.

6 Princeton's "General examination" required students to respond to questions on algebra, real and complex variables, and two advanced topics of their own choosing from different areas of mathematics.

Sigurdur Helgason with his advisor, Salomon Bochner, and his student, Jiri Dadok.

after 1940. It was not directly related to what I had done in Denmark, but it was still within that field. I connected it to another theory, which had become very fashionable at the time, the so-called theory of Banach algebras. So my thesis was on Banach algebras and almost-periodic functions and proved some theorems on both topics. Bochner was very accessible and supportive.

What language did you speak to him in?

Oh, English. It seemed easier. I could read German, but I couldn't speak it. I had English for four years, German for just two. But neither language was comfortable for me. I hadn't spoken a word of English for five years, so it took some effort. But I had no problem understanding lectures: mathematics is easy so far as vocabulary is concerned.

Did you know his story at all?

A little bit. He came to the States, I think in the mid-1930s.

Yes. In April 1933, the Germans kicked everybody with even one Jewish grandparent out of the universities. Bochner was Jewish and Orthodox, so of course he had to leave, and he came to Princeton later that year.

Well! That was good, because many other German mathematicians had very great difficulty finding positions here. He was a big star, no question about that. There has not been much written about his life. There is an article by a colleague of his at Rice, where Bochner went after he retired from Princeton.[7] That colleague was William Veech, who was a former student of Bochner's, and he wrote a little article about his life. He sent it to me, but I am not sure it has been published. Anthony Knapp has an article about him in the biographical memoirs of the National Academy of Sciences, 2004.

Bochner was inspiring to talk to, and he spoke freely about many things.[8] I also liked the fact that although he talked to me about certain problems that he wanted me to get interested in, the fact that I was *not* interested in those problems but started on some other problems instead—he reacted positively to that and was very supportive. So I got into this habit of just following my own taste in the choice of topics to work on. This suited me better.

You met Arthur Mattuck at Princeton. Do you remember that?

Oh, yes. We were good friends there. Frank Peterson also. We were in classes together, as a matter of fact. Frank and I played tennis together, too. Arthur was of course in a different field: he was in algebraic geometry. Frank Peterson was again in another field, in topology. So mathematically, we were separate. But we all had dinner together in Procter Hall in Princeton. We wore black gowns, and at dinner time we talked about mathematics—it was largely shop talk throughout. This was a complete novelty to me, because I lived in a fifty-student dorm in Copenhagen, and there were no mathematicians there except me. They were mostly engineers, but somehow talking shop over meals was something that you just didn't do. In Princeton that was quite common among the mathematicians but frowned upon by others, who even complained to the dean.

Frank and I were in one course together the very first term. That course was given by Ralph Fox. I happened to have met Fox in Copenhagen. He came there for a visit. So he knew that I was going to Princeton, and I contacted him when I got there. Anyway, this was a course in topology, about

7 Bochner was at Princeton 1933–1968 and at Rice University 1968–1982.

8 To his work in analysis Bochner later added two popular books on the history of science, *The Role of Mathematics in the Rise of Science* (Princeton, NJ: Princeton University Press, 1966) and *Eclosion and Synthesis: Perspectives on the History of Knowledge* (New York: Benjamin, 1969).

which I knew next to nothing. It was done partly by the so-called Moore method, where there is no textbook; he simply handed out some notes on theorems that we, the students, were supposed to prove. Moore was a guy from Texas who was famous for that method of teaching.[9] Moore went further in that he forbade the students to look at any sources. No books are allowed; you had to do it all on your own. Fox didn't take that attitude: he welcomed us to consult the literature. And it was a kind of seminar. Fox would say, "Is there anyone that has solved one of these problems here?" And then somebody would say "Yeah." So we would usually be occupied with listening to somebody giving a talk on a problem. But, as I say, there were no exams: there was complete academic freedom. So if you were not interested, or if you were lazy, you didn't have to do anything. But Frank was in the course, and Frank knew the subject already pretty well, so I would talk to him sometimes about topology.

Another person in that course was Gian-Carlo Rota. It was apparent to me that this guy would become quite a scholar; whenever I took a book or journal out of the library, his name was on the list of earlier borrowers. He was a senior, and he had to take this course for credit. He was kind of humble, because he was a senior in a class with mostly graduate students, who by custom did not talk much to undergraduates. So I never talked to him at that time, and I didn't get to know Gian-Carlo until I came to MIT and we became close friends. However, he did an enormous amount of the work in that course, and he was very often the one who was called upon to talk: I would say that he probably gave more problem solutions in that class than anybody else. Later I asked him, "Why did you do that?" He said, "Because unlike you I had to take it for credit!"

Instructor at MIT

After finishing your PhD you came to MIT for the first time in 1954. How did you decide on MIT?

There were these instructorships that you could apply to, so I applied to MIT and I applied to Harvard, both of them. And then a friend of mine, Walter Baily, was driving up to Boston—it was probably early spring 1954—and he

[9] Topologist R. L. Moore (1882–1974) was most famous for his aphorism, "That student is taught the best who is told the least." He wrote his thesis under Oswald Veblen at the University of Chicago and taught mathematics at the University of Texas for 49 years. The "Moore method" of teaching de-emphasizes (or in Moore's case forbade) the use of textbooks in favor of teaching by the students themselves.

said, "Why don't you come along? You can then take a look at MIT, where I studied." So I did. I liked it, so when I was offered the Moore Instructorship, I accepted right away.

Presumably Bochner wrote you a letter of recommendation, especially since he had written a book with Ted Martin called Several Complex Variables—the first American book on the subject.

Oh, yeah, Martin was the chairman then, so I'm sure that Bochner wrote to Martin. The procedure was simpler in those days, because I don't remember asking anyone for a recommendation letter except Bochner. Nowadays the custom is to have several different letters.

Of course, the number of graduate students during those years in the 1950s at MIT was very small, and the offering of courses was relatively limited. But there was a big jump at the end of the 1950s—the Sputnik atmosphere caused a lot of expansion. In earlier days, undergraduate seniors were usually encouraged to go elsewhere for their graduate school, because there wouldn't be enough new to offer them. But that changed in the 1960s, and as a result, we now have some graduate students who were undergraduates at MIT.

Ted Martin was extremely successful in building up the department. Later chairmanships have been rotating, usually for just a five-year period, so people tried to do their research at the same time, whereas in Ted's time, he was a full-time, permanent chairman.[10] He had a monthly get-together at his house on Moon Hill Road in Lexington. He sent a card, "We will be home every Sunday during this month, and feel free to drop in," and people did. Nothing very formal, but it was a positive effort to make people feel comfortable. His house was on a sloping hill, with a wall that was mostly glass, a very large bay window in the living room, and the view from there was great.

He was a talented administrator and *extremely* tactful in dealing with people. I later served on several committees within the department, for example for hiring Moore Instructors, where he was a member. I remember that very well. He would always be willing to accept another point of view. I admired him very much as a chairman. He really deserves tremendous credit for what he did for the math department at MIT. He went on to become Chairman of the Faculty for the whole Institute, which he was for several

[10] William Ted Martin served for 21 years as head of the department (1947–68), the third-longest term in the department's history, after John D. Runkle (1865–1902) and Henry Walter Tyler (1902–1930). After the Martin era the headship alternated between pure and applied faculty. Of the seven heads since then, none has served longer than a decade.

years. He showed up with his family at a celebration of fifty years of the Moore Instructorship, which he started.

Martin died just a couple of years ago.[11] I remember there was an Institute-wide gathering at the Faculty Club, when he was retiring, and he gave a little speech there. He was telling a joke about another guy who was retiring, and who was asked, "What are you going to do now that you're retired?"

"Oh… I think I'll finish my book."

"Oh? I didn't know you were writing a book."

"Oh, I'm not writing one, I'm reading one."

That's the kind of joke that Ted Martin would tell.

Arthur Mattuck remembers that you decided to study Harish-Chandra's work, and that you worked very hard on it at that time.

Yes, that's right. When I was in Princeton I did somehow find out about Harish-Chandra's work. Bochner never talked about it—it was very far from his work. I saw it in the literature. It was unrelated to my thesis, but I was led to it because, as I told you, I got interested in Banach algebras and their relation to abstract harmonic analysis, so it was natural to get involved in Harish-Chandra's work. But that was on a very high level, so the first thing I had to study on my own was Lie group theory.

In the spring of 1954, Harish-Chandra was invited to give a lecture to the physicists. On the faculty were the two physicists Bargman and Wigner, who were among the pioneers of representation theory. I was very happy to go, but there were, I think, no other mathematicians in the audience.

How did you do with the lecture? Did you understand what he was talking about?

Yes, I understood it, because he really wanted to be understood by the physicists. And Wigner, he wasn't afraid of asking simple-minded questions. He was a modest man and might be asking questions where he knew the answer very well but pretended not to, just so that the students would then benefit from the answers. It is not very usual that a professor would play that role. But he did! Bargmann was there too; he wrote a very important paper in the field, and so did Wigner.

I studied Lie theory on my own in Princeton, and I continued after I got to MIT. Lie groups were not covered in a course at MIT at the time. Harvard did have courses on Lie theory, by Gleason and Mackey. Gleason gave a

---

[11] William Ted Martin died on May 30, 2004, at the age of 92.

course on Lie groups the first year I came, just one term, and he based it on Chevalley's book. I followed Gleason's course, I remember. And the next term, Mackey gave a course on Lie algebras where there was of course no analysis involved. Because of the connection with his work on representation theory, Mackey took great interest in Harish-Chandra's work, and he gradually became more interested in applications of representation theory to other fields like physics and number theory as well.

Lie group theory has nowadays become a lot easier. In one year's course you can go much further than you could in the 1950s, because the whole thing has become more streamlined. At the same time, the subject has grown enormously, so reaching the frontier is still quite difficult. Lie groups and representation theory, which is an outgrowth of Lie theory, are strong subjects at MIT now. But as I said, this was not a course given during those two years I was here, when the department and the variety of courses were much smaller. So I would say that even after those two years at MIT, I didn't really have any serious insight into it, partly because nobody else in the area was involved in the subject. Nowadays, students have a much easier time of it.

It seems that every serious research mathematician, at some formative point in his or her career, has to tackle something really difficult in order to push the frontier out further. What did it take for you to master Harish-Chandra's ideas? How did you work on it?

I started systematically reading his papers. But it was not so easy to master Harish-Chandra's work, because although I had gotten some facility with Lie group theory, Harish-Chandra's work was way beyond that. The background for reading his papers was really rather inaccessible. There were some notes from Hermann Weyl at the Institute, which Harish-Chandra always quoted and probably had mastered: Harish-Chandra had been at the Institute, so he had absorbed some of this from Hermann Weyl. While Weyl was a spirited writer, I did not always find him all that clear.

Then there was Élie Cartan's work. But his work was, for one thing, relatively little understood, in spite of its great importance. He was one of the great mathematicians of the period, but his papers were quite a challenge. Hermann Weyl, in reviewing a book by Cartan from 1937, wrote: "Cartan is undoubtedly the greatest living master of differential geometry.... Nevertheless, I must admit that I found the book, like most of Cartan's papers, hard reading."[12]

[12] Weyl, H. (1938). Cartan on groups and differential geometry: A review of *La Théorie des Groupes Finis et Continus et la Géométrie Différentielle traitées par la Méthode*

Geometry has always appealed to me very much. It's just that global differential geometry, at the time I was interested, was rather inaccessible. It is an old subject, but it was rather quiet until the mid-1940s or so. It has now become a very popular subject, but it took a while.

Chevalley's book came out in 1946.[13] He was a professor at Princeton University at the time. His book was very useful, in that it clarified a lot of material that was obscure. It's a book on Lie groups, but it also had a considerable effect in differential geometry. Chevalley's book and *Séminaire "Sophus Lie"* [14] in France, 1955, were the principal background sources for me. Later, in 1956, there appeared a nice little book by Nomizu, *Lie Groups and Differential Geometry*.[15] It did not go very far—eighty pages or so—but it was very useful to me. However, I did not really get seriously involved until I went to Chicago for two years, 1957 to 1959.

In some ways, I was still in graduate school at MIT. I only taught courses that did not take much time. Ted Martin did allow me to give one graduate course, and that required much more preparation. It was called Functional Analysis. That is a very large subject, so I just simply picked out subjects that I liked to talk about.

I picked up a lot of things from communication with other mathematicians. I got acquainted with John Nash right away when I came here, and I was, of course, very impressed with him. I knew of his great work in Riemannian geometry, and he would also often talk about what fields in mathematics were worthwhile and what fields were not. He had plenty of opinions about that, which I took seriously, though with a grain of salt. We were both bachelors, and neither of us liked to cook, so we went to restaurants all the time. Arthur Mattuck was often with us. He was an instructor at Harvard the first year I was at MIT and moved to MIT the following year.

Nash actually started the MIT colloquium. There was a Harvard colloquium, which met once a week on Thursdays. Then Nash felt MIT should have one too, so it was set up on Wednesdays. After every collo-

---

du Repère Mobile, by Élie Cartan. *Bulletin of the American Mathematical Society 44*(1), 598–601.

13 Chevalley, C. (1946). *Theory of Lie groups, I.* Princeton, NJ: Princeton University Press.

14 Lie, S. (1954-1955). *Séminaire "Sophus Lie", 1. Théorie des algèbres de Lie.* Topologie des groupes de Lie. Paris: École normale supérieure. Lie, S. (1955-1956). *Séminaire "Sophus Lie", 2. Hyperalgèbres et groupes de Lie formels.* Paris: École normale supérieure.

15 Nomizu, K. (1956). *Lie groups and differential geometry.* Tokyo: The Mathematical Society of Japan.

quium, at Harvard and MIT, there would be a party, and there you could talk mathematics with everybody—and you did! And you picked up a lot of information that way. I didn't have a car, so I usually relied on Nash or Arthur for transport. So it was very different from how things are now. I mean the community was smaller, but there would be, as I say, a mathematics party twice a week! So it was kind of a graduate school for me.

Did your contact with Nash continue in later years?

No. My period of knowing Nash was 1954 to 1956. After that I saw him very little. Nash had become sick, back in 1959. I saw him later when I was at the Institute in Princeton in 1965. He was not involved in mathematics in any way, and he was under care at a clinic in Princeton. I remember running into him at the library once; we got into a conversation, and I said, "Well, why don't you come for afternoon coffee one day." So he did, but it was very sad. Somehow one didn't know what to do. He was not happy at that clinic, although he talked about sessions with his doctor being productive in some ways. But I didn't follow that up. He was still quite ill at the time.

I didn't see him again until 1998. Then he was completely different, and he had recovered a lot. He was a little subdued compared to his younger days and didn't quite have the same sense of humor as I remembered but was otherwise okay.

Did other figures stand out from your instructorship at MIT?

One of the people I remember from the first years was Wiener. I visited his course once in while, out of curiosity. It was interesting. We did not talk mathematics, but he did talk to me about his stay in Denmark. He told me that he learned to speak Danish and said a few sentences to me in Danish, but he said, "Oh, but I never could master Icelandic."

He came to the common room quite often, where all these graduate students were milling around. He wanted to mix with them a little bit. I played chess with him a few times. I was not a good chess player, but I usually beat him. Chess is usually a rather competitive game. But for Wiener, no! He did it strictly for the fun of it. So we would play a game, and after I won he would say, "Well, you win, let's try another—let's change sides." I found it refreshing: here is this great mathematician, who probably had a considerable competitive streak in him; yet he took chess as strictly a game, nothing more. Philip Franklin, however, was a clever chess player.[16] I was no match for him. He even looked a bit like Lasker, the chess champion.

Princeton, Columbia, Chicago

The MIT mathematics department had a rule that essentially prevented them from hiring instructors, except in very rare instances. You moved around quite a bit in the next few years, but you implied that you were able to continue your work in mastering Lie group theory.

Yes. After two years at MIT, I got a kind of a lectureship in Princeton, and there I worked on applications of Lie theory to differential equations: for example, I generalized Ásgeirsson's mean value theorem to Riemannian homogeneous spaces.

Then from Princeton you went to Chicago.

Princeton was not set to be a one-year job, but I assumed it was just that. Then I got this offer from Chicago for two years, so I jumped on that.

The chairman was Saunders Mac Lane, so he essentially hired me. He was wonderful, a great guy. Mathematically we had nothing in common, really: he was in abstract algebra. But he was a very enterprising chairman. He had actually invited me to a summer session in 1955: they had a special summer on functional analysis, from which I learned a lot. He was visiting MIT, and I think he was looking for younger people to invite to that summer in Chicago. He had some funding for it, so I went there. George Mackey was there, also Kaplansky and Halmos. And I met Irving Segal there. At the time, he was giving a course on group theory in quantum mechanics. So this was a very productive summer.

I met Grothendieck there. He was becoming a star at the time. He had not yet entered algebraic geometry with full steam, but he just delighted in talking about mathematics. I remember we both stayed at the International House at the University of Chicago, and we had our meals there. Once after dinner he said, "Would you like to take a little walk and talk about mathematics?" Sure! Delighted. And I could pick up whatever topic I wanted to talk about, because he knew everything. We discussed a paper by Godement

[16] Philip Franklin (1898–1965) was an analyst who wrote his thesis on the four-color problem under Oswald Veblen at Princeton. He served in the MIT mathematics department from 1924 to 1964.

that I had been reading and that he knew very well.[17] I was also writing a paper, and I was telling him what was in it. And he gave some good suggestions. He was not preaching to me at all: he just *enjoyed* talking about mathematics.

This was before he entered algebraic geometry and changed that subject. He had written his thesis, a very strong one, in topological vector spaces. At the time, he worked with Laurent Schwartz and Jean Dieudonné. But then he changed fields completely.

Mathematics is such a highly specialized field. How do mathematicians who are "talking shop" bridge the gaps between their different interests?

The various branches of mathematics in the last, let's say, 60 or 70 years are much less isolated than they used to be. To some extent it was Bourbaki's project to make the communication easier between different fields. That was of course not the only project, but they certainly made a big contribution in that direction. For example, in the subject of topology, they certainly had a role in making that subject useful for the general mathematician. This is an important feature of modern mathematics, that with some effort you *can* relate to people in other fields. For an example, take representation theory and differential geometry: these would be very distinct fields if you go back a hundred years. But nowadays they are related: Representation theory is an outgrowth of Lie group theory. Lie group theory relates to differential geometry. So they are connected in a significant way.

Did you return to MIT after your two years at Chicago?

No, I spent one year at Columbia in between. I shared an office with Harish-Chandra. At that time, I was teaching a course on Lie groups, and I began my book on the subject. He himself gave a course on [Carl Ludwig] Siegel's work on quadratic forms, but I followed that too. He was very pleasant and enjoyed discussing mathematics.

At that time, he had never taught Lie groups at Columbia; he had been there quite a few years but somehow never lectured there about his own work or even the background of it. That would have involved quite a bit of effort for him, because this material was available mainly from Weyl's

---

17 Roger Godement was a student of Henri Cartan and member of Bourbaki. He wrote a very enthusiastic review of Harish-Chandra's first paper on infinite dimensional representations, a paper that was later awarded the Cole prize.

seminar notes from the Institute, but he probably realized that it would be so much work to make this comprehensible to graduate students that he had never done it. In fact, Weyl had been lecturing on material where lots of stuff was taken for granted because Cartan had done it. Later, Harish-Chandra did give a whole year's course on his own work and used my 1962 book as background material for the students.

Steele Prize Material

It seems like this demonstrates a common pattern of mathematical progress. Weyl was based on Cartan, Harish-Chandra on Weyl, but it sounds like many gaps remained to be filled in—and that such spadework is motivated not by the discussions of the experts, but by the need to place it before students.

You published Differential Geometry and Symmetric Spaces in 1962—quite an early stage in an academic career to take time off to write a research-level textbook. The bet paid off, you might say, as the book and its revision were two of the three books for which you later won the Steele Prize, in August, 1988.[18] How did you develop the material for that first book, and what made you decide to write it?

It was a project that I became committed to in Chicago: I already started planning it there in detail. Chicago was on the quarter system, so I gave a quarter course on topics in Lie theory. It was very small, very few people were there, but it was okay. I remember discussing the project with Chern, who was a major mover in the field of differential geometry. He was very encouraging.

We had a seminar in Chicago that went on probably a whole quarter. I was getting interested in not just Lie groups but symmetric spaces, an outgrowth of Lie group theory, which was the title of the seminar. Joe Wolf,[19] who later became a big contributor to the subject, was a graduate student, but he was the one that instigated that seminar. Participants were Chern, Spanier, Palais,[20]

---

[18] Helgason, S. (1962). *Differential geometry and symmetric spaces.* San Diego, CA: Academic Press. Helgason, S. (1978). *Differential geometry, Lie groups, and symmetric spaces.* San Diego, CA: Academic Press. Helgason, S. (1984). *Groups and geometric analysis.* San Diego, CA: Academic Press.

[19] Joseph Wolf received his PhD from Chicago in 1959 under S.-S. Chern for a dissertation entitled "On the Manifolds Covered by a Given Compact, Connected Riemannian Homogeneous Manifold."

[20] Richard Palais received his PhD from Harvard in 1956 from advisors Andrew Gleason and George Mackey. His dissertation was entitled "A Global Formulation of the Lie Theory

Rinehart,[21] Lashof,[22] and me, and a couple of graduate students. So that was quite stimulating. But it was unsystematic. People picked out what they could talk about, and some notes materialized. It was not something where one lecture was a continuation of the previous one. You jumped around at will.

I gave a more systematic course in Lie groups for a whole year at Columbia. That was on a lower level, but still at graduate level, and at the same time I was working on this book: I would say one-third of the book, the differential geometry part, was done that year. Then I continued after I came to MIT: the first year, 1960, I gave an undergraduate course on differential geometry. I had the notes already written out, and I simply tested out the first chapter of my book on the students.

The book was in two parts. Differential geometry was the first half of it and some of the second part, but the main topic was symmetric spaces, which was a natural continuation. The first part dealt with things that were classical but where the proofs were rather hard to dig out. So in some ways the first part was more difficult and a little frustrating, because the main results were known, but the proofs were often either not rigorous or were in Élie Cartan's difficult style.

Now, Ambrose was a serious expert at MIT in differential geometry at the time. He had also studied Cartan's work very much, and a book by Bishop and Crittenden was in part based on his lectures and his collaboration with Singer. But his taste was a little different. His differential geometry was more in the direction of fiber bundles, which included lots of machinery that I did not want to include, because it wasn't quite needed for what I had in mind. Fiber bundles came about in topology, but they fit into differential geometry in a very natural way. But my main aim was an exposition of these so-called symmetric spaces, and there this machinery of fiber bundles wasn't really necessary.

---

of Transformation Groups." He was an Instructor at the University of Chicago from 1956–58 and spent most of his career at Brandeis.

[21] Bruce Reinhart received his PhD from Princeton 1956, with Donald Spencer as his advisor.

[22] Richard Lashof received his PhD from Columbia for a dissertation entitled "Topological Group Extensions and Lie Algebras of Locally Compact Groups" and spent most of his career at the University of Chicago.

Sigurdur Helgason with his wife, Artie, in Santorini.

Writing an advanced textbook at that age was a little risky, because that doesn't weigh very highly in the job market. I wrote some short papers in the 1950s and a long one in 1959, that is true. The last one was related to the book without overlapping it. Of course, it was an advanced book, containing some new results, and I hoped it would be a useful one. But still I thought, better get it done quickly! So I was working very hard on it. I finished it in 1961, just before my son was born. My wife Artie typed it all on ditto masters, and our hands were always blue from the ink when making corrections. A later book of mine, in 1984, just went to the printer handwritten. That was a useful shortcut.

During that year, 1960–1961, I was a kind of a hermit: I didn't go anywhere, in spite of the place being full of old friends! Arthur Mattuck and Frank Peterson and Gian-Carlo Rota, they were all colleagues. It was very nice. I don't think there were parties twice a week at that time—at least, maybe there were, but I didn't go to them twice a week. I was busy. But it was worth it, because the book became reasonably useful. I revised it thoroughly back in 1978,[23] for example taking into account very important contributions by my colleague Victor Kac. He had not published this material in full before and had not been able to take it out of Russia, so he had to reconstruct it from memory.

Writing is not the main thing that mathematicians are trained to do. What is good mathematical writing? What do you try to do in your books?

Well, clarity is the main thing for me. When you read some old mathematics, let's say Cartan's work, you are completely baffled by the seeming informality of his proofs, which came from his experience, combined with uncanny geometric insight. You realize: here's some really good stuff, but you also see that it *does* need to be explained better. So clarity becomes an overriding issue.

Then you want to do something more: you want to somehow say something that ties the various theorems together in a natural way. I don't know how well I do that, although I try. In the 1962 book, I wanted to quickly get to the second part, so the first part is a bit condensed. So the writing is nothing to brag about. But I think it's clear, at any rate.

August, 1988, must have been a good month for the MIT mathematics department, as both you and Gian-Carlo Rota won the Steele Prize that year. Do you remember that?

I do, I do [laughs]. There was a centennial meeting of the American Mathematical Society, so as a result, there were a lot of historical talks, and it attracted a lot of people. It's in Providence, only an hour drive from Cambridge, so I just drove down there, and I gave Gian-Carlo a ride back, because he was also at the meeting. So I asked him, when I was about to take him home, "Are you going back there tomorrow?"

"Yah, I am going back there tomorrow, I—I have to go, because I am getting the Steele Prize."

I said, "That's curious, because I am, too!"

They tell you in writing, but they keep it confidential, so nobody knows until the time. Neither of us knew that the other had received it. So I said, "Well, look, we are near to a very nice restaurant"—it was Le Bocage, then on Huron Avenue in Cambridge—"why don't we celebrate?" So we had a fancy dinner together, right there.

---

[23] Helgason, S. (1978).

Radon Transforms

You did a lot of work on Radon transforms.

Yah, that became a hobby of mine. That was an interest that developed from a book by a mathematician named Fritz John.[24] He died a few years ago. That book of his came out in 1955, and I got acquainted with it partly because an Icelandic mathematician, Leifur Ásgeirsson, who knew John very well, lent me the galley proofs. It starts with an account of a work by Radon—who was an Austrian mathematician, at the time, although he was born in Germany[25]—who did this work back in 1917.

Radon's 1917 paper was about reconstructing a function from its various different line integrals.

Exactly. More generally, the explicit reconstruction of a function from its hyperplane integrals. This paper was relatively little known because it was published in a very obscure journal. But Fritz John knew about it and used it as a kind of foundation for his book, and I found this combination of geometry and analysis absolutely fascinating. I was amazed that this was so little known. This has changed drastically in recent years.

In my paper in 1959, I have a generalization of Radon's inversion formula to Riemannian manifolds of constant curvature. That was my first inroad. Then I had a seminar on it at MIT. At that time, there wasn't much material available, and the seminar was rather short. Then I proved another theorem in the subject and gave that in a course in the fall term of 1963. I always remember that because when I just finished the class where I lectured on that theorem, now called the support theorem, I heard that President Kennedy had been assassinated. A colleague told me about it as I was returning to my office. Most people remember what they were doing when Kennedy died, and so do I.

Radon transforms became popular because of the application in medicine through CAT scanning, which started around 1963 with Cormack's idea: in normal X-ray picture taking, you use X-rays just in one direction. But Cormack discovered that if we use X-rays in all directions through some

24 John, F. (1955). *Plane waves and spherical means applied to partial differential equations.* New York: Interscience Publishers, Inc.

25 Johann Radon (1887–1956) was born in an area of Bohemia that is now in the Czech Republic and lived in both Germany and Austria.

inhomogeneous material, you can get full determination of the variable density of the material through an explicit inversion formula. Hounsfield computerized this. So the two of them together made CAT scanning possible.[26]

The support theorem that I mentioned fits in here. There was an AMS meeting in Boulder in 1963, and that's when I stated it first. Cormack actually found out that I had worked on this subject, and he wrote me a letter concerning the support theorem and its medical applications, of which I had no idea. I was just interested in the theorem for mathematical reasons. There was nothing of that nature in John's book, but the practical application comes about in the following way: the support theorem tells you that if there is some part—say, the heart—that you wish to avoid sending X-rays through, you can still determine the density outside that critical area by only using X-rays that avoid that area. So that is medically useful.

You've been dealing with this for some time. You wrote papers on it as early as 1959, as you said. You published a book on it in 1980.[27] *And then in 1990 and 1994, you and David Vogan got grants to continue researching this.*

That's right. David Vogan always seems to have the answer when I consult him on something involving Lie groups. Victor Guillemin is a big Radon transform expert. He came in the 1960s. I think his PhD is from the early 1960s.[28]

So the interest in this is driven by these practical applications, X-rays, tomography, and there were some geological uses as well.

Yes, there is a theoretical side to it, and then an applied side. Now, I have only thought about the pure side, I have to admit that. And I have combined that with my interest in symmetric spaces. So there is a Radon transform on symmetric spaces, which is kind of my specialty. But I have listened to lectures from the applied side, because there are conferences where both sides appear, so there is very often at AMS meetings a session on Radon

---

[26] Allan M. Cormack and Godfrey N. Hounsfield shared the 1979 Nobel Prize in Physiology or Medicine.

[27] Helgason, S. (1980).. *The Radon transform (Progress In Mathematics).* Boston: Birkhäuser.

[28] Victor Guillemin received his PhD from Harvard University in 1962 and joined the faculty of MIT in 1966. He received the 2003 Leroy P. Steele Prize for Lifetime Achievement from the AMS for his "critical role in the development of a number of important areas in analysis and geometry."

transforms and tomography. There are several people who have interests on both sides. A very prominent one is Todd Quinto at Tufts. He is a PhD from MIT.[29] He is a student of Victor Guillemin's, although we talked a lot when he was writing his thesis, so I know him well; and he is active in both sides of the subject. He is now, I think, a chairman at Tufts. Another professor at Tufts, Fulton Gonzalez, is very active in the field. He wrote his thesis with me in 1984, entitled "Radon Transforms on Grassmann Manifolds."

Thus it's an attractive subject, because it combines geometry and analysis and, in the case of symmetric spaces, combines this with Lie group theory. So I have in fact two books related to this in the works. They are both extensions of earlier books. One is called *Geometric Analysis on Symmetric Spaces*, which includes chapters on the Radon transform on symmetric spaces; but then I have another one in the works, *Integral Geometry and Radon Transforms*, which is a continuation of the one that you mentioned. It's a new edition of it, updated and less technical, so it's meant for a wider audience.

Mathematics in Iceland

You took on a few administrative tasks in the department as well, am I right?

Yes, a few. I have served on various ad hoc and standing committees and was chairman of the Graduate Committee for seven years, sometime in the eighties. Phyllis Ruby was a great help to me at that time. She was the first graduate administrator and, with George Thomas, set up the Graduate Office in the form it is today. She served in the department for 41 years and was a helpful friend of the graduate students.

I gather that there were more graduate students from Iceland around the time that you were graduate chairman. [Laughs]

There were two. That has nothing to do with me. One of them, Gísli Másson, was appointed by Nesmith Ankeny, who was my predecessor. And the other one, Einar Steingrímsson, I think I recused myself in that case. Their fields were different from mine (algebra and combinatorics, respectively) and neither took any classes with me. After them came Kári Ragnarsson, who wrote

29 E. Todd Quinto wrote his thesis, "On the Locality and Invertibility of Radon Transforms," under Victor Guillemin (1978). He was interim mathematics department head (2007–2008) at Tufts University.

Sigurdur Helgason with Phyllis Ruby, with whom he worked closely when he chaired the Graduate Committee.

a thesis in algebraic topology. I had him in several courses. Altogether there have been four PhDs from Iceland at the department in past years, and one is coming as a graduate student this year. There is also one faculty member in the electrical engineering department who was born in Iceland.

Has mathematics in Iceland become better since you left?

There was essentially just one mathematician in Iceland when I came to the States: that was Leifur Ásgeirsson. Leifur was a good friend of mine, and I did some work on subjects that interested him, just because he told me about them. Since then, there has been a math department at the University of Iceland so there are several people there now, and the program leads to a good bachelor's degree. But the department is still quite small. There are several Icelandic mathematicians in this country, actually.

Do you retain contact with people in Iceland?

Oh, yes. I do. I go back there most summers; I just was there in June. It's not very far—same distance as to California, because you fly directly from Boston now. I have, of course, several relatives that I keep in touch with.

There are reunions where we go and see old gymnasium classmates, every ten years certainly, even now. What is left of the class is actually only about fifty percent, but they get together quite regularly. But then I have good contact with the mathematicians, too, although there is nobody there who works directly in my field.

They had an international conference there a year ago, which attracted people in my area.[30] David Vogan and a couple of Icelandic mathematicians did a great job in organizing it. It dealt with three topics: harmonic analysis, Radon transforms, and representation theory. So it was all Lie group–oriented. And it was not too large for Iceland to organize, maybe at most 90 participants, so very easy to manage. The unpredictable Icelandic weather cooperated nicely, so country sightseeing during half a day was quite successful. The Icelandic landscape is unusual in its absence of trees: you have a view towards infinity in all directions.

~ August 4, 2008

---

[30] "International Conference on Integral Geometry, Harmonic Analysis and Representation Theory" in honor of Sigurdur Helgason on the occasion of his 80th birthday (Reykjavik, Iceland, August, 2007).

WITHDRAWN
Carleton College Library